Directed Molecular Evolution of Proteins

Edited by S. Brakmann and K. Johnsson

Related Titles from Wiley-VCH

Kellner,R.; Lottspeich, F.; Meyer, H. E.

Microcharacterization of Proteins

1999
ISBN 3-527-30084-8

Bannwarth,W.; Felder, E.; Mannhold, R.; Kubinyi, H.; Timmermann, H.

Combinatorial Chemistry. A Practical Approach

2000
ISBN 3-527-30186-0

Gualtieri,F.; Mannhold, R.; Kubinyi, H.; Timmermann, H.

New Trends in Synthetic Medicinal Chemistry

2001
ISBN 3-527-29799-5

Clark,D. E.; Mannhold, R.; Kubinyi, H.; Timmermann, H.

Evolutionary Algorithms in Molecular Design

2000
ISBN 3-527-30155-0

Directed Molecular Evolution of Proteins

or How to Improve Enzymes for Biocatalysis

Edited by
Susanne Brakmann and Kai Johnsson

 **WILEY-VCH**

The Editor of this volume

Dr. Susanne Brakmann
AG „Angewandte Molekulare Evolution"
Institut für Spezielle Zoologie
Universität Leipzig
Talstraße 33
04103 Leipzig, Germany

Prof. Dr. Kai Johnsson
Institute of Molecular
and Biological Chemistry
Swiss Federal Institute of
Technology Lausanne
CH-1015 Lausanne, Switzerland

Cover Illustration   Recent advances in automation
and robotics have greatly facilitated the high –
throughput screening for proteins with desired
functions. Among other devices liquid handling
tools are integral parts of most screening robots.
Depicted are 96-channel pipettors for the microliter-
and submicroliter range (illustrations kindly
provided by Cybio AG, Jena).

Library of Congress Card No.:
applied for

British Library Cataloguing-in-Publication Data
A catalogue record for this book is available from
the British Library.

**Die Deutsche Bibliothek – CIP Cataloguing-in-Pub-
lication Data**
A catalogue record for this publication is available
from Die Deutsche Bibliothek.

© Wiley-VCH Verlag GmbH, Weinheim 2002

Printed on acid-free paper.

Printed in the Federal Republic of Germany.

Composition   Mitterweger & Partner
Kommunikationsgesellschaft mbH, Plankstadt
Printing   betz-druck GmbH, Darmstadt
Bookbinding   Großbuchbinderei J. Schäffer
GmbH & Co. KG, Grünstadt

ISBN 3-527-30423-1

Contents

List of Contributors

Prof. Dr. Frances H. Arnold
Chemical Engineering 210–41
California Institute of Technology
1201 East California Boulevard
Pasadena, California 91125, USA

Prof. Dr. Stephen Benkovic, Dr. Stefan Lutz
Department of Chemistry
The Pennsylvania State University
414 Wartik Laboratory
University Park, Pennsylvania 16802, USA

Prof. Dr. Uwe Bornscheuer
Institut für Chemie und Biochemie
Ernst-Moritz-Arndt-Universität
Soldmannstraße 16
17487 Greifswald

Prof. Dr. Virginia W. Cornish
Columbia University
Department of Chemistry
3000 Broadway, MC 3167
New York, NY 10027-6948, USA

Dr. Rolf Daniel
Institut für Mikrobiologie und Genetik
Georg-August-Universität
Grisebachstraße 8
37077 Göttingen

Prof. Dr. Jacques Fastrez
Laboratoire de Biochimie Physique
et de Biopolymères
Université Catholique de Louvain
Place L. Pasteur, 1.Bte 1B
B-1348 Louvain-la-Neuve, Belgium

Prof. Dr. Donald Hilvert
ETH Hönggerberg
Laboratorium für Organische Chemie
HCI, F339
CH-8093 Zürich, Schweiz

Prof. Dr. Lawrence A. Loeb
Department of Pathology
School of Medicine
University of Washington
Box 357705
Seattle, Washington 98195–7705, USA

Prof. Dr. Alfred Pingoud
Institut für Biochemie
Justus-Liebig-Universität
Heinrich-Buff-Ring 58
35392 Giessen

Prof. Dr. Manfred T. Reetz
Max-Planck-Institut für Kohlenforschung
Kaiser-Wilhelm-Platz 1
45740 Mülheim

Prof. Dr. Peter K. Schuster
Institut für Theoretische Chemie und Strahlenchemie
Universität Wien
Währingerstraße 17
A-1090 Wien, Österreich

Dr. Andreas Schwienhorst
Institut für Mikrobiologie und Genetik
Georg-August-Universität
Grisebachstraße 8
37077 Göttingen

Prof. Dr. K. Dane Wittrup
Dept. Chemical Engineering & Div. Bioengineering
and Environmental Health
Massachusetts Institute of Technology
Cambridge, MA 02139, USA

1

Introduction

Kai Johnsson, and Susanne Brakmann

The application of evolutionary and combinatorial techniques to study and solve complex biological and chemical problems has become one of the most dynamic fields in chemistry and biology. The book presented here is a loose collection of articles aiming to provide an overview of the current state of the art of the directed evolution of proteins as well as highlighting the challenges and possibilities in the field that lie ahead.

Although the first examples of directed molecular evolution date back to the pioneering experiments of S. Spiegelman *et al.* and of M. Eigen and W. Gardiner, who proposed that evolutionary approaches be adapted for the engineering of biomolecules [1, 2], it was the success of methods such as phage display for *in vitro* selection of peptides and proteins as well the selection of functional nucleic acids using the SELEX procedure (Systematic Evolution of Ligands by Exponential enrichment) that brought the power of this concept to the attention of the general scientific community [3, 4]. In the last decade, directed evolution has become a key technology for biomolecule engineering. The success of the evolutionary approach, however, not only depends on the potency of the method itself but is also a result of the limitations of alternative approaches, as our lack of understanding of the structure-function relationship of proteins in general hinders the rational design of biomolecules with new functions. What are the prerequisites for a successful directed evolution experiment? In its broadest sense, (directed) evolution can be considered as repeated cycles of variation followed by selection. In the first chapter of the book, the underlying principles of this concept and their application to the evolutionary design of biomolecules are reviewed by P. Schuster – one of the pioneers in the field of molecular evolution.

Naturally, the first step of each evolutionary project is the creation of diversity. The most straightforward approach to create a library of proteins is to introduce random mutations into the gene of interest by techniques such as error-prone PCR or saturation mutagenesis. The success of random mutagenesis strategies is witnessed by their ample appearances in the different chapters of this book describing case studies of particular classes of proteins and enzymes. In addition, recombination of mutant genes by DNA shuffling or related techniques can be used to create additional diversity and to accumulate rapidly beneficial and additive point mutations [5]. This is a key technique that also surfaces in the majority of the chapters. The sequence space searched by these approaches is, however, quite limited. DNA shuffling between homologous genes, which has also been called family shuffling, allows yet unexplored regions of sequence space to be accessed [6]. In the chapter by S. Lutz and S. J. Benkovic, an approach to create chimeras even between non-homologous genes and its application in protein engineering is described.

An interesting alternative to the generation of libraries with *in vitro* methods is the generation of so-called environmental libraries, described by R. Daniel. Here, advantage is taken of natural microbial diversity by isolating and cloning environmental DNA and by using the resulting libraries to search for novel biocatalysts.

After the creation of diversity, i.e. the generation of a library of different mutants, the protein(s) with the desired phenotype (function or activity) have to be selected from the library. This can be achieved by either selection or screening procedures. The principal advantage of selection is that much larger libraries can be examined: the number of clones that can be subjected to selection is, in general, five orders of magnitudes above those that can be sorted by advanced screening methods. Impressive examples for the power of true selection, where the survival of the host is directly coupled to the desired phenotype, can be found in the chapters written by D. Hilvert *et al.* and J. F. Davidson *et al.*. The major challenge of most selection approaches is to couple the desired phenotype, such as the catalysis of an industrially important reaction, to the survival of the host. But what can be done if the desired phenotype cannot provide a direct selective advantage to a given host organism? Different approaches appear feasible: if the desired property binds to a given molecule, display systems for the protein of interest such as phage display, ribosomal display or mRNA display, and the subsequent *in vitro* selection of binders by so-called panning procedures are established technologies [3, 7, 8]. A recent publication by the group of J. W. Szostak describes the employment of *in vitro* selection of functional proteins from libraries of completely randomized 80mers (actual library size $\sim 10^{13}$) using mRNA display. This work highlights the power of *in vitro* selection, and is a striking example of an experiment that would simply be impossible to perform using screening procedures [9]. In the chapter written by P. Soumillion and J. Fastrez, an interesting extension of this approach, the *in vitro* selection of novel enzymatic activities using phage display, is reviewed. Here, clever selection schemes link the immobilization of the phage to the desired reactivity.

Another approach to the selection of biomolecules with novel functionalities, i.e. binding, or even enzymatic activity, is based on the yeast two- and three-hybrid system. The potential and limitations of these and related approaches are reviewed in the chapter contributed by the group of V. W. Cornish *et al.*

Despite their inferiority in terms of number of clones examined, screening procedures have become increasingly important over the last years. One important reason for this is the enormous technological progress that has been achieved in automation and miniaturization, allowing up to 10^6 different mutants to be screened in a reasonable timeframe. An overview of advanced screening strategies is given in the article of A. Schwienhost. In the chapter written by K. D. Wittrup a discussion of the prerequisites for a successful screening process is given, analyzing the outcome of the directed evolution of proteins displayed on cell surfaces as a function of the screening conditions. The power of intelligently designed screening processes is demonstrated in the following contributions: M. T. Reetz and K.-E. Jaeger describe screening techniques to engineer the enantioselectivity of enzymes; T. Lanio *et al.* present their approaches for the evolutionary generation of restriction endonucleases, U. T. Bornscheuer reports on the functional optimization of lipases, and last but not least, P. C. Cirino and F. H. Arnold give an overview of directed evolution experiments with heme enzymes.

Clearly, there are various developments and applications in the field of directed evolution that are not covered by any of the articles published in this book. Nevertheless, we hope to provide a snapshot of this rapidly developing field that will inspire and support scientists with different backgrounds and intentions in planning their own experiments.

Finally, we would like to thank all authors for their contributions, and P. Gölitz and K. Kriese of Wiley-VCH for their continuous motivation and help in getting this book published.

References

[1] S. Spiegelman, I. Haruna, I. B. Holland, G. Beaudreau, D. Mills, *Proc. Natl. Acad. Sci. USA* **1965**, *54*, 919–927.

[2] M. Eigen, W. Gardiner, *Pure Appl. Chem.* **1984**, *56*, 967–978.

[3] G. P. Smith, *Science* **1985**, *28*, 1315–1317.

[4] a) C. Tuerk, L. Gold, *Science* **1990**, *249*, 505–510; b) A. D. Ellington, J. W. Szostak, *Nature* **1990**, *346*, 818–822.

[5] W. P. Stemmer, *Nature* **1994**, *370*, 389–391.

[6] A. Crameri, S. A. Raillard, E. Bermudez, W. P. Stemmer, *Nature* **1998**, *391*, 288–291.

[7] J. Hanes, A. Plückthun, *Proc. Natl. Acad. Sci. USA* **1997**, *91*, 4937–4942.

[8] R. W. Roberts, J. W. Szostak, *Proc. Natl. Acad. Sci. USA* **1997**, *94*, 12297–12302.

[9] A. D. Keefe, J. W. Szostak, *Nature* **2001**, *410*, 715–718.

2
Evolutionary Biotechnology – From Ideas and Concepts to Experiments and Computer Simulations

Peter Schuster

Research on biological evolution entered the realm of science in the 19th century with the centennial publications by Charles Darwin and Gregor Mendel. Molecular models for evolution under controlled conditions became available only in the second half of the twentieth century after the initiation of molecular biology. This chapter presents an account of the origins of molecular evolution and develops the concepts that have led to successful applications in the evolutionary design of biopolymers with predefined properties and functions.

2.1
Evolution *in vivo* – From Natural Selection to Population Genetics

Nature is the unchallenged master in design by variation and selection and since Charles Darwin's epochal publication of the "Origin of Species" [1, 2] the basic principles of the mechanism behind natural selection have become known. Darwin deduced his principle of evolution from observations "in the field" and compared species adapted to their natural habitats with the results achieved through artificial selection by animal breeders and in nursery gardens. Natural selection introduces changes in populations by differential fitness, which is tantamount to the instantaneous differences in the numbers of decedents between two competing variants. In artificial selection the animal breeder or the gardener interferes with the natural selection process by discarding the part of the progeny with undesired properties. Only shortly after the publication of Darwin's "Book of the Century" the quantitative rules of genetics were discovered by Gregor Mendel [1, 2]. It took, nevertheless, about seventy years before Darwin's theory was united successfully with the consequences of Mendel's results in the development of population genetics [2, 3].

The differential equations of population genetics are commonly derived for sexually replicating species and thus deal primarily with recombination as the dominant source of variation. Mutation is considered as a rather rare event. In evolutionary design of biopolymers the opposite is true: Mutation is the common source of variation and recombination occurs only with special experiments, "gene shuffling" [4], for example. In the formulation of the problem we shall consider here the asexual case exclusively. The mathematical expression dealing with selection through differential fitness is then of the form

$$\frac{dx_k}{dt} = x_k \left(f_k - \Sigma_{j=1}^n f_j x_j\right) = x_k(f_k - \Phi); \quad k = 1, 2, \ldots, n. \tag{1}$$

The fraction of variant I_k is denoted by x_k with $\sigma_k x_k = 1$; f_k is its fitness value. Accordingly, we introduced $\phi = \sigma_k f_k x_k$ as the mean fitness of the population. The mathematical role of ϕ is to maintain the normalization of variables. The interpretation of Eq. (1) is straightforward: Whenever the differential fitness, f_k-ϕ, of a variant I_k is positive or its fitness is above average, $f_k > \phi$, dx_k/dt is positive and this variant will increase in frequency. The opposite is true if $f_k < \phi$, then the fraction of the corresponding variant will decrease and ultimately approach zero: The variant has died out. Selection thus chooses the variant I_m with the highest fitness value, $f_m = \max\{f_k, k = 1,2,...,n\}$, and after sufficiently long time only this variant will be present in the population, $\lim_{t\to\infty} x_m = 1$. In other words, if we wait long enough, all less fit variants will have died out, and the population becomes homogeneous.

The typical evolutionary scenario considered by population genetics is characterized by low mutation rates. Then the arrival of a new variant by mutation, I_λ in a currently optimized population (containing exclusively I_m) is a rare event and the dynamics of Eq. (1) is visualized in response to such an instant. Apart from a stochastic initial phase, during which the new species is in danger of dying out by accident, the course and the outcome of the selection process is determined exclusively by the difference in fitness values: $s = f_\lambda - f_m$. The value of s is reflected by the number of generations that are required to select the advantageous mutant (see Fig. 2.1). In nature selective advantages of emerging mutants are commonly very small and hence thousands of generations are required before a new variant can take over in the population.

Population genetics saw a major extension by Motoo Kimura [5] who suggested that adaptive mutations were extremely rare, most mutants were selectively neutral, and the predominant role of evolution was the elimination of deleterious variants. Kimura's view was strongly supported by the data obtained from comparative sequence analysis of proteins and nucleic acids [6], which became the basis of current molecular phylogeny. Genotypes are changing steadily and this also during epochs of phenotypic stasis. Despite overwhelming indirect hints for neutral evolution from molecular data, the first direct proof came only recently from experiments on bacterial evolution under controlled conditions: The change in phenotypic properties, like cell size, shows clear

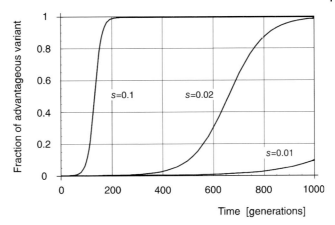

Fig. 2.1. Selection of advantageous variants. The individual curves show selection and fixation of mutants in populations of $N = 10000$ individuals according to the equation $x(t) = x_0 / \{x_0 + (1 - x_0)\exp(-st)\}$. Time t is measured in generations or replication steps, x_0 is the initial frequency of the new variant in the population, and $s = f' - f$ is its selective advantage. The curves shown above use initial conditions of a single copy in the population, $x = 0.0001$.

punctuation [7] whereas genomic DNA sequences continue to change during phenotypic stasis at the same pace or even faster than during the adaptive periods [8]. Kimura's approach is based on a stochastic description of the selection process: Every newly formed variant has a certain probability to reach fixation that increases with its s value, which is measured relative to the fitness of the currently dominant type in the population. In the neutral case, $s = 0$, populations migrate through sequence space in a random walk like manner. The random walk is modeled by diffusion in a continuous space of genotypes (which is an approximation to the sequence space concept discussed below). Computer simulations of neutral evolution performed in the 1990s [9 ,10] confirmed Kimura's view.

Populations genetics, although successful in its own right, suffers from two major problems when confronted with present-day molecular biology: (i) Mutation is handled as some rare external event, which is not part of the regularly considered dynamics, and (ii) the phenotype is represented only by its fitness value, which is assigned as a parameter to the corresponding genotype.

2.2
Evolution *in vitro* – From Kinetic Equations to "Magic Molecules"

The undeniable efficiency and beauty of Nature's solutions to often exceedingly complex problems has, nevertheless, raised the desire to make use of similar evolutionary techniques in order to solve problems in technology through exploitation of the natural recipe. In the area of biotechnology the problem is to design molecules for predefined purposes and, starting in the 1980s [11], this goal has been pursued with great success.

The idea of mimicking evolution by suitable experiments in the test tube was born in the 1960s by Sol Spiegelman and his coworkers at Columbia University [12]. The setup of such serial transfer experiments is shown in Fig. 2.2. RNA molecules of viral origin were transferred into a test tube containing a medium suitable for viral RNA replication. This replication assay contained a virus-specific replication enzyme, $Q\beta$-replicase, as well as activated monomers in form of nucleoside tri-phosphates. Spiegelman was able to show that natural selection in the sense of Fig. 2.1 occurs whenever there are entities, cellular organisms or molecules, which multiply and, occasionally, produce modified progeny because of imperfect reproduction. Indeed the rate of RNA synthesis increased roughly by one order of magnitude over some 70 serial transfers in the setup sketched in Fig. 2.2. In addition we show the increase in replication rate during the first 27 transfers: The rate rises by a factor of three within only six transfer steps (no.8 – no.13), and we notice a clear indication of stepwise optimization of replication rate. Thus, occurrence of evolution in the Darwinian sense as the interplay of variation and natural selection is not bound to the existence of cellular life.

Molecular biologists have discovered and are currently still revealing a true wealth of data on the nature of the genetic machinery, the processing of biological information, and regulation and control of cells and organisms. After the molecular structure of nucleic acids had been correctly derived by James Watson and Francis Crick [13], the nature of the space in which the evolving populations travel was clear: Sequence space is a discrete space of all DNA (or RNA) sequences with a distance defined by mutation and/or recombination (Fig. 2.3). Leaving aside recombination and assuming the point mutation as the elementary process or basic move in the creation of new genotypes, sequence space is a generalized hypercube with the Hamming distance as metric [14] (It is worth mentioning that by initially neglecting mutation, recombination spaces were also successfully defined [15]). Two properties of sequence space are highly important: (i) It is a high-dimensional object with the lengths of the genome, λ measured in nucleotides, being the dimension and hence distances are short, and (ii) all points in sequence space, i.e. all sequences, are equivalent.

In his seminal paper on the evolution of molecules, Manfred Eigen [16] combined the knowledge of molecular biology and chemical reaction kinetics and formulated a

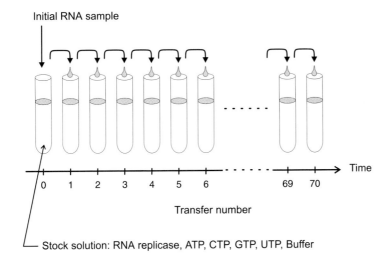

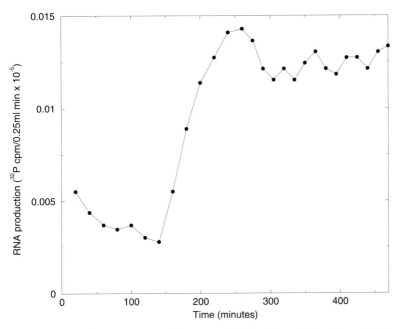

Fig. 2.2. RNA evolution experiments. The upper part shows the technique of serial transfer applied to evolution of RNA molecules in the test tube. After a given time interval a small sample is transferred into the next test tube containing fresh stock solution. Thereby the materials, which were consumed during RNA synthesis, are replenished. The stock solution contains an enzyme required for replication, for example Qβ-replicase, and activated monomers (ATP, UTP, GTP, and CTP), which are the building blocks for polynucleotide synthesis. The rate of RNA synthesis (lower part) is measured through incorporation of radioactive GTP into the newly produced RNA molecules. The rate of replication shows stepwise increase. An early decrease is observed, because first a quasi-species is formed by the master sequence through production of mutants of lower fitness. The figure is redrawn from the data in [12].

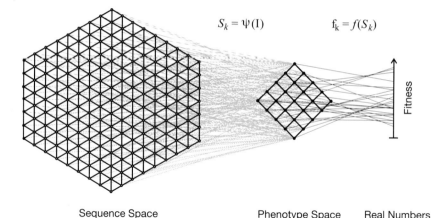

$$S_k = \psi(I) \qquad\qquad f_k = f(S_k)$$

Fitness

Sequence Space   Phenotype Space   Real Numbers

Fig. 2.3. Sequence space and genotype – phenotype mappings. Mapping genotypes onto phenotypes and into fitness values. The sketch shows a map from sequence (or genotype) space onto phenotype space, as described in the text, and further into the real numbers resulting in fitness values assigned in two steps to the individual genotypes. The second map is a "landscape", which could also be illustrated by a three-dimensional plot. Both mappings are usually many-to-one and thus non-invertible. Sequence and phenotype space are high-dimensional objects; they are sketched here by two-dimensional illustrations.

model which describes replication, mutation, and selection by means of a network of kinetic equations:

$$\frac{dx_k}{dt} = x_k \left(Q_{kk} f_k - \Phi \right) + \Sigma_{j=1,\, j \neq k}^{n} Q_{kj}\, f_j\, x_j; \quad k = 1, 2, \ldots, n. \tag{2}$$

It is straightforward to interpret Eq. (2) as an extension of Eq. (1): The replication process is a network of parallel reactions leading to the correctly copied product, $I_k \rightarrow 2I_k$, with probability Q_{kk}, and to a variant, $I_k \rightarrow I_k + I_j$, with probability Q_{jk}. The two production terms in Eq. (2) describe correct reproduction of I_k and its production from other genotypes through mutation, $I_j \rightarrow I_j + I_k$, and the third term containing ϕ is identical to the fitness weighting term in Eq. (1). Since Eq. (2) is intended to refer to an experimental setup for studying the evolution of molecules, we cannot be content with a mathematical interpretation of $\phi(t)$; what we need now is a physical process defining it. It is indeed straightforward to identify this term as a dilution flux whose effect is to control the total number of replicating molecules. A flow reactor that could, in principle, serve this purpose is shown in Fig. 2.4. A non-neutral replicating ensemble contains the fittest genotype called the *master sequence*. Commonly, this is also the most frequent type (Fig. 2.5). If mutation is a sufficiently frequent event the master sequence is surrounded by a cloud of mutants consisting of either close relatives or more distant variants of sufficiently high fitness. Under suitable conditions the master

Stock Solution ⟶   Reaction Mixture ⟶

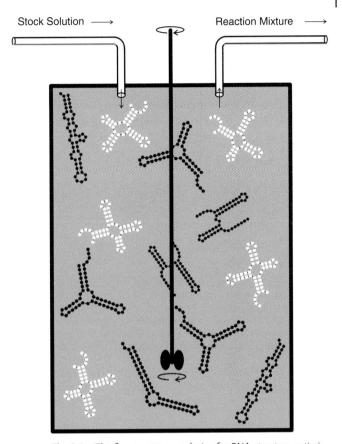

Fig. 2.4. The flow reactor as a device for RNA structure optimization. RNA molecules with different shapes are produced through replication and mutation. New sequences obtained by mutation are folded into minimum free energy secondary structures. Replication rate constants are computed from structures by means of predefined rules (see text). For example, the replication rate is a function of the distance to a target structure, which was chosen to be the clover leaf shaped tRNA shown above (white shape) in the reactor. Input parameters of an evolution experiment *in silico* are: the population size N, the chain length λ of the RNA molecules as well as the mutation rate p.

sequence and its mutant cloud approach a steady genotype distribution called *quasi-species*. The concept of quasi-species was found useful and important for understanding virus evolution [17].[1] In addition, quasi-species of RNA molecules *in vitro* were

1) Often it is very difficult to find out whether or not an experimentally observed distribution of genotypes is stationary. The notion of virus quasi-species has been coined to characterize a distribution of variants around a fittest genotype irrespectively of its closeness to a steady state.

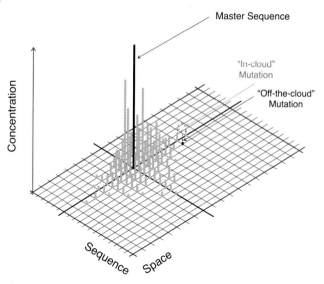

Fig. 2.5. A quasi-species-type mutant distribution around a master sequence. The quasi-species is an ordered distribution of polynucleotide sequences (RNA or DNA) in sequence space. A fittest genotype or master sequence I_m, which is commonly present at highest frequency, is surrounded in sequence space by a "cloud" of closely related sequences. Relatedness of sequences is expressed (in terms of error classes) by the number of mutations which are required to produce them as mutants of the master sequence. In case of point mutations the distance between sequences is the Hamming distance. In precise terms, the quasi-species is defined as the stable stationary solution of Eq. (2) [16, 19, 20]. In reality, such a stationary solution exists only if the error rate of replication lies below a maximal value called the error threshold. In this region, i.e. below the often sharply defined mutation rate of the error threshold, the population is structured as shown in the figure. Above the critical error rate the stationary solution of Eq. (2) is (practically) identical with the uniform distribution. The uniform distribution, however, can never be realized in nature or *in vitro* since the number of possible nucleic acid sequences, 4^λ, exceeds the number of individuals by many orders of magnitude even in the largest populations. Then the actual behavior is determined by incorrect replication leading to random drift: populations migrate through sequence space. We distinguish two classes of mutations: "In-cloud" mutations (gray), which lead to an already existing variant, and "off-the-cloud" mutations (black), which produce a new genotype.

studied in replication assays consisting of a virus-specific RNA replicase, in particular Qβ-replicase [18].

An important feature of the replication-mutation kinetics of Eq. (2) is its straightforward accessibility to justifiable model assumptions. As an example we discuss the uniform error model [18, 19]: This refers to a molecule which is reproduced sequentially, i.e. digit by digit from one end of the (linear) polymer to the other. The basic assumption is that the accuracy of replication is independent of the particular site and the nature of the monomer at this position. Then, the frequency of mutation depends exclusively on the number of monomers that have to be exchanged in order to mutate from I_k to I_j, which are counted by the Hamming distance of the two strings, $d(I_j, I_k)$:

$$Q_{jk} = q^{\lambda} \varepsilon^{d(I_j, I_k)} = q^{\lambda} \left(\frac{1-q}{q} \right)^{d(I_j, I_k)}. \tag{3}$$

Within this model all mutation rates can be expressed in terms of only three quantities; the chain length of the polymer, λ, the single-digit accuracy of replication, q, often expressed as mutation rate per site and replication, $p = 1 - q$, and the Hamming distance, $d(I_j, I_k)$. Finally, the (dependent) parameter, $\varepsilon = (1 - q)/q$ is the ratio between single digit mutation rate and accuracy.

Equation (2) sustains a stationary state that can be characterized as a mutation equilibrium provided the replication process is sufficiently accurate, $q > q_{min}$. This minimal accuracy of replication is readily obtained from a straightforward estimate that is based on the condition of non-vanishing frequency of the master sequence

$$Q_{min} = q_{min}^{\lambda} = \sigma_m^{-1} = \frac{\sum_{k=1, k \neq m}^{n} x_k f_k / (1 - x_m)}{f_m} \tag{4}$$

The minimum accuracy of replication is tantamount to a maximal tolerable mutation rate, $p_{max} = 1 - q_{min}$, that has been called the error threshold. At mutation rates which are higher than threshold, the structured quasi-species is replaced by the uniform distribution. In other words, all variants including the master sequence occur with the same probability, when the replication accuracy is too low. A simple and straightforward estimate shows that a uniform distribution cannot exist with biopolymers: The number of possible variants of chain length λ built from κ classes of monomers, λ^{κ}, is hyper-astronomically large and no population size on Earth can ever come close to such values. Consequently, the populations drift randomly through sequence space, and this phenomenon of highly error-prone reproduction might be characterized as *random replication*.

It is worth considering Eq. (4) from a different point of view. The replication accuracy q is assumed to be determined by the replication machinery and therefore cannot be varied. Then, the error threshold restricts the chain length and defines an upper value for sufficiently faithful replication:

$$Q_{min} = q^{\lambda_{max}} \quad \text{with} \quad \lambda_{max} \approx \frac{\ln \sigma}{1 - q} = \frac{\ln \sigma}{p}. \tag{5}$$

Closely related to this equation is an interesting observation: The product of chain length and mutation rate is approximately constant for many classes of organisms [21 – 23]. This constant is close to one for lytic RNA viruses, close to 0.1 for RNA retroviruses, and approximately 1/300 for DNA-based microbes. Smaller but still nearly constant values were found for higher organisms.

An interesting detail of the quasi-species concept was predicted more than twelve years ago [24] and has been observed recently with virus populations [25] and computer simulations [26]: We assume two genotypes of high fitness, each one surrounded by a specific mutant cloud (Fig. 2.5). Genotype I_{m1} has higher fitness compared to I_{m2} but less efficient mutants in the sense of a mutant cloud with lower mean fitness. The quasi-species considered as a function of the mutation rate p may show a rearrangement reminiscent of a phase transition at some critical replication accuracy $q_{cr} = 1 - p_{cr}$. At low mutation rates, $p > p_{cr}$, the difference in fitness values determines selection and hence, the master sequence with higher fitness, I_{m1}, dominates. Above the critical mutation rate, $p > p_{cr}$, however, mutational backflow to the master is decisive and then I_{m2} is selected.

Replication-mutation kinetics *in vitro* and its major result, the concept of molecular quasi-species, set the stage for a new kind of biotechnology that is based on variation and selection [11]. The application of artificial evolution to produce molecules binding to given targets started in 1990 independently in two research laboratories [27, 28]. The results of about a decade of evolutionary biotechnology were summarized in many reviews (examples are [29 – 31]). The essential idea of the evolutionary design of molecules for predefined purposes consists in the application of consecutive selection cycles, where each comprises the three phases: (i) amplification, (ii) diversification, and (iii) selection (Fig. 2.6). Amplification and diversification of nucleic acid molecules have now become routine methods in molecular biology. DNA and RNA can be multiplied by many different assays; we mention here only PCR and the 3SR reaction. Variation and diversity can be achieved in two different ways: (i) replication with enhanced mutation rates, and (ii) chemical synthesis of random sequences. The selection process, nevertheless, requires intuition and ingenuity. As an example we consider the production of optimal binders to predefined targets called *aptamers*. A universally applicable method for the evolutionary design of ligands is the SELEX (**s**ystematic **e**volution of **l**igands by **ex**ponential enrichment) technique [27, 28]. Here, the selection criterion is retention on a chromatographic column with covalently attached target molecules. Changing the solvent allows selection constraints to be tuned. Commonly, some twenty to thirty selection cycles are sufficient to obtain optimally binding molecules. In favorable cases it is possible to obtain aptamers with binding constants in the nanomolar range [30].

Design of RNA molecules with novel catalytic functions called *ribozymes* (**ribo**nucleotide en**zymes**) started out from the reprogramming of naturally occurring molecules to accept unnatural substrates [32, 33]: A specific RNA cleaving ribozyme, a class I (self-splicing) intron, was modified through variation and selection until it operated efficiently on DNA. The evolutionary path of such a transformation of catalytic activity has been recorded in molecular detail [34]. The basic problem in the evolutionary design of new catalysts is the availability of appropriate analytical tools for the detection of activity. The technique of chemical tagging, for example, uses a covalently attached detectable marker, which is cleaved off in the molecules with the desired catalytic activity. Inactive molecules are unable to split off the tag. They can be detected by the presence of the tag after reaction, and they are excluded from further selection rounds. It is particularly interesting that molecules with novel catalytic activities were selected from pools of random sequences. The first successful experiment of

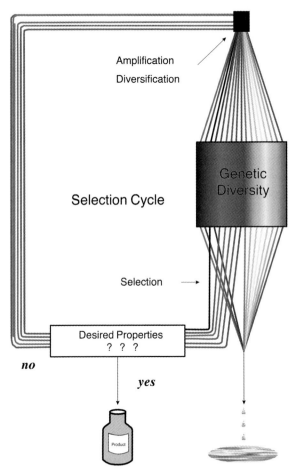

Fig. 2.6. Evolutionary design of biopolymers in selection cycles. Properties of biomolecules, for example binding to a target or catalytic function, are optimized iteratively through selection cycles. Each cycle consists of three phases: (i) amplification, (ii) diversification by replication with problem adjusted error rates (or random synthesis), and (iii) selection. Amplification and diversification are carried out by well established methods in molecular biology. Examples are the polymerase-chain-reaction (PCR) and the self sustained sequence replication reaction (3SR reaction). Both allow for enhanced mutation rates. Selection still requires ingenious concepts. Examples are the SELEX method and chemical tagging as discussed in the text.

this kind was reported already in 1993 from Jack Szostak's research laboratory [35]. Meanwhile many different new ribozymes have been created and selected in this way [30].

A particularly fascinating property of RNA is its capability to combine the properties of information carrier and catalyst in a single molecule: The information is carried in the sequence and the function is a property of the molecular structure. This peculiar feature of RNA makes it predestined for evolution experiments as well as for the selection of functional molecules from random pools. A molecule, once detected by its function and isolated from the reaction mixture, can be amplified and diversified instantaneously. Since sequence and function are two features of the same molecule, the tricky problem to design a link between the information carrier and the functional molecule, as encountered in the case of messenger RNA and enzyme, does not exist. These unique properties make RNA a kind of "magic molecule" and gave rise to the idea of an RNA world that might have played an important role at the origin of life (For a collection of articles on this subject see [36]).

Concluding this section it is worth comparing *in vitro* evolution and artificial selection of molecules with predefined properties from a wider perspective. Evolution in the test tube, like evolution *in vivo*, is based on natural selection through differential fitness. In other words what counts for survival is exclusively fertility or replication rate, eventually modified by the progeny's probability of survival to the reproductive age or the mutation rate, respectively. Artificial selection of molecules is like animal breeding or the creation of new plant variants in nursery gardens: Molecules or individual organisms carrying the desired properties are picked out at will by the experimenter. Human intervention defines fitness and the only limitation is a non-zero number of descendants. Evolutionary design of molecules may be considered as "breeding of molecules". While it is simple and straightforward for the animal breeder to pick out the black puppies and to discard the brown ones, when black dogs are the predefined goal, selection of molecules clearly requires knowledge of physics and chemistry and creative ingenuity in the design of suitable equipment.

2.3
Evolution *in silico* – From Neutral Networks to Multi-stable Molecules

Two or more genotypes are neutral in evolution when the selection constraint is unable to distinguish between them. Early sequence comparison data [6] apparently confirmed Motoo Kimura's idea of neutral drift in population genetics [5]. Accordingly, many different genotypes could give rise to the same phenotype, and depending on the conditions, different phenotypes can share the same fitness value. Direct evidence for neutral evolution under controlled conditions came only two years ago: Serial transfer experiments with *Escherichia coli* bacteria over several ten thousands of generations show punctuated or stepwise rather than continuous increase in fitness [7]. Changes in genotypes occur during the epochs of phenotypic stasis with the same frequency as or even faster than in the adaptive periods [8]. Occurrence of neutral evolution is a fact but still many questions remain open: What does neutrality in phenotype space really mean? How important is the degree of neutrality in evolution, and how can we measure it? Given one particular phenotype as an initial condition, which phenotypes are accessible through mutation and/or recombination? What determines the efficiency and the success of evolutionary processes? How important is population size? Given that the evolutionary process is a sequence of more or less random events, are there invariant features or "constants of evolution"?

In order to answer these open questions, at least on the level of a sufficiently simple and experimentally testable system, a model based on the properties of RNA molecules has been conceived (for a recent review of the RNA model see [37]): Genotypes are represented by RNA sequences and the phenotypes are modeled by RNA secondary structures[2] (Fig. 2.7, for a recent review on RNA secondary structures see [38]). Then, the rules of folding RNA sequences into secondary structures of minimal free energies determine the relations between RNA genotypes and phenotypes. The investigations of global sequence-structure relations comprise two aspects: (i) the mapping of RNA sequences into RNA structures is analyzed as a typical and mathematically accessible example of a (simple) genotype-phenotype map, and (ii) evolution is studied by simulating replication-mutation kinetics in populations of molecules under the conditions of a flow reactor (Fig. 2.4).

Formally, sequence-structure relations are considered as mappings from sequence space onto a discrete space of structures (for a review see [39]):

$$\psi : \{I; d_{ij}^h\} \Rightarrow \{S; d_{ij}^s\} \text{ or } S_k = \psi \, (I). \tag{6}$$

The equation expresses that the space of all genotypes, the sequence space I, is a discrete space with the Hamming distance as metric. It is mapped onto a discrete space of structures called shape space with the structure distance as metric (We use I rather than I_k in order to indicate different numbering schemes used for sequences and structures). The evolutionarily relevant quantity, the fitness value f_k as shown in Fig. 2.3, is derived from the phenotype S_k through evaluation, which can be understood as another mapping, a map from shape space into the positive real numbers including zero, $f_k = f(S_k)$. Both maps need not be invertible in the sense that more than one phenotype may have the same fitness value, and more than one sequence may lead to the same structure. We shall study here neutrality induced by the first map, ψ in Eq. (6).

---

2) The idea to represent the phenotype in RNA evolution experiments by the RNA structure was formulated by Sol Spiegelman [12].

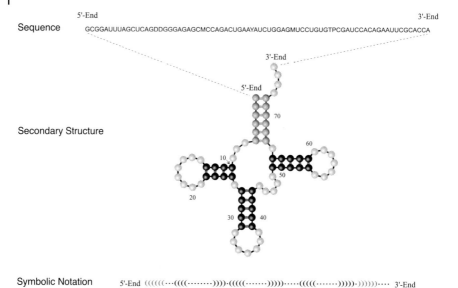

Sequence

5'-End — GCGGAUUUAGCUCAGDDGGGAGAGCMCCAGACUGAAYAUCUGGAGMUCCUGUGTPCGAUCCACAGAAUUCGCACCA — 3'-End

Secondary Structure

Symbolic Notation   5'-End (((((···(((·········))))·(((········)))))····((((·······)))))·))))))···· 3'-End

Fig. 2.7. RNA secondary structures. The nucleotide sequence of tRNA[phe] (shown in the upper string) is presented together with the secondary structure of minimal free energy and the symbolic notation (lower string). The sequence contains several modified nucleotides (**D, M, P, T, Y**) in addition to the conventional bases (**A, U, G, C**). Individual nucleotides in the secondary structure are shown as light gray (single bases), dark gray and black pearls (base pairs). The tRNA structure is a clover leaf with three hairpin loops (adjacent stacks are shown in black) and a closing stack (shown in dark gray), which completes the central multi-loop. The symbolic notation assigns one of the three symbols ".", "(", and ")" to each nucleotide depending on its binding state, "unpaired", "paired downstream", and "paired upstream", respectively. Downstream and upstream refer here to the conventional direction in polynucleotide sequences going from the 5'-end to the 3'-end. The three stacks of the hairpin loop (black) are embraced by the base pairs of the closing stack (gray).

Systematic, but computationally extremely demanding explorations of sequence-structure mappings determine the secondary structures of all κ^{λ} different sequences of given chain length λ by means of a fast folding algorithm. Properties of interest are derived by exhaustive enumeration [40]. These studies, although limited to small molecules ($\lambda \leq 30$), provide the basis for general insights into genotype-phenotype mappings (as sketched in Fig. 2.3). The results obtained can be summarized in four major findings: (i) The number of sequences (genotypes) exceeds the number of secondary structures (phenotypes) by many orders of magnitude, (ii) relatively few common structures are opposed by a great number of rare structures, (iii) a sphere with a radius much smaller than $\lambda/2$, the radius of sequence space, around an arbitrarily chosen reference sequence contains one or more sequences for each structure (*shape space covering* [41]), and (iv) the pre-images of common structures are extended *neutral networks* in sequence space [41], which, in principle, span the entire space. Because of their general importance, we shall discuss here only neutral networks and their role in evolutionary processes.

Neutrality in sequence space implies that two or more sequences form identical structures. It is straightforward to verify that the sketch in Fig. 2.3 gives a correct impression: As mentioned above there are many more sequences than structures, and a very large number of structures are rare. Evolutionarily important structures are common, since rare structures often formed by a handful of sequences are often very hard to find through evolutionary searches. All sequences forming a given structure S_k represent the pre-image of S_k in sequence space, and we shall call them the neutral set G_k of structure S_k. The neutral set is converted into a graph, the neutral network G_k, by connecting all pairs of sequences with Hamming distance one through an edge (Fig. 2.8). Generic properties of neutral networks were studied by means of a mathematical model based on random graph theory [42]. The mean degree of neutrality of structure S_k, measured as the relative frequency of neutral nearest neighbors, $\overline{\lambda_k}$, is the quantity that determines the properties of neutral networks in sequence space. How are the sequences forming the same structure distributed in sequence space? The extension of neutral networks reminds of a percolation phenomenon:

$$G_k \text{ is connected, if } \quad \overline{\lambda_k} > \lambda_{cr} = 1 - \kappa^{-1/\kappa-1}, \text{ and} \tag{7a}$$

$$G_k \text{ is partitioned, if } \quad \overline{\lambda_k} < \lambda_{cr} = 1 - \kappa^{-1/\kappa-1}. \tag{7b}$$

Connectedness of a neutral network, implying that it consists of a single component, is important for evolutionary optimization. Populations usually cover a connected area in sequence space and they migrate (commonly) by the Hamming distance moved. Accordingly, if they are situated on a particular component of a neutral network, they can reach all sequences of this component. If the single component of the connected neutral network of a common structure spans all sequence space, a population on it can travel by random drift through whole sequence space.

Neutral networks connect sequences forming the same secondary structure of minimum free energy. Every sequence, however, forms a great number of sub-optimal structures, which are also computable by suitable algorithms. Seen from a given structure S_k, the neutral set G_k is surrounded by the set of compatible sequences C_k. This set contains all sequences which form S_k as sub-optimal or minimum free energy structure. By taking two structures at random, say S_j and S_k, and considering the two sets of compatible sequences, C_j and C_k, it was proven [42] that the intersection is always non-empty: $C_j \cup C_k \neq \emptyset$. In other words, this *intersection theorem* can be expressed by: Given an arbitrary pair of structures, there will be at least one sequence that can adopt both structures[3].

---

3) It is important to stress that the intersection theorem cannot be extended to three or more structures: For three or more structures there may but need not exist a sequence that can form all of them [42].

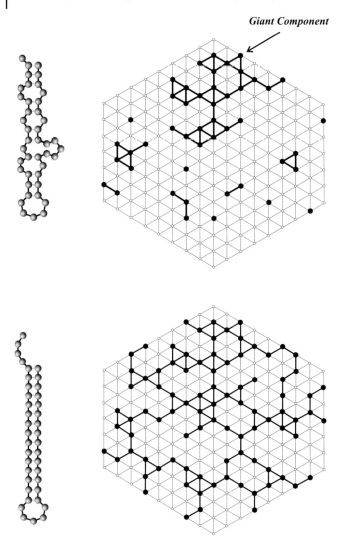

Fig. 2.8. Neutral networks in sequence space. The pre-image of the structure in the lower part of the figure is a connected neutral network spanning whole sequence space. Networks of this class are typical for frequent structures. The upper part of the figure shows an example of a parti- tioned network, which consists of one giant component and many small islands. Connectivity is determined by the mean fraction of neutral neighbors, $\overline{\lambda_k}$, of the pre-image of the corresponding structure, S_k, in sequence space.

The existence of extended and connected neutral networks in RNA sequence space was proven by an elegant experiment recently published by Erik Schultes and David Bartel [43]. At the starting point for their work were two ribozymes of known structures with chain length $\lambda=88$: (i) an RNA ligase evolved in the laboratory [44], and (ii) a natural cleavage ribozyme isolated from hepatitis delta virus RNA [45]. The two structures have no base pair in common and apparently no common phylogenetic history. Then, an RNA sequence was designed and synthesized at the intersection of the two neutral networks of the reference structures. This means that a chimeric sequence was synthesized which was compatible with both structures. The chimera did form both structures on folding and showed both activities, although they were substantially weaker than those of the reference ribozymes, the ligase and the cleavage ribozyme, respectively. Only two or three selected point mutations or base pair exchanges are required, however, to reach full catalytic efficiency. Still, the two optimized RNA molecules have a Hamming distance of about forty from their reference sequences. Next, Schultes and Bartel explored further the mutational neighborhoods and found neutral paths of Hamming distance around 40, by preparing and analyzing series of RNA sequences, along which neighboring sequences differing in a single base or base pair only. Without interruption these two neutral paths lead from the chimeric RNA with both catalytic activities to the two reference ribozymes. This result presents a direct proof for a sequence space-wide extension of the two neutral networks as well as an experimental confirmation of the existence of a non-empty intersection of the two compatible sets. The existence of multi-stable RNA molecules has been derived also by means of a recently developed kinetic folding algorithm [46], which resolves the folding process to elementary steps involving single base pairs. Application to sequences at the intersection of structures allows the design of molecules switching between two or more conformations with predefined rate constants [47].

Computer simulations of evolution in sequence space through replication and mutation in populations of RNA molecules under the conditions of a flow reactor (Fig. 2.4) were carried out first in the 1980s [48]. Typical sustainable population sizes are between one thousand and one hundred thousand molecules. The mutation rate, p, is adjusted to the chain lengths of the molecules so that the majority of mutation events leads to single point mutations and double mutations in a single replication event are very rare. Basic to these *in silico* studies is a straightforward introduction of phenotypes, represented by molecular structures, into the model (Fig. 2.9). Every newly formed genotype produced in the population by an off-the-cloud mutation (Fig. 2.5) is folded into its minimum free energy structure and the resulting structure is evaluated to yield the replication rate of fitness value of the new molecular variant. These early studies of evolution *in silico* provided already clear evidence for the punctuated nature of the optimization process and neutral drift during the epochs of phenotypic stasis, independent of whether the simulations were conceived to aim at one

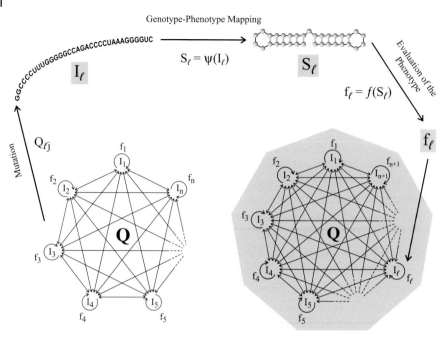

Fig. 2.9. Evolutionary dynamics with phenotypes. The sketch shows a sequence of events following an "off-the-cloud" mutation and leading an innovation, which consists in the incorporation of a new mutant into the replication-mutation ensemble: (i) A new variant sequence, I_λ, is created through a mutation, $I_j \rightarrow I_\lambda$, (ii) the sequence is converted into a structure, $S_\lambda = \Psi(I_\lambda)$, and (iii) the fitness of the new phenotype is determined by means of the mapping $f_\lambda = f(S_\lambda)$. Eventually, the new variant is fully integrated into the replication-mutation ensemble.

particular target structure or at some property shared by several classes of structures. Later on, further studies on neutral evolution were performed with the goal to check the diffusion approximation of random drift [10]. A more recent investigation [49, 50] explored and revealed the mechanism of punctuated evolution. A typical plot of the course of the mutation-selection process is shown in Fig. 2.10: The mean distance to target of the population (which is a measure of fitness in these simulation experiments) is plotted against time and shows pronounced punctuation. Adaptive periods are interrupted by long epochs of stasis with respect to fitness. Evolution in genotype space, however, neither slows down nor stops on the mean fitness plateaus [51]. Inspection of the sequence distribution of the population provides new insights into the process.

An evolutionary trajectory leading from an initial population to the final state is characterized by a uniquely defined time-ordered series of phenotypes, called the *relay series* [49]. It can also be understood as a series of transitions between pairs of consecutive phenotypes in the relay series. Transitions are off-the-cloud mutations leading to new phenotypes and fall into two classes: (i) minor or continuous transitions and

(ii) major or discontinuous transitions.[4] Minor transitions between structures occur with high frequency and involve changes that are easy to accomplish with a single point mutation, like opening or closing of single base pairs adjacent to stacks. Opening of stacks with marginal stability also falls into this class. The sequence constraint is low: Almost every sequence forming the initial structure yields the final structure of a minor transition on one or a few different single point mutations. Major transitions between structures require simultaneous changes in several adjacent and/or distant base pairs and occur at single point mutations with low probability only. Major transitions are characterized by strong constraints on initial sequences. In other words, they require special initial sequences and thus occur with low probability when averaged over the entire neutral network.

Analysis of the dynamics on the plateaus of constant fitness falls into one of two different scenarios: (i) Neutral evolution in the conventional sense consisting of changing genotypes that give rise to the same phenotype or *phenotypic stasis* expressed by a single phenotype on the relay series, and (ii) a *neutral random walk* on a subset of closely related *phenotypes* of identical fitness, which are accessible from each other through minor transitions, that manifest itself by a sometimes large number of steps in the relay series with frequent repetitions of particular phenotypes. Very rarely, fitness neutral major transitions are also observed inside fitness plateaus. As we shall see below the two scenarios are not very different in reality: Scenario (ii) is readily converted into scenario (i) by an increase in population size. Each quasi-stationary epoch ends with a major transition that is accompanied by a gain in fitness. A straightforward interpretation of this finding suggests that the population undertakes a random search during the epochs of phenotypic stasis until a mutant sequence is produced that initiates a fitness improving major transition. A cascade of fitness improving minor transitions commonly follows the major transition, and the close neighborhood of the new variant is thereby instantaneously explored.

The explanation given above is strongly supported by the dynamics observed in genotype space. When the population enters a fitness plateau the distribution of genotypes is very narrow (Fig. 2.10). Then, while the population diffuses on a neutral subspace of sequence space, the width of the mutant cloud increases steadily and seems to approach a saturation phase. Instantaneously, when the population reaches the end of the fitness plateau, the width of the distribution drops as the population passes a bottleneck in genotype space. This picture of population dynamics on the neutral subset, slow spread and fast contraction, is complemented by a recording of the migration of the population center through sequence space. On the plateau, during the spread of the distribution, the center is almost stationary or drifts very

---

4) The choice of the adjectives "*continuous*" and "*discontinuous*" points to topological relations between the pre-images of the corresponding structures in sequence space [52].

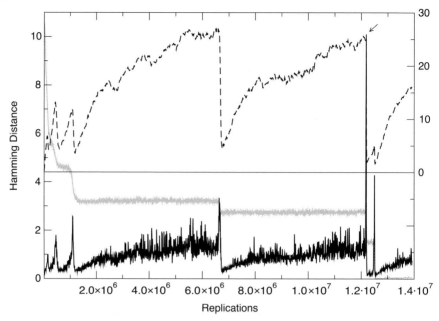

Fig. 2.10. Variability in genotype space during punctuated evolution. Shown are the results of a simulation of RNA optimization towards a tRNA target with population size N = 3000 and mutation rate p = 0.001 per site and replication event: (i) The trace of the underlying trajectory recording the average distance from target, $<d_r^s(t)>$ (gray, left ordinate scaled by 0.22, or full length is 50) and (ii) two plots of different measures of evolution in genotype space, the migration of the population center $<\delta d_c^h/\delta t>$ (with $\delta t = 8000$ replications) and the width of the population $<d_p^h>$, against time expressed as the total number of replications performed until time t. The upper plot is a measure of genotype diversity and shows the mean Hamming distance within the population ($<d_p^h>$, dotted line, right ordinate). The lower curve presents the Hamming distance between the centers of the population at times t and t+δt ($<\delta d_c^h/\delta t>$, full line, left ordinate) and measures the drift velocity of the population center. The arrow indicates a sharp peak of $<\delta d_c^h/\delta t>$ at the end of the second long plateau, which reaches a height of Hamming distance ten.

slowly. At the end of the quasi-stationary epoch, however, the velocity of the population center shows a sharp peak corresponding to a jump in sequence space. Major transitions lead to genotypes, which represent bottlenecks for evolutionary optimization.

Individual trajectories of evolution in the flow reactor are not reproducible in detail. Relay series of different computer runs under identical conditions[5] involve different structures and the corresponding genotypes have sequences that diverge from initial conditions. Almost all quantities, for example the number of replications required to reach the target or the number of minor transitions, show widely scattered distribu-

---

[5] Identical conditions here means that everything was chosen to be the same except the seeds of the random number generator.

tions. Population size effects on the evolutionary processes are pronounced. The number of replications increases with population size, a dramatic effect is seen with the number of minor transitions: It decreases by a factor of about four in the range between $N = 1000$ and $N = 100000$ molecules. The number of major transitions, however, shows only small scatter and is remarkably constant in this range of population sizes. Modeling of neutral evolution by means of a birth-and-death process provides a straightforward interpretation of this result: Minor transitions have a sufficiently high probability of occurrence such that frequent variants, once formed, stay in a larger population and do not reappear in further steps of the relay series. The low sensitivity of the numbers of major transitions to both population size and sequence of random events, however, makes them candidates for constants of evolution: They represent essential innovations and their number appears to depend only on initial and final state.

2.4
Sequence Structure Mappings of Proteins

In this section we do not aim at a presentation of the current state of the art in the design of proteins by variation and selection. This will be done in great detail in the other chapters of this volume. What we shall try to do instead is a comparison of results derived for proteins and RNA molecules to point out common features as well as differences.

The experimental results of selection and evolution of molecules derived here came mainly from investigations on RNA molecules and this simply because RNA is better suited for studies, since (i) RNA unites the properties of genotype and phenotype in one and the same molecule, and (ii) the bases in the base pairs of the stacking regions of RNA are complementary (AU, GC, and sometimes GU). These relations are fundamental for the simple logic for secondary structure formation, and have no counterpart in proteins. In addition, RNA secondary structures play almost always the role of an intermediate in the kinetic folding process and thus have a physical meaning. A third problem with the evolutionary design of proteins is the problem to link messengers and function carriers. This can be solved elegantly by the various "display techniques": phage, bacterial, ribosomal display and others. Another elegant method based on a covalent link between RNA and protein has been used in a paper discussed below [53]. Although variation – selection methods are available for proteins, they cannot compare successfully with the ease of selection procedures when both properties are contained in the same molecule like in the case of RNA.

Protein sequence space was postulated as a useful tool for discussing protein evolution already in 1970 [54]. Later on most extensive model studies were more or less confined to rather simple lattice models [55]. Systematic studies on random sequence model proteins [56] gave two important results: (i) more sequences than structures, and (ii) a few common folds compared to a great variety of rare folds. The second finding was also obtained by different stability considerations [57]. It is worth noticing that the frequency distribution of protein lattice fold is remarkably similar to that of RNA molecules with random sequences of the same chain length [40]. Shape space covering as observed with RNAs does not hold for lattice proteins [41].

Neutral networks [41] represent more or less the basic and most important feature of genotype-phenotype mappings. Although protein structure and function has been discussed with respect to neutrality for a very long time, direct evidence for neutrality and neutral networks came only recently from empirical potentials and neural network studies [58, 59]. Other investigations on protein "foldability landscapes" are in general agreement with the existence of extended neutral networks too [60, 61].

It is worth mentioning in this context that there seems to be a general difference between RNA and protein landscapes: Certain amino acid composition ratios between hydrophobic and hydrophilic amino acids presumably give rise to insoluble aggregates and this may lead to "holes" in protein sequence space. Perhaps, the concept of "holey adaptive landscapes" as favored in a series of recent papers on models of evolution [62] might be useful in this context.

Finally, two experimental results are highly relevant in this context: The first study on true random sequence proteins [53] revealed that the occurrence of function in protein sequence space has approximately the same probability, 10^{-12}, as in RNA sequence space. The second remarkable finding showed that very different structures of proteins, with no sequence homology, of course, gave rise to the binding affinity to AMP, the target molecule. More studies following along the line of this elegant experiment will provide the desired insight in protein sequence-structure mapping. The second experiment was done four years ago [63]: Two protein molecules with 50 % sequence homology have entirely different structures. A fully β-sheet structure was turned into an α-helix bundle by changing only half of the amino acid residues. Entirely different structures can be found at not too large Hamming distances in sequence space.

2.5
Concluding Remarks

What distinguishes the evolutionary strategy from conventional or rational design? The primary and most important issue is that we need not know the structure that yields the desired function. It is sufficient to derive an assay that allows for testing whether or not a candidate molecule has the desired property. At the current state of the art, *de novo* rational design of biopolymers gives very poor results and as long as this deficiency in structure prediction methods cannot be overcome, evolutionary search for function will be superior.

Variation and selection turns out to be an enormously potent tool for improvement also *in vitro*. Why this is so, does not trivially follow from the nature of random searches. The efficiency of Monte-Carlo methods may work very poorly as we know from other optimization problems. The intrinsic regularities of genotype-phenotype mappings with high degrees of neutrality and very wide scatter of the points in sequence space, which lead to the same or very similar solutions, are the clues to evolutionary success.

Acknowledgements

The work reported here was supported financially by the Austrian *Fond zur Förderung der wissenschaftlichen Forschung* (FWF), Projects P-13093-GEN, P-13887-MOB, and P-14898-MAT as well as by the *Jubiläumsfond der Österreichischen Nationalbank*, Project No.7813.

References

[1] K. Sander, *Biologie in unserer Zeit*, **1988**, 18, 161–167 (in German).

[2] G. de Beer, *Notes and Records of the Royal Society of London* **1964**, 19, 192–226.

[3] R. A. Fisher, *The genetical theory of natural selection*, Oxford University Press, Oxford (UK), **1930**.

[4] W. P. C. Stemmer, *Proc. Natl. Acad. Sci. USA*, **1994**, 91, 10747–10751.

[5] M. Kimura, *The neutral theory of molecular evolution*, Cambridge University Press, Cambridge (UK), **1983**.

[6] J. L. King, T. H. Jukes, *Science*, **1969**, 788–798.

[7] S. F. Elena, V. S. Cooper, R. E. Lenski, *Science* **1996**, 272, 1802–1804.

[8] D. Papadopoulos, D. Schneider, J. M. Meier-Eiss, W. Arber, R. E. Lenski, M. Blot, *Proc. Natl. Acad. Sci. USA*, **1999**, 96, 3807–3812.

[9] B. Derida, L. Peliti, *Bull. Math. Biol.*, **1991**, 53, 355–382.

[10] M. A. Huynen, P. F. Stadler, W. Fontana, *Proc. Natl. Acad. Sci. USA* **1996**, 93, 397–401.

[11] M. Eigen, W. C. Gardiner, *Pure Appl. Chem.* **1984**, 56, 967–978.

[12] S. Spiegelman, *Quart. Rev. Biophys.*, **1971**, 4, 213–253.

[13] H. F. Judson, *The eighth day of creation*, Jonathan Cape, London,**1979**.

[14] R. W. Hamming, *Coding and information theory*, 2nd ed., Prentice Hall, Englewood Cliffs, NJ, **1989**.

[15] P. F. Stadler, G. P. Wagner. *Evol. Comp.*, **1998**, 5, 241–275.

[16] M. Eigen, *Naturwissenschaften*, **1971**, 58, 465–523.

[17] E. Domingo, J. J. Holland, *Annu. Rev. Microbiol.*, **1997**, 51, 151–178.

[18] C. K. Biebricher, W. C. Gardiner, *Biophys. Chem.*, **1997**, 66, 179–192.

[19] M. Eigen, P. Schuster, *Naturwissenschaften*, **1977**, 64, 541–565.

[20] M. Eigen, J. McCaskill, P. Schuster, *Adv. Chem. Phys.*, **1989**, 75, 149–263.

[21] J. W. Drake, *Proc. Natl. Acad. Sci. USA*, **1991**, 88, 7160–7164.

[22] J. W. Drake, *Proc. Natl. Acad. Sci. USA*, **1993**, 90, 4171–4175.

[23] J. W. Drake, B. Charlesworth, D. Charlesworth, J. F. Crow. *Genetics*, **1998**, 148, 1667–1686

[24] P. Schuster, J. Swetina, *Bull. Math. Biol.*, **1988**, *50*, 635–660.

[25] C. L. Burch, L. Chao, *Nature*, **2000**, *406*, 625–628.

[26] C. O. Wilke, J. L. Wang, C. Ofria, R. E. Lenski, C. Adami, *Nature*, **2001**, *412*, 331–333.

[27] A. D. Ellington, J. W. Szostak, *Nature*, **1990**, *346*, 818–822.

[28] C. Tuerk, L. Gold, *Science*, **1990**, *249*, 505–510.

[29] A.Watts, G. Schwarz, *Biophys. Chem.*, **1997**, *66 (2/3)*, 67–284.

[30] D. S. Wilson, J. W. Szostak, *Ann. Rev. Biochem.*, **1999**, *68*, 611–147.

[31] L. Gold, C. Tuerk, P. Allen, J. Binkley, D. Brown, L. Green, S. MacDougal, D. Schneider, D. Tasset, S. R. Eddy. In: R. F. Gestland, J. F. Atkins, eds. *The RNA world*. Cold Spring Harbor Press, Plainview, NY, **1993**, pp. 497–509.

[32] A. A. Beaudry, G.F. Joyce, *Science*, **1992**, *257*, 635–641.

[33] R. R. Breaker, *Chem. Rev.*, **1997**, *97*, 371–390.

[34] N. Lehman, G. F. Joyce, *Current Biology*, **1993**, *3*, 723–734.

[35] D. P. Bartel, J. W. Szostak, *Science*, **1993**, *261*, 1411–1418.

[36] R. F. Gesteland, J. F. Atkins, eds. The RNA world. Cold Spring Harbor Press, Plainview, NY, **1993**.

[37] P. Schuster, *Biol. Chem.*, **2001**, *382*, in press.

[38] P. Higgs, *Quart. Rev. Biophys.*, **2000**, *33*, 199–253.

[39] P. Schuster, P. F. Stadler, In: M. J. C. Crabbe, M. Drew, A. Konopka, *Handbook of Computational Chemistry*, Marcel Dekker, New York, **2001**, in press.

[40] W. Grüner, R. Giegerich, D. Strothmann, C. Reidys, J. Weber, I. L. Hofacker, P. F. Stadler, P. Schuster, *Mh. Chem.*, **1996**, *127*, 355–389.

[41] P. Schuster, W. Fontana, P. F. Stadler, I. L. Hofacker, *Proc. Roy. Soc. London B*, **1994**, *255*, 279–284.

[42] C. Reidys, P. F. Stadler, P.Schuster, *Bull. Math. Biol.*, **1997**, *59*, 339–397.

[43] E. A. Schultes, D. P. Bartel, *Science*, **2000**, *289*, 448–452.

[44] E. H. Ekland, J. W. Szostak, D. P. Bartel, *Science*, **1995**, *269*, 364–370.

[45] A. T. Perotta, M. D. Been, *J. Mol. Biol.*, **1998**, *279*, 361–373.

[46] C. Flamm, W. Fontana, I. L. Hofacker, P. Schuster, *RNA*, **2000**, *6*, 325–338.

[47] C. Flamm, I. L. Hofacker, S. Maurer-Stroh, P. F. Stadler, M. Zehl, *RNA* **2001**, *7*, 254–265.

[48] W. Fontana, P. Schuster, *Biophys. Chem.*, **1987**, *26*, 123–147.

[49] W. Fontana, P. Schuster, *Science*, **1998**, *280*, 1451–1455.

[50] P. Schuster, W. Fontana, *Physica D*, **1999**, *133*, 427–452.

[51] P. Schuster, A. Wernitznig, *Is there a constant number of evolutionary innovations required to reach a given target?* Preprint, **2001**.

[52] B. M. Stadler, P. F. Stadler, G. P. Wagner, W. Fontana, *J. Theor. Biol.*, **2002**, in press.

[53] A. D. Keefe, J. W. Szostak, *Nature*, **2001**, *410*, 715–718.

[54] J. Maynard Smith, *Nature* **1970**, *225*, 563–564.

[55] K. Yue, K. M. Fiebig, P. D. Thomas, H. S. Chan, E. I. Shakhnovich, K. A. Dill, *Proc. Natl. Acad. Sci. USA*, **1993**, *90*, 1942–1946.

[56] H. Li, R. Helling, C. Tang, N. Wingreen, *Science*, **1996**, *273*, 666–669.

[57] S. Govindarajan, R. A. Goldstein, *Proc. Natl. Acad. Sci. USA*, **1996**, *93*, 3341–3345.

[58] A. Babajide, I. L. Hofacker, M. J. Sippl, P. F. Stadler, *Folding & Design*, **1997**, *2*, 261–269.

[59] A. Babajide, R. Farber, I. L. Hofacker, J. Inman, A. S. Lapedes, P. F. Stadler, *J. Theor. Biol*, **2001**, *212*, 35–40.

[60] S. Govindarajan, R. A. Goldstein, *Biopolymers*, **1997**, *42*, 427–438.

[61] S. Govindarajan, R. A. Goldstein, *Proteins*, **1997**, *29*, 461–466.

[62] S. Gavrilets, *Trends in Ecology and Evolution*, **1997**, *12*, 307–312.

[63] S. Dalal, S. Balasubramanian, L. Regan, *Nat. Struct. Biol.*, **1997**, *4*, 548–552.

3
Using Evolutionary Strategies to Investigate the Structure and Function of Chorismate Mutases[1)]

Donald Hilvert, Sean V. Taylor, and Peter Kast

3.1
Introduction

Evolution is the slow and continual process by which all living species diversify and become more complex. Through recursive cycles of mutation, selection and amplification, new traits accumulate in a population of organisms [1]. Those that provide an advantage under prevailing environmental conditions are passed from one generation to the next. Since ancient times, man has exploited evolution in a directed way to produce plants and animals with useful characteristics. Crossbreeding individuals with favorable traits successfully harnesses sexual recombination, one of the most powerful evolutionary strategies to generate new variants. From these crossings, progeny with improved features are chosen for additional breeding cycles, thus channeling the course of development.

Biologists and chemists have recently begun to use evolutionary strategies to study and tailor the properties of individual molecules rather than whole organisms. An array of methods has been developed to generate diversity in populations of molecules. Depending on the experiment, mutagenesis might entail degenerate oligonucleotide-directed or error-prone DNA synthesis [2, 3], shuffling of mutant DNA fragments [4, 5], or combinatorial syntheses of chemical compound libraries [6]. From the resulting molecular ensembles, desirable members must be identified by selection or screening procedures. In the amplification step, the self-replicating properties of the evolving molecules can be exploited (e.g. PCR amplification of nucleic acids). Alternatively, more of the desired compound can be prepared by large-scale chemical synthesis.

In practice, the sorting step is the most critical part of any laboratory evolution experiment. How can rare but useful variants be efficiently isolated from complex mix-

1) This article was adapted from a review that appeared in Angew. Chem. **2001**, *113*, 3408–3436; Angew. Chem Intl. Ed. **2001**, *40*, 3310–3335

tures of less desirable molecules? A chemist seeking the optimal octameric peptide inhibitor for a particular enzyme might start by making systematic substitutions at a single site in a known peptide. If allowed substitutions are restricted to the 20 standard amino acids, assaying the 20 different octameric peptides individually to find the highest affinity inhibitor would be relatively trivial. Simultaneously randomizing two positions in the peptide would generate 400 (20^2) different variants. Screening and testing such a library would be substantially more involved, but still manageable. However, it would be a daunting task indeed to identify the best inhibitor in a combinatorial library of fully randomized octameric peptides having 20^8 (2.6×10^{10}), individual members – unless an extremely efficient and sensitive sorting procedure were available.

Genetic selection is perhaps the most powerful technique currently available for analyzing and directing the evolution of large populations of biomacromolecules. In this chapter, we focus on its use as a tool in studying the structure and function of chorismate mutases. The reader is also referred to other more general reviews [7 – 17].

3.2
Selection versus Screening

Identifying interesting variants in large combinatorial libraries can be accomplished either by assaying all members individually or by applying conditions that allow only variants of interest to appear. Unfortunately, the terms "screening" for the former strategy and "selection" for the latter are often confounded in the literature. The basic difference between these approaches is illustrated schematically in Fig. 3.1.

In general, screening strategies require an active search of all variants in the library. This is depicted in panels A and B of Fig. 3.1 from the point of view of a bumblebee. Desirable nectar-producing flowers must be detected against a large background of

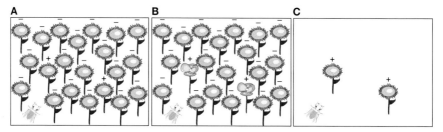

Fig. 3.1. Different search modes for finding nectar-producing flowers (marked with +). (A) Random screening. The bumblebee must check each flower individually to differentiate producers from the large number of non-producers (marked with −). (B) Facilitated screening. The bee can distinguish nectar-producing from non-producing flowers on the basis of their distinct phenotypes. (C) Selection. The bee only encounters the desired flowers because nectar production was a prerequisite for plant growth in this case.

flowers unable to produce nectar (undesired library members). If desired and undesired variants are difficult to distinguish, screening must be conducted one at a time in
"random" or "blind" fashion (Fig. 3.1A). A "facilitated screening" strategy is sometimes possible if the desired variants have a distinctive appearance or "phenotype"
(Fig. 3.1B). Nevertheless, irrelevant library members are always present, lowering
the signal-to-noise ratio and competing for resources.

In contrast, selection strategies exploit conditions favoring exclusive survival of desired variants, thereby mimicking the culling process associated with true Darwinian
evolution. Panel C of Fig. 3.1 shows how the bee easily encounters nectar-producing
flowers under selective conditions that eliminate the flowers that do not produce
nectar.

3.2.1
Classical solutions to the sorting problem

Selection and/or screening techniques are widely used in molecular biology, particularly for the cloning of DNA fragments or genes into circular DNA molecules called
vectors, which can be replicated inside bacteria by the cellular DNA synthesis machinery [18, 19]. Cloning involves enzymatic ligation of the ends of the insert fragment with
the corresponding complementary ends of an appropriately cut vector [20]. The resulting circular molecule, a plasmid, is then transferred into bacteria (usually the species
Escherichia coli) by a process called transformation. During transformation, only a
small percentage of cells successfully take up DNA. To select for these transformed
bacteria, the cell suspension is plated onto agar medium in petri dishes that contain a
particular antibiotic. Since the vector also carries a gene that specifies resistance
against this antibiotic, only those cells that have received the original vector or the
new plasmid survive, multiply, and form a colony on the agar medium. Because every
one of the approximately 10^7 cells in the colony contains identical DNA, they all belong
to the same clone.

After this very effective initial selection step, the molecular biologist is left with the
problem of finding those clones that have the desired plasmid rather than the original
vector molecule lacking the inserted gene [19]. Depending on the complexity of the
cloning strategy and the efficiency of the ligation step, relatively few of the antibio-
tic-resistant clones may actually possess the desired DNA fragment. The examples
in Fig. 3.2 show how an *in vivo* selection approach, employing a positive-selection
cloning vector, is superior to either random or facilitated screening in detecting clones
with the desired genetic makeup (or "genotype").

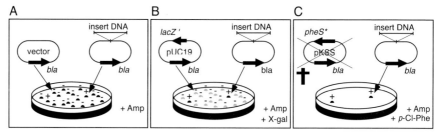

Fig. 3.2. Search strategies for differentiating colonies with plasmids containing an inserted gene (marked with +) from those with only vector DNA (marked with −) after plating on nutrient agar plates following transformation [20]. The *bla* gene encodes β-lactamase, which renders transformants resistant to the antibiotic ampicillin. (A) Random screening. The phenotype of all colonies is identical. DNA from many randomly picked clones must be isolated and analyzed to find the desired plasmid. (B) Facilitated screening. The cloning vector pUC19 [150] carries a fragment of the *lacZ* gene encoding β-galactosidase. The encoded polypeptide (the α-fragment of the enzyme) can associate with the product of a cellular *lacZ* fragment to form functional β-galactosidase [151], which cleaves 5-bromo-4-chloro-3-indolyl β-D-galactopyranoside (X-gal) resulting in the precipitation of a blue indigo derivative [152]. Vector-containing colonies consequently have a blue color on agar plates containing X-gal. In contrast, the desired plasmid-containing clones form col-

orless colonies because the cloned gene was inserted into the coding region for the α-fragment, thereby abolishing β-galactosidase activity. The clones with a white colony phenotype (shown here as black spots) must be picked from the background of blue colonies (shown here as gray spots) to find the correct genotype. (C) Selection. The positive-selection vector pKSS carries a mutant *phES* gene, which confers relaxed substrate specificity to the encoded phenylalanyl-tRNA synthetase α-subunit [153]. In the presence of the phenylalanine analog p-Cl-phenylalanine, bacteria expressing the mutant *phES* on pKSS cannot survive because incorporation of the analog into cellular protein causes general failure of protein function. However, transformed cells can form colonies on nutrient agar plates containing ampicillin and p-Cl-phenylalanine if the conditionally lethal *phES* on pKSS is destroyed by insertion of a cloned DNA-fragment. Thus, all colonies growing after plating the transformation mixture have the correct genotype.

3.2.2
Advantages and limitations of selection

The main advantage of selection over screening is that many more library members can be analyzed simultaneously. This is because uninteresting variants are never seen. As a consequence, surveying libraries is much faster and can be carried out with higher throughput. In the best screening protocols currently available, which take advantage of fluorogenic or chromogenic substrates [21], the maximum number of library members that can be assayed is about 10^5 [22]. In contrast, up to 10^{10} clones can be assessed in a single experiment using genetic selection *in vivo* in *E. coli* cells. This number is an upper limit based on transformation efficiencies obtainable with conventional protocols [23]. The absolute limit, at least in principle, is only dependent on the scale of transformation.

Nevertheless, it is challenging to develop suitable selection strategies for every desired catalytic task. In particular, *in vivo* selection schemes may be difficult or even impossible to devise if enzymes that work in non-natural environments, such as organic solvents, are desired [24, 25]. Coupling of the target reaction to survival in the selection step may require development of complex, non-trivial and intelligent assays. With every selection system, and in particular with very complex ones, there always exists the possibility that viable but unanticipated or undesired solutions to the posed survival problem will surface [26, 27]. If these false positives become too abundant, an efficient screening step may be necessary, or the system may need to be redesigned to eliminate this background by introduction of an additional selection step.

It is in the realm of very large combinatorial libraries that selection rather than screening gains crucial importance. As the focus shifts from randomizing an eight-residue peptide to a 100 amino acid protein (the typical size of a small functional domain, for example a chorismate mutase domain), the number of sequence permutations rises to an astronomical 20^{100}. The ability to assay even a tiny fraction of this sequence space in directed molecular evolution experiments demands selection, even though initial development of an appropriate system may be considerably more involved than the setup of a screening procedure.

3.3
Genetic Selection of Novel Chorismate Mutases

The utilization of evolutionary strategies in the laboratory can be illustrated with proteins that catalyze simple metabolic reactions. One of the simplest such reactions is the conversion of chorismate to prephenate (Fig. 3.3), a [3,3]-sigmatropic rearrangement. This transformation is a key step in the shikimate pathway leading to aromatic amino acids in plants and lower organisms [28, 29]. It is accelerated more than a million-fold by enzymes called chorismate mutases [30].

Chorismate mutases have been intensively studied because of the scarcity of other enzyme-catalyzed pericyclic reactions. Mutases from different organisms exhibit similar kinetic properties, though they may share little sequence similarity [30]. This dissimilarity extends to their tertiary and quaternary structures, as shown by comparison of chorismate mutases from *Bacillus subtilis* (BsCM) [31, 32], *E. coli* (EcCM) [33], and the yeast *Saccharomyces cerevisiae* (ScCM) [34, 35] (Fig. 3.4). BsCM, a member of the AroH class of chorismate mutases, is a symmetric homotrimer, packed as a pseudo-*a*/*β*-barrel. It has three identical active sites located at the subunit interfaces (Fig. 3.4A). In contrast, EcCM (Fig. 3.4B) and ScCM (Fig. 3.4C) are distantly related homodimeric members of the all-*a*-helical AroQ class of chorismate mutases [36, 37]. Homology searching in sequence databases has yielded other proteins that belong to each of these structural classes [36 – 38]. Catalytic antibodies with modest chorismate mutase activity have also been generated [39 – 41]. They have typical immunoglobulin folds, as shown in Fig. 3.4D for antibody 1F7 [42].

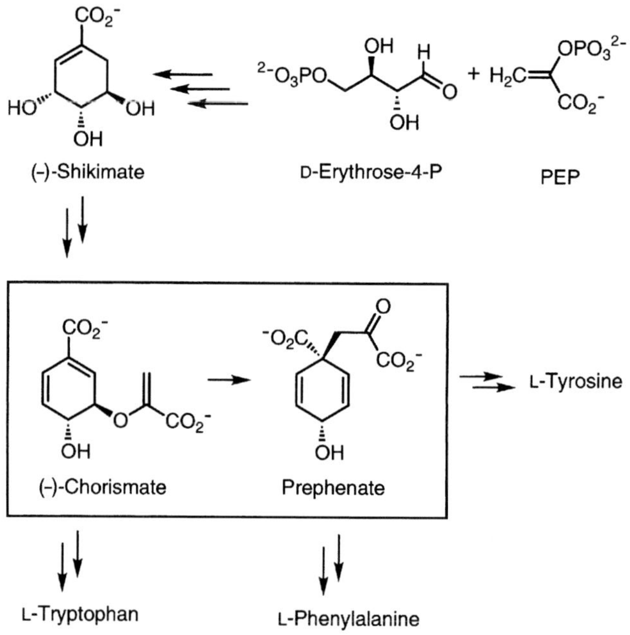

Fig. 3.3. Shikimate pathway for the biosynthesis of aromatic amino acids in plants and lower organisms. The [3,3]-sigmatropic rearrangement of chorismate into prephenate is shown in the box. PEP, phosphoenolpyruvate.

The architectural diversity that nature exploits to promote this relatively simple chemical transformation is impressive and raises many intriguing questions. How do the individual enzymes work? What similarities are present and what differences? Why are natural chorismate mutases 10^4 times more effective than the catalytic antibody? Can mechanistic insights be used to improve the latter? More generally, what are the underlying chemical determinants of each structure? The individual catalytic domains of the BsCM, EcCM and antibody proteins are each approximately 100 residues in length. Why does the polypeptide adopt a mixed α/β conformation in one case, an all-α-helical structure in another, and an all-β topology in the third? Using random mutagenesis and genetic selection *in vivo*, the relationship between sequence, structure and function can be rapidly and effectively examined.

A   **B**

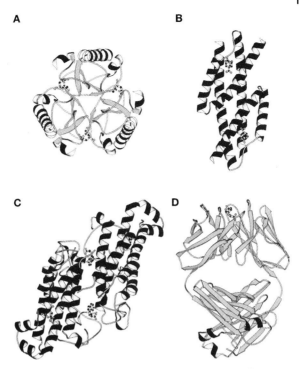

C   **D**

Fig. 3.4. Ribbon diagrams of three naturally occurring chorismate mutases and a catalytic antibody. Helices are shown in dark gray and β-sheets in light gray. A transition state analog inhibitor (**1**, see Fig. 3.6) is shown in ball-and-stick representation at each active site. (A) Monofunctional chorismate mutase from *B. subtilis* (BsCM) [31, 32]. (B) Chorismate mutase domain of the bifunctional chorismate mutase-prephenate dehydratase from *E. coli* (EcCM) [33]. (C) Allosterically regulated chorismate mutase from the yeast *S. cerevisiae* (ScCM) [35]. Tryptophan, which activates the enzyme, is shown bound at the allosteric sites located at the dimer interface. (D) Catalytic antibody 1F7 [42], which was generated using **1** as a hapten [39].

3.3.1
The selection system

Over the past decade, several strains of yeast [43, 44] and *E. coli* [45, 46] have been engineered that lack chorismate mutase. A typical bacterial selection system is depicted schematically in Fig. 3.5. It is based on *E. coli* strain KA12 [45], which has deletions of the chromosomal genes for both bifunctional chorismate mutases (chorismate mutase-prephenate dehydrogenase and chorismate mutase-prephenate dehydratase). Monofunctional versions of prephenate dehydratase [47] and prephenate dehydrogenase [48] from other organisms are supplied by the plasmid pKIMP-UAUC, leaving the cells deficient only in chorismate mutase activity [45].

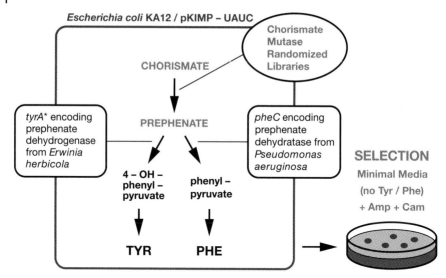

Fig. 3.5. An engineered *E. coli* selection system lacking endogenous chorismate mutase activity [45]. The genes encoding the bifunctional enzymes chorismate mutase-prephenate dehydrogenase and chorismate mutase-prephenate dehydratase were deleted, and monofunctional versions of the dehydrogenase and dehydratase were supplied on plasmid pKIMP-UAUC. Poten- tial chorismate mutases from random gene libraries are evaluated on the basis of their ability to complement the genetic defect and allow the cells to grow in the absence of added tyrosine (Tyr) and phenylalanine (Phe). The library plasmid and pKIMP-UAUC carry genes for ampicillin (Amp) and chloramphenicol (Cam) resistance, respectively.

As a consequence of the missing enzyme, KA12/pKIMP-UAUC bacteria are unable to grow in the absence of exogenously added tyrosine and phenylalanine. Their metabolic defect can be thought of as an interrupted circuit. Supplying the cells with an additional plasmid encoding a natural chorismate mutase can repair this genetic defect. Alternatively, a mutant of the natural enzyme or even an unrelated polypeptide can be introduced into the cell. If the protein is able to catalyze the conversion of chorismate to prephenate, the cells will be able to produce their own tyrosine and phenylalanine and grow under selective conditions (i.e. without tyrosine and phenylalanine supplementation). If the protein does not accelerate the rearrangement, the cells will not grow at all. As a consequence, attention can be focused on functional clones without wasting time and resources on non-productive members of the library. Additionally, the linkage of the desired phenotype (catalyst-enabled growth under selective conditions) with the responsible genotype (the gene encoding the catalyst) is automatic in this system, allowing easy isolation, diversification and further evolution of interesting library variants.

Because misplacement of catalytic residues by even a few tenths of an angstrom can mean the difference between full activity and none at all, enzymatic activity represents an extremely stringent criterion for selection. Given this, insightful answers to many of the questions posed in the previous section can be obtained by searching intelligently designed protein libraries for active chorismate mutases.

3.3.2
Mechanistic studies

3.3.2.1 Active site residues

Our understanding of the chorismate rearrangement derives largely from mechanistic studies of the uncatalyzed reaction [49 – 52] and from computation [53 – 59]. In order to rearrange, chorismate, which normally has a pseudo-diequatorial conformation in solution [51], must first adopt a pseudo-diaxial conformation (Fig. 3.6). It then reacts exergonically [60] via a transition state with a chairlike geometry [50]. The process appears to be concerted but asynchronous, with C–C bond formation lagging behind C–O bond cleavage [49].

The transition state for the enzymatic reaction has been shown to have a chairlike geometry as well [61], and conformationally constrained compounds that mimic this structure, such as the oxabicyclic dicarboxylic acid 1 (Fig. 3.6), are good inhibitors of chorismate mutase enzymes [62 – 64]. How a protein might stabilize this high-energy species has been a matter of some debate. Recently, heavy atom isotope effects were used to characterize the structure of the transition state bound to BsCM [65]. A very

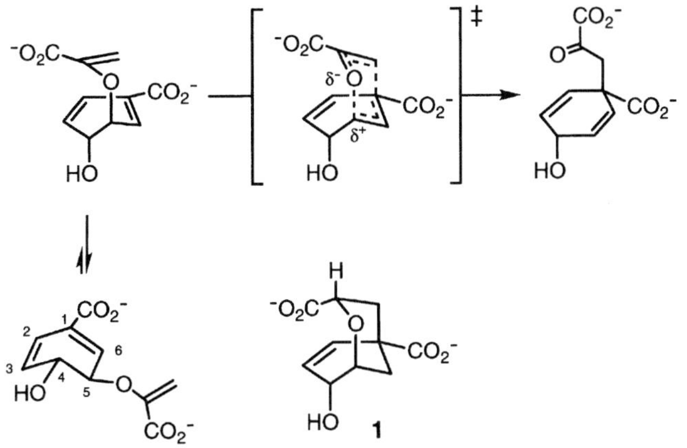

Fig. 3.6. Chorismate prefers a pseudodiequatorial conformation in solution. It must adopt a disfavored pseudodiaxial conformation to reach the pericyclic transition state. The conformationally constrained oxabicyclic dicarboxylic acid 1, which mimics the transition state, is a potent inhibitor of natural chorismate mutases [62]. Antibodies raised against this compound also catalyze the reaction, albeit 100 to 10,000-times less efficiently than their natural counterparts [39, 41].

large ^{18}O isotope effect at O(5) (ca. 5 %), the site of bond cleavage, and a small, normal ^{13}C isotope effect at C(1) (ca. 0.6 %), the site of bond formation, show that the rearrangement step is significantly rate determining for this enzyme. These findings also confirm that the enzymatic reaction proceeds through a concerted but asymmetric transition state. Comparison with theoretical isotope effects obtained by Becke3LYP/6–31G* calculations [65] suggests, however, that the enzymatic transition state might be more highly polarized than its solution counterpart, with more C–O bond cleavage and less C–C bond formation.

Examination of the residues that line the active site of BsCM (Fig. 3.7) [31, 32] suggests a plausible explanation for the greater polarization of the bound transition state. The positively charged guanidinium group of Arg90 is located within hydrogen-bonding distance of the ether oxygen of the bound substrate, where it can stabilize the partial negative charge that builds up at this site in the transition state. The contribution of such a residue to catalysis would normally be investigated by site-directed mutagenesis. For instance, Arg90 might be replaced with lysine or methionine to examine what happens when the guanidinium is replaced with a different cation or when the positive charge is removed completely. But because a selection system is available, all 20 natural amino acids can be evaluated at position 90 simultaneously. Those amino acids that yield a functional enzyme can be quickly identified by sequencing the genes able to complement the chorismate mutase deficiency. When this experiment was performed, the results were dramatic [45]. Only arginine at position 90 yielded an active enzyme. Even the conservative replacement of arginine with lysine gave an enzyme that was dead *in vivo*. Kinetic characterization of representative mutants showed that removal of the arginine side chain reduces catalytic efficiency by five to six orders of magnitude [38].

If the role of Arg90 is electrostatic stabilization of the developing negative charge on the ether oxygen of chorismate in the transition state, one might wonder why complementation was not observed when arginine was replaced with another positively charged residue like lysine. One possibility is that the Arg90Lys variant is produced poorly in *E.*

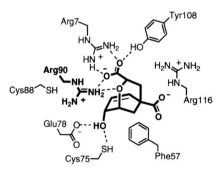

Fig. 3.7. Schematic representation of the BsCM active site. An extensive array of hydrogen bonding and electrostatic interactions is used to bind transition state analog **1** (shown in bold). The side chain of Arg90 (also in bold) is oriented so that its guanidinium group is placed within hydrogen-bonding distance of the ether oxygen of the ligand.

coli. The growth phenotype depends on the total amount of chorismate mutase activity present in the cell, which is determined by the specific activity of the catalyst and also its concentration. If the catalyst is active, but present only at extremely low concentrations, no growth will be observed. While the BsCM mutants appear to be produced at comparable levels within the selection system [45], exceptions are possible.

A potentially more interesting explanation is that the active site of the enzyme places intrinsic structural constraints on possible substitutions. The lysine side chain is shorter than that of arginine and it is conceivable that it cannot reach far enough into the active site to place its ε-ammonium group within hydrogen-bonding distance of the critical ether oxygen. If this is true, additional mutations at other positions within the active site might allow replacement of Arg90 without loss of function.

To test this idea, two residues within the binding pocket of BsCM were simultaneously mutated – Arg90 and Cys88 [45]. The side chain of Cys88 is about 7 Å distant from bound ligand, but it lies against the side chain of Arg90 and might influence the conformation of the latter (Fig. 3.7). Combinatorial mutagenesis of these two residues gives 400 variants, which can be rapidly evaluated in the chorismate mutase-deficient selection strain. The results of this experiment proved quite informative (Fig. 3.8) [45]. They showed first that Cys88 is not essential for catalysis. It can be replaced by large, medium-sized and small amino acids; as long as an arginine is present at position 90, an active enzyme is obtained. If a small residue (glycine or alanine, for example) replaces Cys88, active double mutants with an additional Arg90Lys mutation were found, showing that the guanidinium group is not crucial for catalysis. Perhaps the most interesting result to emerge from this study is that a cation at position 90 can be eliminated altogether, provided that Cys88 is replaced with a lysine.

Kinetic studies on the Cys88Lys/Arg90Ser double mutant show that lysine at position 88 restores three of the five to six orders of magnitude in catalytic efficiency lost upon removal of the arginine at position 90 [38]. Crystallographic studies of this double mutant show that the side chain of Lys88 extends into the active site and orients its ammonium group like the guanidinium group of Arg90 in the wild-type enzyme, within hydrogen-bonding distance of the ether oxygen of bound ligand (Fig. 3.9) [38]. Thus, combinatorial mutagenesis and selection resulted in a redesigned active site, affording a novel solution to the chemical problem of catalyzing the chorismate mutase reaction.

The results of these selection experiments are mechanistically significant insofar as they support a critical role for a cation in the mechanism of chorismate mutase. No active catalysts were found that lacked a cation in the vicinity of the substrate's ether oxygen. The conclusion that a cation is crucial for high chorismate mutase activity can thus be made with much greater confidence than would have been possible following a conventional mutagenesis experiment in which only single substitutions were considered [66]. Of course, we cannot exclude the possibility that other, equally effective

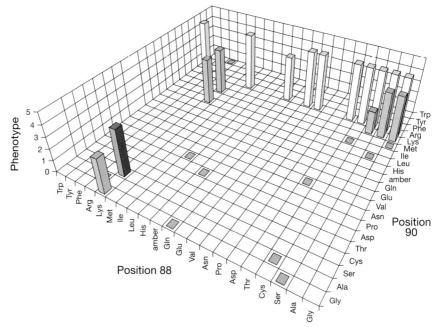

Fig. 3.8. Outcome of a combinatorial muta-genesis and selection experiment involving si-multaneous randomization of positions 88 and 90 in BsCM [45]. Amino acids are ordered according to sidechain volume. Active variants that were sequenced and a few inactive controls are shown with their *in vivo* activity, or phenotype. The phenotype is reflected by colony size, which is rated from 0 (no growth) to 5 (wild-type growth). White columns show variants with wild-type Arg90. The dark gray column is variant Cys88Lys/Arg90Ser, which was analyzed in more detail (see Fig. 3.9).

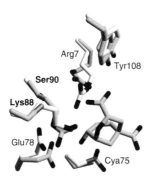

Fig. 3.9. Superposition of residues from unliganded C88K/R90S (carbon atoms in white, residues labeled) and wild-type BsCM complexed with inhibitor **1** (gray carbons) [38]. Cys75 is partially oxidized to the sulfinate (Cya75) in the crystal of the mutant. Nitrogen and oxygen atoms are shown in black, and sulfur atoms are shown in dark gray.

arrangements of catalytic groups lacking a cation might be found in the future through more extensive mutagenesis. However, it is worth noting that both E. coli [33] and yeast [34] chorismate mutases have a cationic lysine in an equivalent position in their respective active sites (Fig. 3.10), even though their tertiary structures are otherwise unrelated to BsCM. This lysine has been shown by mutagenesis to be catalytically essential in EcCM [67, 68].

In this context, it is also interesting that the 10^4-fold less efficient catalytic antibody 1F7 [39, 40] lacks a comparable strategically placed cation (Fig. 3.11) [42]. This fact is not terribly surprising given the structure of the transition state analog used to generate this catalyst (see Fig. 3.6) [39]. Nevertheless, it does suggest possible strategies for producing antibodies with much higher activity. For example, additional negative charges could be designed into a hapten to elicit new antibodies that contain the cat-

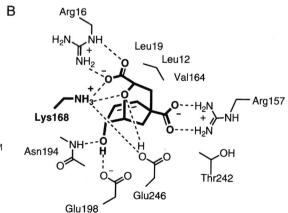

Fig. 3.10. Schematic representation of the EcCM (A) and ScCM (B) active sites complexed with transition state analog **1**. The transition state analog and the lysine counterparts of Arg90 in BsCM are shown in bold.

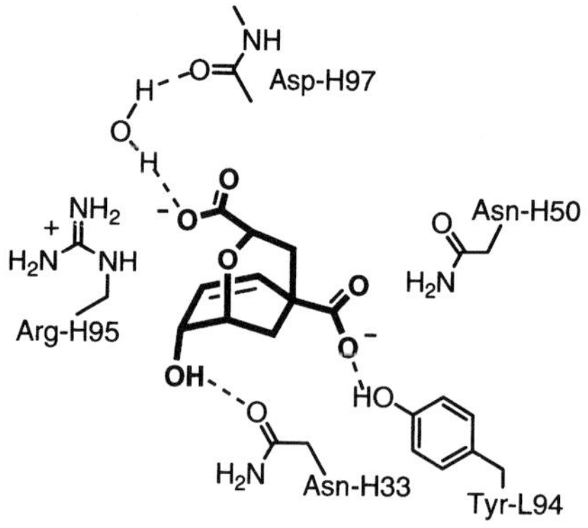

Fig. 3.11. Schematic representation of the active site of catalytic antibody 1F7 with bound **1**. ArgH95 is not within hydrogen-bonding distance of the ligand's ether oxygen [42].

alytically essential cation. Alternatively, a cation might be directly engineered into the 1F7 active site using site-directed mutagenesis. A third approach utilizing selection may ultimately be the most effective strategy for optimizing antibody activity, however.

The Fab fragment of 1F7 has already been shown to function in the cytoplasm of a chorismate mutase-deficient yeast strain [43, 44]. When produced at a sufficiently high level, the catalytic antibody is able to replace the missing enzyme and weakly complement the metabolic defect. Conceivably, therefore, it can be placed under selection pressure to identify variants that have higher catalytic efficiency. Preliminary results from such experiments appear quite promising [69].

3.3.2.2 Random protein truncation

When structural information for a protein is lacking, identifying appropriate residues for mutational analysis is difficult. In such cases, random mutagenesis coupled with selection can provide a powerful means of analysis. We have used such an approach to examine how the seventeen C-terminal residues of BsCM contribute to enzyme efficiency [70].

BsCM is 127 amino acids long. The first 115 residues adopt a well-defined structure, ending in a 3_{10} helix near the active site (Fig. 3.4A) [31, 32]. Residues beyond this point are not seen in X-ray structures of the unliganded enzyme [31], implying that they are highly disordered or have multiple accessible conformations. Upon ligand binding to BsCM, some ordering of the C-terminus is observed crystallographically [31]. In addition, ligand-induced conformation changes have been inferred from Fourier transform infrared (FTIR) spectroscopic studies [71]. These observations raise the possibility that the flexible C-terminal tail might play some role in enzyme function. For example, it might serve as a lid to the active site that sequesters the substrate from bulk solution and/or contributes stabilizing interactions to the transition state.

To address these questions, we developed a strategy involving random C-terminal truncation of the enzyme followed by selection of functional clones in the KA12/

pKIMP-UAUC system [70]. Using a PCR procedure with a partially randomized oligonucleotide, we constructed two gene libraries in which the BsCM codons corresponding to residues 116 – 127 or residues 111 – 127 (including the crystallographically observed 3_{10} helix) were randomized. The libraries were designed to optimize the frequency of stops at each mutagenized codon. In this way, a nested set of randomly truncated proteins was created. Since only functional genes survive selection, dispensable portions of the encoded proteins can be rapidly and directly identified. Non-native sequences are also explored by this method, so residues that might fulfill a crucial structural or mechanistic role can be discovered as well.

After construction and transformation into the bacterial selection strain, both libraries contained $10^4 – 10^5$ members. Essentially all members of the library in which the 12 C-terminal residues were mutagenized were viable on selection medium. These results demonstrate that none of the 12 C-terminal residues is absolutely required for enzymatic activity. They can be mutated and the last 11 residues can even be removed without significantly impairing activity *in vivo*. In contrast, when the last 17 residues were mutagenized, only ~25 % of the library was able to grow on selection plates. The additional five randomized amino acids form a well-defined 3_{10} helix and provide contacts with the rest of the protein and its ligands. Sequencing data show that mutations are tolerated at these positions, although a clear preference for wild-type Lys111, Ala112, Leu115 and Arg116 is observed. Moreover, active enzymes shorter than 116 amino acids were not found. Apparently, residues in the 3_{10} helix cannot be removed without killing the enzyme.

Kinetic characterization of several selected BsCM variants shows that truncation or mutation of the C-terminal tail has little effect on the turnover number (k_{cat}) of the enzyme (Tab. 3.1). When chorismate is bound to the active site of the variants, it is converted to prephenate nearly as efficiently as with wild-type BsCM. However, a substantial reduction in the k_{cat}/K_m value is evident (Tab. 3.1). This finding indicates that the C-terminus, while not directly involved in the chemical transformation of bound ligand, does contribute to enzymatic efficiency by uniform binding of substrate and transition state.

Tab. 3.1. Catalytic parameters of truncated BsCM variants [70]

BsCM variant	C-terminal amino acid sequence	k_{cat} (s^{-1})	k_{cat}/K_m (M^{-1} s^{-1})
5-8	KAVVLR	26	2.8×10^3
1-3	NSNVLRP	30	2.0×10^3
V-7	KAVVLLT	26	1.6×10^3
5-11	KAVVLRPN	23	5.6×10^3
wild-type	KAVVLRPDLSLTKNTEL	46	6.9×10^5

3.3.3
Structural studies

Selection methods are ideally suited for examining factors that influence protein structure and stability. The informational content of protein sequences is notoriously non-uniform, with many residues being highly tolerant to a wide range of substitutions, whereas others cannot be altered without dramatic consequences for folding or function [72, 73]. Protein structure analysis is thus a combinatorial problem requiring efficient methods for simultaneous evaluation of many different alternative sequences. An experimentally accessible population of 10^{10} molecules is miniscule compared with the 20^{100} ($= 10^{130}$) possibilities for a 100 amino acid long polypeptide. Nevertheless, libraries of this size can often provide statistically meaningful insights into intrinsic secondary structural preferences in proteins [74, 75], constraints on segments that link secondary structural elements [76 – 78], and optimal packing arrangements of residues in the interior of a protein [79 – 82].

3.3.3.1   Constraints on interhelical loops

The role of interhelical turns in determining protein structure has been explored in relatively simple four-helix-bundle proteins by generating combinatorial libraries and screening them for functional variants. For example, Hecht and coworkers mutagenized a three-residue interhelical turn in cytochrome *b-562* [76]. All 31 variants isolated from the random library adopted stable native-like structures as judged by the characteristic red color associated with heme complexes of cytochrome *b-562*. Similar results have been obtained with mutagenized turns in Rop [77, 83], a dimeric RNA-binding four-helix-bundle protein, although detailed analysis of individual mutants showed some variation in thermostability [84]. Collectively, these studies suggest that there are relatively few constraints on sequence or length in interhelical segments.

How general are such conclusions? Do they apply to more complex proteins? Are there more elusive features that influence turn preferences? Selection for catalytic activity, an extremely sensitive probe of structural integrity, has the potential to reveal more subtle sequence preferences than simple screening protocols. We tested this premise using the EcCM enzyme (Fig. 3.4B) as a template [85]. The two identical polypeptides that make up this helical bundle protein consist of three helices joined by two loop segments (Fig. 3.12A) [33]. The L2 loop connecting the H2 and H3 helices lies farthest from the active site and was targeted for analysis.

Initially, three solvent-exposed residues in the turn (Ala65 – His66 – His67, Fig. 3.12B) were randomized and catalytically active variants identified by selection *in vivo*. The starting ensemble contained 8000 (20^3) distinct members, of which more than 63 % were found to complement the chorismate mutase deficiency of the bacterial host strain, albeit with widely varying growth rates. The high percentage of com-

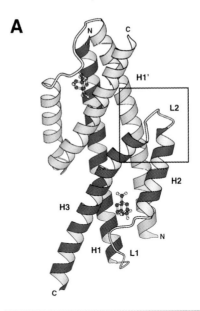

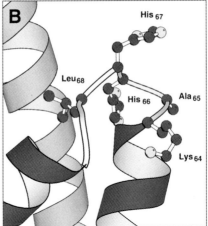

Fig. 3.12. Ribbon structure (A) and L2 loop (B) of EcCM. Random mutagenesis of the L2 loop followed by selection for chorismate mutase activity *in vivo* showed little sequence constraint on solvent exposed turn residues, aside from a modest bias in favor of hydrophilic amino acids [85]. In contrast, long-range tertiary contacts impose a strict requirement for hydrophobic aliphatic amino acids at position 68.

plementing clones clearly confirms that there are few restrictions on this loop segment. Nevertheless, the range of growth rates observed for complementing clones suggests that substitutions in the turn can affect either protein production, stability or catalytic activity. In fact, careful sequence comparison uncovered a statistically meaningful preference for hydrophilic residues at these solvent-exposed positions in the variants affording the fastest growth. Although proline-containing turns with alternative backbone conformations are also tolerated, such sequences were only found in partially active clones.

Randomization of the same Ala65–His66–His67 tripeptide in combination with Lys64, which is part of the adjacent H2 helix (Fig. 3.12), enabled investigation of the extent to which proximal secondary structure influences the allowable loop substitution patterns. A library containing all possible 160,000 (20^4) members was constructed and subjected to selection. Again, a large fraction of these sequences (>50 %) could functionally replace the native sequence, showing that the secondary structure at position 64 does not limit the range of side chains allowed at this site and that its randomization does not affect residue preferences in the neighboring turn region.

Quite different results were obtained when the buried turn residue Leu68 was randomized together with Ala65–His66–His67. Unlike the other turn residues, the side chain of Leu68 is buried at the interface between L2 and H1' (helix H1 of the other subunit) and thus sequestered from solvent (Fig. 3.12). In this case, only 6 – 7 % of the 160,000 possible sequences complemented the chorismate mutase deficiency, and sequence analysis indicated an extremely strict requirement for hydrophobic aliphatic amino acids at position 68. Thus, in contrast to the weak influence of a residue involved in a secondary structure element, long-range tertiary interactions between residues in an interhelical turn and residues elsewhere in the protein can impose strong constraints on allowable sequence substitutions.

The EcCM study points out the key advantages of genetic selection over screening methods. First, much larger libraries can be evaluated with little extra effort. Screening 10^5 individual clones in the absence of a convenient spectroscopic assay would be an arduous undertaking, but "survival of the fittest" selection readily sorts the entire mixture into active, partially active and inactive enzymes. Second, the greater stringency imposed by the requirement that selected variants be catalytically active allows nonobvious features to be discerned, such as the preference for hydrophilic residues at solvent-exposed positions, or the identification of turns with altered backbone conformations. These finely tuned details of protein structure might be missed with less stringent assay criteria.

3.3.4
Altering protein topology

All proteins are composed of a limited set of secondary structure elements. Yet, we are just beginning to learn how such building blocks can be combined to yield well-defined tertiary and quaternary structures. Evolutionary strategies that allow simultaneous evaluation of huge numbers of alternatives on the basis of a selectable phenotype, such as catalytic activity, can greatly aid such efforts.

3.3.4.1 **New quaternary structures**

Reengineering of existing protein scaffolds represents a manageable first step toward the larger goal of designing functional enzymes *de novo*. Alteration of the oligomerization state of various proteins, for instance converting monomers to multimers [86] or multimers to monomers [87 – 92], has received particular attention. Such systems may shed light on the evolutionary origins of multimeric proteins [93, 94].

Converting an oligomeric protein into a catalytically functional monomer is particularly difficult when the individual polypeptides must be untangled, as in the intricately entwined homodimeric EcCM [33]. Nevertheless, it seemed possible that insertion of a flexible "hinge-loop" into the long H1 helix that spans the EcCM dimer would alter the enzyme's quaternary structure. The important role of hinge-loops in determining the oligomerization state of a variety of proteins has been noted previously [93, 94]. Although the active sites of EcCM are constructed from residues contributed by both polypeptide chains, such a segment might allow the N-terminal half of H1 to bend back on itself, displacing the second polypeptide, to yield a monomeric and catalytically

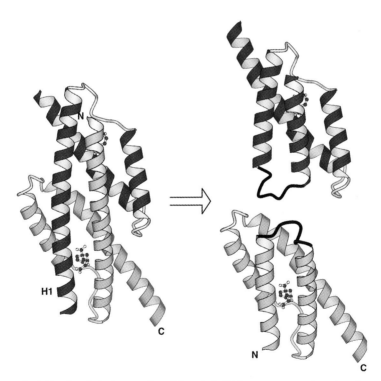

Fig. 3.13. Topological conversion of dimeric EcCM into a monomer. Insertion of a flexible loop (black) into the dimer-spanning H1 helix would allow the N-terminal portion of the helix to bend back on itself, displacing the other polypeptide to form a functional four-helix-bundle catalyst.

active four-helix-bundle (Fig. 3.13). Depending on loop length and composition, other oligomeric states could also conceivably be populated.

To test these possibilities, random segments of four to seven residues were inserted into the middle of the EcCM H1 helix [95]. The individual libraries (designated L4, L5, L6 and L7) have a maximum theoretical diversity of 160,000 (20^4), 3.2×10^6 (20^5), 6.4×10^7 (20^6), and 1.28×10^9 (20^7) distinct members, respectively. In each case, transformation of chorismate mutase-deficient bacteria yielded roughly 10^7 clones, giving fully diverse and redundant coverage of the L4 and L5 libraries, ~10 % sequence coverage of the L6 library, and ~1 % coverage of the L7 library.

No functional clones were found in the L6 library under selection conditions, but 0.05 %, 0.002 %, and 0.5 % of the L4, L5, and L7 libraries, respectively, complemented the genetic defect. Several active enzymes from the libraries were characterized biochemically (Fig. 3.14). Insertion of four or seven amino acid segments into the middle of the H1 helix yielded monomeric enzymes with near wild-type activity, but these proteins were unstable and tended to aggregate. In contrast, insertion of a five amino acid segment (Cys-Phe-Pro-Trp-Asp) yielded a well-behaved hexameric chorismate mutase (as judged by analytical ultracentrifugation) that is about 200-fold less active than the wild-type protein.

The properties of the hexameric species, when compared with those of the monomers, suggest that protein stability is an important driving force in the evolution of oligomeric proteins [95]. However, additional work is needed to elucidate how the subunits are organized and how quaternary structure influences function.

On a more general note, loop length appears to play a more important structural role than might have been anticipated from other experiments [83]. The low percentage of

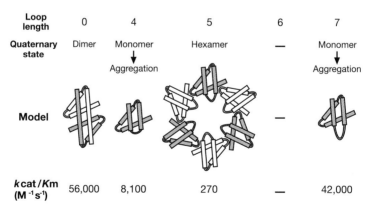

Loop length	0	4	5	6	7
Quaternary state	Dimer	Monomer ↓ Aggregation	Hexamer	—	Monomer ↓ Aggregation
Model				—	
k_{cat}/K_m (M^{-1}s^{-1})	56,000	8,100	270	—	42,000

Fig. 3.14. Models for the structural organization of hinge-loop variants of EcCM showing the number of residues in the inserted hinge-loop, the oligomeric state of active variants, and experimentally determined k_{cat}/K_m values [95].

active proteins recovered in these selection experiments indicates that relatively few loop sequences permit a change in quaternary structure without affecting active site structure. Moreover, the scarcity of functional clones (1 out of every 50,000 L5 library members, for example) also means that it would have been extremely difficult to find active enzymes by screening with an *in vitro* chorismate mutase assay alone.

3.3.4.2 Stable monomeric mutases

Although the topological conversion of the EcCM homodimer to a monomer preserves the enzyme active site, it yields a much less stable protein. A cluster of hydrophobic residues between the two active sites appears to stabilize the dimer [37], but these interactions would be lost in the monomer. Exposure of these buried apolar groups to solvent presumably favors aggregation as well. Because the highly charged binding pocket is located in the interior of the helical bundle [33], the monomer itself lacks a conventional hydrophobic core and, hence, the usual driving force for protein folding [96].

A thermostable dimer was therefore considered as an alternative starting point for the design of a stable monomeric mutase. A large number of EcCM sequence homologues, some from thermophilic organisms, are known. For example, the hyperthermophilic archaeon *Methanococcus jannaschii* produces a chorismate mutase (MjCM) that is 25 °C more stable than EcCM [37]. Despite only 21 % sequence identity, six prominent residues that line the active site are strictly conserved and the two enzymes have comparable activities. Since the hydrophobic core of MjCM is very similar to that of EcCM, interactions distant from the dimer interface must be responsible for its additional stability. These same interactions were expected to stabilize the desired monomer.

Topological variants of MjCM were prepared by inserting a flexible loop into the middle of the H1 helix [97], as described in the previous section for EcCM (see Fig. 3.13). In this case, though, the turn segment was modeled on a known helix-turn-helix motif in *E. coli* seryl-tRNA synthetase [98]. Two residues in helix H1 (Leu22 and Lys23, numbered according to the homologous EcCM sequence) were duplicated and six residues of random sequence were inserted between the repeated amino acids. Two point mutations (Leu22a to Glu and Ile79 to Arg) were also included as an element of negative design to disfavor dimer formation and to minimize aggregation of monomers by reducing the exposed hydrophobic surface area. This design yields a library with 6.4×10^7 (20^6) individual members. The library was constructed at the genetic level, used to transform the chorismate mutase-deficient *E. coli* strain ($>10^8$ transformants), and evaluated by selection.

In contrast to the selection experiments described above which were carried out on solid media, the library was grown in liquid culture lacking tyrosine and phenylalanine. This allows a direct competition for resources and amplification of the small

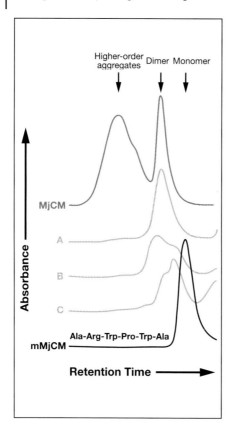

Fig. 3.15. Analytical size-exclusion chromatography distinguishes between dimeric and monomeric chorismate mutases [97]. Wild-type MjCM (top trace) was found to be a mixture of dimer and higher-order aggregates. Most selected variants (e.g. traces A, B and C) were dimers or mixtures of dimers and monomers. Only one of 26 variants tested (mMjCM, bottom trace) eluted as a monomer. It had the six amino acid insert shown. Analytical ultracentrifugation confirmed that this protein is monomeric in solution.

fraction of the original library containing a functional chorismate mutase (0.7 %). After three days of selection, >80 % of the clones in the library complemented the genetic defect, which corresponds to a >100-fold enrichment.

By eliminating inactive or weakly active clones, selection greatly facilitates the search for a tractable monomer. However, function, not topology, is the basis for selection. For this reason, representative clones were subsequently screened by size-exclusion chromatography. Only one of the 26 clones analyzed in this way had a retention time expected for a monomer (Fig. 3.15). This protein had the six-residue insert Ala-Arg-Trp-Pro-Trp-Ala. Its monomeric nature was confirmed by analytical ultracentrifugation. It is also highly helical, undergoes cooperative thermal and chemical denaturation [$\Delta G_U(H_2O)$ = 2.7 kcal mol^{-1}], and has significant catalytic activity. Its k_{cat} value for the rearrangement of chorismate is identical to that of MjCM, while its K_m value is elevated only three-fold [97].

The combination of selection (>100-fold enrichment) and screening (1 in 26) shows that fewer than 0.05 % of the possible turn sequences are capable of yielding well-behaved, monomeric proteins. This result again contradicts the simple expectation that most interhelical turn sequences are functionally equivalent [76]. It also underscores in dramatic fashion the advantage of using evolutionary approaches. Individual screening of 1000 – 10,000 proteins to find the one with the desired activity and topology would represent a major experimental undertaking.

Interestingly, cells harboring the monomer grow much more effectively in the absence of tyrosine and phenylalanine than cells containing the wild-type MjCM dimer [97]. Although the precise reasons for this difference are unknown, it may reflect the fact that the MjCM dimer forms catalytically inactive aggregates when overproduced in *E. coli*, whereas the monomer shows no tendency to do so (Fig. 3.15). Not only does the process of genetic selection maximize catalytic activity, but problems associated with protein misfolding, aggregation or toxicity are simultaneously minimized. This feature may be of value in the production of biocatalysts for commercially practical applications, where stable, easily produced enzymes are essential.

3.3.5
Augmenting weak enzyme activity

True Darwinian evolution involves multiple cycles of mutation and selection. This process can be mimicked in a laboratory setting to optimize the properties of an inefficient enzyme. The hexameric but weakly active chorismate mutase [95] described in Section 3.3.4.1 has been improved in this way [99]. Mutations were introduced into the gene encoding the hexamer subunit by DNA shuffling (Fig. 3.16) [5, 100], which mimics sexual recombination *in vitro*. Improved variants were selected, as before, by their ability to complement the chorismate mutase deficiency in bacteria. Plasmid DNA was isolated from the fastest growing cells and the entire procedure was repeated.

After two rounds of mutation and selection, several clones were found that grew at wild-type levels. The corresponding enzymes were isolated and characterized biochemically [99]. One variant, containing three mutations (Ser15Asp/Leu79Phe/Thr87Ile), had a nine-fold improvement in k_{cat} and a 35-fold improvement in k_{cat}/K_m. Size-exclusion chromatography also showed that it is still considerably larger than the wild-type dimer, but more detailed biophysical studies will be needed to determine its precise oligomerization state and subunit organization. This catalyst was subjected to a further round of mutagenesis and selection and additional clones with somewhat improved k_{cat}/K_m values have been identified. Interestingly, many of these are missing the C-terminal histidine tag, which had been introduced to facilitate purification, plus an additional five to eight amino acids. Eliminating the C-terminal extension of the H3 helix could conceivably improve packing interactions between the subunits in the higher order oligomer.

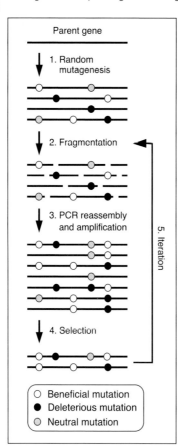

Fig. 3.16. DNA shuffling [5, 100]. Techniques such as error-prone PCR can be used to mutagenize a parent gene. The resulting pool of homologous genes contains beneficial, deleterious or neutral point mutations. DNaseI treatment of this pool gives rise to a set of random fragments. Novel recombinations accompany reassembly of the random fragments into full-length genes. Selection for variants with improved function provides the starting point for another round of shuffling. Point mutations arise during the recombination process itself, making every round a genuine Darwinian evolution cycle.

In principle, directed evolution procedures could be repeated indefinitely until any desired activity has been attained. At some point, however, the catalyst will be sufficiently active that the host cell grows like the wild-type strain, making selection for further improvement difficult. This is true for the modified hexamer even though it is still an order of magnitude less efficient than the homodimeric MjCM [37]. Since total activity depends on the catalyst concentration as well as specific activity, reducing the available catalyst concentration can further increase selection pressure. In practice, intracellular protein concentrations can be lowered in a variety of ways, including the use of low copy plasmids [101], weak promoters [102] and inefficient ribosome binding sites [103].

3.3.6
Protein design

The *de novo* design of *functional* proteins represents a far more ambitious goal than topological redesign of an existing enzyme. A number of small protein scaffolds have been successfully designed from first principles [104 – 110], often with the aid of computer algorithms [111, 112]. Nevertheless, conferring function to these molecules – the ability to recognize another molecule with high selectivity or to catalyze a chemical reaction – remains an enormous challenge.

In principle, the same evolutionary approaches that have been successfully used to characterize the mechanistic and structural determinants of enzymes and to alter their topology should aid the design of new protein catalysts. For example, one could imagine directly selecting active catalysts from large random protein libraries [113]. However, sequence space is infinitely vast. A library with the mass of the earth (5.98×10^{27} g) would contain at most 3.3×10^{47} different sequences. This number represents but a miniscule fraction of the total diversity available to even a small protein, yet far exceeds what is accessible to experimental investigation. Unless catalysts are present in ensembles of random variants at an unexpectedly high frequency, even the most sensitive and efficient selection procedure will be fruitless.

An attainable compromise between these two approaches might utilize basic structural information, such as the sequence preferences of helices and sheets or the tendency of hydrophobic residues to be buried in the protein interior, to intelligently design random sequence libraries from which functional catalysts can be selected. Binary patterning of polar and nonpolar amino acids [75, 114, 115], for example, is a potentially general strategy for protein design that exploits the nonrandom distribution of amino acids in folded structures (Fig. 3.17). It has been successfully exploited to create compact helical bundles that share many of the properties of native proteins, including protease resistance and cooperative folding [116 – 118].

To explore the feasibility of such an approach for the design of active catalysts, we have systematically replaced the secondary structural elements in the homodimeric helical bundle chorismate mutase (Fig. 3.18) with binary-patterned units of random sequence. Genetic selection was then used to assess the catalytic capabilities of the proteins in the resulting libraries, providing quantitative information about the robustness of this particular protein scaffold and insight into the subtle interactions needed to form a functional active site [119].

In the first stage of our experiment, modules that correspond to the H1 and H2/H3 helices were combined individually with the appropriate complementary wild-type helical segments of the thermostable MjCM', a protease-resistant version of the dimeric *M. jannaschii* chorismate mutase (Fig. 3.18) [37]. This amounted to randomizing 37 % and 42 % of the entire protein for the H1 and H2/H3 replacement modules,

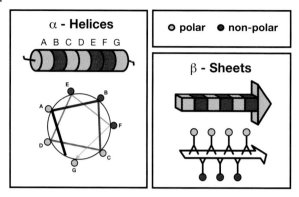

Fig. 3.17. Amphiphilic secondary structures. By specifying the locations of polar (light gray) and nonpolar (dark gray) residues
through simple binary patterns, amphiphilic helical or β-sheet
structures can be designed [154].

respectively. At each randomized position, only four polar (asparagine, aspartate, glu-
tamate and lysine) or four nonpolar (isoleucine, leucine, methionine and phenylalanine) residues were permitted. The loops and conserved catalytic residues were kept
constant in the initial design.

The chorismate mutase-deficient bacterial strain was transformed with the binary-
patterned module libraries and the ability of the partially randomized proteins to fold
into ordered, catalytically active structures was assessed by complementation of the
chorismate mutase deficiency. Roughly 0.02 % of the H1 library clones were active. In other words, about 1 out of 4500 binary-patterned enzymes is a viable catalyst. A somewhat smaller fraction of the H2/H3 library, 0.006 %, was found to complement the genetic deficiency (i.e. 1 out of 17,500 variants). Given the apparent abundance of catalysts, it would have been possible to find active enzymes even if the three
active site residues in each of the two modules (Arg11, Arg28 and Lys39 in H1, and
Arg51, Glu52 and Gln88 in H2/H3) had been additionally mutagenized in these libraries. Assuming that only a single amino acid is tolerated at each active site position,
the probability of finding a catalyst would only have decreased by a factor of 20^3. Thus,
active chorismate mutases would still have been isolable from experimentally accessible libraries containing as few as 10^8 members ($20^3 \times 10^4$). Binary patterning is
clearly useful in a directed evolution approach for catalyst design.

Catalysts from both the H1 and H2/H3 libraries were isolated and characterized
[119]. Despite the restricted set of building blocks used in the randomized segments, the enzymes are remarkably active, with modestly elevated K_m values and
k_{cat} values in the range 0.2 to 2.3 s^{-1}, which compare favorably with the value of
5.7 s^{-1} for wild-type MjCM [37]. While most positions in the protein are relatively
tolerant to substitution, sequence analysis has identified a few sites as quite restric-

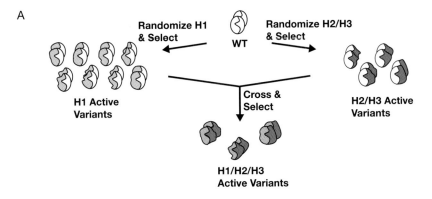

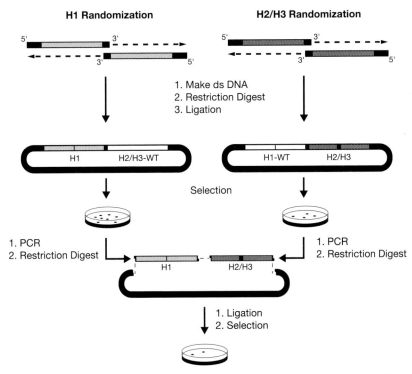

Fig. 3.18. Generation of large binary-patterned libraries of AroQ chorismate mutase genes. (A) A two-stage strategy was adopted involving separate randomization and selection of functional variants of the H1 (light gray) and H2/H3 helices (dark gray), followed by combination of functional binary-patterned segments from the initial libraries and reselection. In the final constructs, ~80 % of the protein was randomized. (B) The binary-patterned genes were constructed from synthetic random oligonucleotides corresponding to the segments encoding the H1 or H2/H3 helices and assembled in vector fragments (black) as shown. In the initial step, the randomized segments were combined with their wild-type counterparts H2/H3–WT and H1-WT (white). Selection was carried out using the system shown in Fig. 3.5. WT, wild-type MjCM'.

tive. For instance, all active clones have a strictly conserved Asn-Lys dyad at positions 84 and 85; such a preference is not evident in unselected clones. In retrospect, this finding can be rationalized by examination of the EcCM structure [33]. The homologue of Asn84 contributes a hydrogen bond to one of the carboxylate groups of bound ligand, and the methylene groups of the adjacent lysine probably make van der Waals interactions with the ligand. Apparently, none of the other amino acids allowed at these positions would have been able to make such interactions.

The preselected H1 and H2/H3 modules were combinatorially crossed in much the same way as immunoglobulin heavy and light chains are combined in the immune system to create diverse antibody structures [120]. In the resulting H1/H2/H3 binary-patterned libraries, ~80 % of the MjCM' residues were randomized and ›90 % of the protein was constructed from only eight amino acids (Fig. 3.18A). Interestingly, only about 1 out of every 10,000 possible combinations yields a functional catalyst, even though each of the preselected H1 and H2/H3 components is individually functional in a wild-type context. The native MjCM' H1 and H2/H3 helices are obviously much more effective than the selected segments in templating correctly folded, active enzymes, presumably because they are better able to tolerate destabilizing changes in the varied portions of the protein.

As observed for the individual H1 and H2/H3 libraries, certain positions are highly conserved in active H1/H2/H3 variants. In addition to the Asn84/Lys85 dyad seen previously, the majority of complementing clones contain the residues Ile14/Asp15/Asp18. These amino acids, which are not as highly conserved in the H1 library, probably help stabilize the active site. In EcCM, the residue at position 14 interacts directly with bound ligand through van der Waals contacts, whereas residues 15 and 18 are involved in second-sphere interactions with the catalytically important residues Arg51 and Arg28.

Biochemical characterization of one of the selected H1/H2/H3 variants confirms that it is a helical homodimer that undergoes cooperative thermal denaturation (T_m ~44 °C). It catalyzes the chorismate to prephenate rearrangement with a k_{cat} value only 15-fold lower than that of the *M. jannaschii* enzyme and a K_m value that is 40-fold higher [119]. By combining binary patterning with two evolutionary cycles, we have thus created a novel helix-bundle protein from a simplified set of building blocks that has biophysical and functional properties similar to those of natural chorismate mutases. Optimization of this and related catalysts from the library through additional rounds of directed evolution will be useful for identifying key interactions that influence folding, function and stability.

Experiments of this kind can provide valuable insights into the chemical determinants of protein structure. The results support speculations that ancient proteins might have been constructed from small numbers of amino acids [121 – 123]. Nevertheless, the relatively low abundance of active enzymes in the combinatorial libraries underscores the difficulty of generating helical scaffolds capable of catalysis – even when both the position and identity of all critical active site residues are specified in advance.

We estimate [119] that finding a catalyst in a library of binary patterned but fully randomized AroQ templates would have required a diversity of at least 10^{24} members. Experience suggests that other folds will be substantially more difficult to engineer [124]. Since libraries of this size are experimentally inaccessible, incremental approaches will undoubtedly be needed for the *de novo* design of tailored enzymes. Initially, a stable scaffold might be developed, binding and catalytic groups subsequently introduced, and the entire ensemble finally optimized for the desired activity. Indeed, recent studies have shown that ATP-binding polypeptides can be isolated from random libraries containing as few as $10^{11} - 10^{12}$ members [125]. An efficient selection system will be invaluable, if not absolutely essential, for extending the capabilities of such molecules to catalysis.

3.4
Summary and General Perspectives

Nature has solved the problem of protein design through the mechanism of Darwinian evolution. Every one of the proteins in our cells, from enzymes and receptors to structural proteins, has arisen by this process. As the examples presented here with chorismate mutase amply demonstrate, evolutionary strategies can also be successfully exploited in the laboratory to study the structure and function of existing proteins and to engineer new ones.

Site-directed mutagenesis, by which one amino acid in a protein is substituted for another, is an established and invaluable tool in protein chemistry. However, the size and complexity of proteins often precludes unambiguous interpretation of the results of such experiments. This is due to the non-uniform distribution of structural information within a protein sequence – the fact that some residues are tolerant to substitution, whereas others cannot be replaced without deleterious consequences – and to the many energetically similar conformational states available to any polypeptide. These factors, and the extreme sensitivity of catalytic activity to seemingly modest structural perturbation, make characterization and (re)design of enzymes in the laboratory difficult.

In contrast to methods involving single substitutions, combinatorial strategies are ideally suited for evaluating the complex network of interactions in proteins that ultimately determine form and function [73]. As shown in the mechanistic and structural investigations of chorismate mutases, degenerate protein ensembles can illuminate subtle trends that might be missed in analyses of individual amino acid replacements. This information can be used to verify and refine our understanding of protein structure and function. Evolutionary strategies can also be used in conjunction with data on structure and mechanism to support design efforts [8], as shown by the creation of novel monomeric and binary patterned chorismate mutases. The iterative nature of evolutionary approaches is particularly advantageous in this regard, allowing rapid optimization of the population of molecules under study toward some desirable goal. Even in the absence of detailed structural information or a clear understanding of how a system works, directed evolution can be used to tailor the properties of proteins for specific practical applications.

Of course, successful application of evolutionary methods requires an efficient method for sifting through the many library members to identify those that have desirable properties, be they recognition of a particular ligand or catalysis of an interesting chemical transformation. For small libraries, finding the useful variants is readily accomplished by screening. Even with large libraries, depending on the difficulty of the task at hand – and the distance the starting structure is from the desired endpoint in sequence space [126] – screening can often be exploited with great effectiveness [127 – 132]. Moreover, advances in automation and miniaturization can be expected to extend the utility of screening methods for high-throughput analysis of reasonably large molecular ensembles [10, 11, 133].

For problems where statistically meaningful correlations are sought or solutions are extraordinarily rare, selection is likely to be the method of choice. *In vivo* selection schemes allow exhaustive searches of libraries of 10^{10} protein variants, and significantly larger libraries can be handled with *in vitro* methods [134 – 140]. The principal challenge in developing an effective selection protocol is the linkage of the macromolecular property of interest to a biological readout, thereby allowing facile recovery of the responsible encoding gene. In principle, any metabolic process can be exploited for this purpose, as can transformations involving the creation or destruction of vital nutrients or toxins [141, 142]. Clever *in vitro* schemes further expand the range of possibilities for selecting molecules with interesting binding and catalytic properties [143–145].

Although *in vivo* directed evolution experiments are now well established, improvements in several areas could significantly enhance the utility of such approaches for diverse practical applications. Improved protocols for mutagenesis [146 – 148] and for transforming microorganisms [23, 149], for instance, might allow the creation of better-designed and still larger protein libraries. Selection schemes that are more versatile and general than those currently available, permitting selection for any desired function, would also clearly be of great value. Furthermore, we can dream of extending evolutionary principles to a wider range of molecules than nucleic acids and proteins. Realization of such an ambitious goal would require developing novel encoded combinatorial libraries as well as strategies for extracting desirable molecules and their blueprints from such ensembles. To that end, new methods for "autocatalytic" detection and selective amplification of the "best" molecules after each evolutionary cycle will be needed.

Evolutionary approaches will never supplant "rational" approaches to the study of molecules. Rather, they powerfully augment the classical strategies of dissection and design. In fact, directed evolution is likely to be most effective in combination with detailed structural and mechanistic information for exploring and engineering the properties of proteins, nucleic acids and other encoded molecules. In the post-genomic era, these methods are likely to become increasingly valuable in analyzing and integrating the flood of data that is emerging for so many biological systems. In the future, we can expect that intelligent application of evolutionary methods will help to clarify the links between sequence and structure, structure and function, and between interacting macromolecules in the cell. Novel applications in medicine and industry will not be far behind.

Acknowledgement

The authors wish to thank Kinya Hotta and Gavin MacBeath for assistance in the preparation of graphics. The financial support of the ETH Zürich, the Schweizerischer Nationalfonds, and Novartis Pharma are gratefully acknowledged.

References

[1] R. Dawkins, *The Blind Watchmaker*, Longham, London, **1986**.
[2] M. K. Trower, in *Methods Mol. Biol.*, Humana Press, Totowa, NJ, **1996**, p. 390.
[3] R. C. Cadwell, G. F. Joyce, *PCR Meth. Appl.* **1992**, *2*, 28.
[4] W. P. C. Stemmer, *Bio/Technology* **1995**, *13*, 549.
[5] W. P. C. Stemmer, *Proc. Natl. Acad. Sci. USA* **1994**, *91*, 10747.
[6] J. N. Abelson, in *Methods Enzymol.*, Vol. 267, Academic Press, New York, NY, **1996**.
[7] P. Kast, D. Hilvert, *Pure Appl. Chem.* **1996**, *68*, 2017.
[8] P. Kast, D. Hilvert, *Curr. Opin. Struct. Biol.* **1997**, *7*, 470.
[9] R. T. Sauer, *Fold. Des.* **1996**, *1*, R27.
[10] F. H. Arnold, *Acc. Chem. Res.* **1998**, *31*, 125.
[11] F. H. Arnold, A. A. Volkov, *Curr. Opin. Chem. Biol.* **1999**, *3*, 54.
[12] D. C. Demirjian, P. C. Shah, F. Moris-Varas, *Topics Curr. Chem.* **1999**, *200*, 1.

[13] U. Kettling, A. Koltermann, M. Eigen, *Topics Curr. Microb. Immun.* **1999**, *243*, 173.
[14] A. Koltermann, U. Kettling, *Biophys. Chem.* **1997**, *66*, 159.
[15] L. P. Encell, D. M. Landis, L. A. Loeb, *Nat. Biotechnol.* **1999**, *17*, 143.
[16] A. Skandalis, L. P. Encell, L. A. Loeb, *Chem. Biol.* **1997**, *4*, 889.
[17] J. Fastrez, *Mol. Biotechnol.* **1997**, *7*, 37.
[18] B. Lewin, *Genes VII*, Oxford University Press, London, **2000**.
[19] J. D. Watson, M. Gilman, J. Witkowski, M. Zoller, *Recombinant DNA: A short course*, 2 ed., W. H. Freeman, San Francisco, **1992**.
[20] J. Sambrook, D.W. Russell, *Molecular Cloning. A Laboratory Manual*, 3rd ed., Cold Spring Harbor Laboratory Press, Cold Spring Harbor, New York, **2001**.
[21] P. B. Fernandes, *Curr. Opin. Chem. Biol.* **1998**, *2*, 597.
[22] A. Qureshi-Emili, G. Cagney, *Nat. Biotechnol.* **2000**, *18*, 393.

[23] D. Hanahan, J. Jessee, F. R. Bloom, *Methods Enzymol.* **1991**, *204*, 63.

[24] L. You, F. H. Arnold, *Protein Eng.* **1996**, *9*, 77.

[25] F. H. Arnold, J. C. Moore, *Adv. Biochem. Eng. Biotechnol.* **1997**, *58*, 1.

[26] C. M. Gates, W. P. C. Stemmer, R. Kaptein, P. J. Schatz, *J. Mol. Biol.* **1996**, *255*, 373.

[27] R. R. Breaker, A. Banerji, G. F. Joyce, *Biochemistry* **1994**, *33*, 11980.

[28] U. Weiss, J. M. Edwards, *The Biosynthesis of Aromatic Amino Compounds*, Wiley, New York, **1980**.

[29] E. Haslam, *Shikimic Acid: Metabolism and Metabolites*, Wiley, New York, **1993**.

[30] B. Ganem, in *Comprehensive Natural Products Chemistry, Vol. 5: Enzymes, Enzyme Mechanisms, Proteins and Aspects of NO Chemistry* (Ed.: C. D. Poulter), Elsevier, Oxford, UK, **1999**, pp. 343.

[31] Y. M. Chook, J. V. Gray, H. Ke, W. N. Lipscomb, *J. Mol. Biol.* **1994**, *240*, 476.

[32] Y. M. Chook, H. Ke, W. N. Lipscomb, *Proc. Natl. Acad. Sci. USA* **1993**, *90*, 8600.

[33] A. Y. Lee, P. A. Karplus, B. Ganem, J. Clardy, *J. Am. Chem. Soc.* **1995**, *117*, 3627.

[34] N. Sträter, G. Schnappauf, G. Braus, W. N. Lipscomb, *Structure* **1997**, *5*, 1437.

[35] Y. Xue, W. N. Lipscomb, R. Graf, G. Schnappauf, G. Braus, *Proc. Natl. Acad. Sci. USA* **1994**, *91*, 10814.

[36] W. Gu, D. S. Williams, H. C. Aldrich, G. Xie, D. W. Gabriel, R. A. Jensen, *Microb. Comp. Genomics* **1997**, *2*, 141.

[37] G. MacBeath, P. Kast, D. Hilvert, *Biochemistry* **1998**, *37*, 10062.

[38] P. Kast, C. Grisostomi, I. A. Chen, S. Li, U. Krengel, Y. Xue, D. Hilvert, *J. Biol. Chem.* **2000**, *275*, 36832.

[39] D. Hilvert, S. H. Carpenter, K. D. Nared, M.-T. M. Auditor, *Proc. Natl. Acad. Sci. USA* **1988**, *85*, 4953.

[40] D. Hilvert, K. D. Nared, *J. Am. Chem. Soc.* **1988**, *110*, 5593.

[41] D. Y. Jackson, J. W. Jacobs, R. Sugasawara, S. H. Reich, P. A. Bartlett, P. G. Schultz, *J. Am. Chem. Soc.* **1988**, *110*, 4841.

[42] M. R. Haynes, E. A. Stura, D. Hilvert, I. A. Wilson, *Science* **1994**, *263*, 646.

[43] K. Bowdish, Y. Tang, J. B. Hicks, D. Hilvert, *J. Biol. Chem.* **1991**, *266*, 11901.

[44] Y. Tang, J. B. Hicks, D. Hilvert, *Proc. Natl. Acad. Sci. USA* **1991**, *88*, 8784.

[45] P. Kast, M. Asif-Ullah, N. Jiang, D. Hilvert, *Proc. Natl. Acad. Sci. USA* **1996**, *93*, 5043.

[46] P. Kast, M. Asif-Ullah, D. Hilvert, *Tetrahedron Lett.* **1996**, *37*, 2691.

[47] G. Zhao, T. Xia, R. S. Fischer, R. A. Jensen, *J. Biol. Chem.* **1992**, *267*, 2487.

[48] T. Xia, G. Zhao, R. S. Fischer, R. A. Jensen, *J. Gen. Microbiol.* **1992**, *138*, 1309.

[49] L. Addadi, E. K. Jaffe, J. R. Knowles, *Biochemistry* **1983**, *22*, 4494.

[50] S. D. Copley, J. R. Knowles, *J. Am. Chem. Soc.* **1985**, *107*, 5306.

[51] S. D. Copley, J. R. Knowles, *J. Am. Chem. Soc.* **1987**, *109*, 5008.

[52] J. J. Gajewski, J. Jurayj, D. R. Kimbrough, M. E. Gande, B. Ganem, B. K. Carpenter, *J. Am. Chem. Soc.* **1987**, *109*, 1170.

[53] O. Wiest, K. N. Houk, *J. Org. Chem.* **1994**, *59*, 7582.

[54] O. Wiest, K. N. Houk, *J. Am. Chem. Soc.* **1995**, *117*, 11628.

[55] P. D. Lyne, A. J. Mulholland, W. G. Richards, *J. Am. Chem. Soc.* **1995**, *117*, 11345.

[56] M. M. Davidson, I. R. Gould, I. H. Hillier, *J. Chem. Soc. Perkin Trans.* **1996**, *2*, 525.

[57] M. M. Davidson, I. H. Hillier, *J. Chem. Soc. Perkin Trans.* **1994**, *2*, 1415.

[58] D. L. Severence, W. L. Jorgensen, *J. Am. Chem. Soc.* **1992**, *114*, 10966.

[59] H. A. Carlson, W. L. Jorgensen, *J. Am. Chem. Soc.* **1996**, *118*, 8475.

[60] P. Kast, Y. B. Tewari, O. Wiest, D. Hilvert, K. N. Houk, R. N. Goldberg, *J. Phys. Chem. B* **1997**, *101*, 10976.

[61] S. G. Sogo, T. S. Widlanski, J. H. Hoare, C. E. Grimshaw, G. A. Berchtold, J. R. Knowles, *J. Am. Chem. Soc.* **1984**, *106*, 2701.

[62] P. A. Bartlett, C. R. Johnson, *J. Am. Chem. Soc.* **1985**, *107*, 7792.

[63] P. A. Bartlett, Y. Nakagawa, C. R. Johnson, S. H. Reich, A. Luis, *J. Org. Chem.* **1988**, *53*, 3195.

[64] J. V. Gray, D. Eren, J. R. Knowles, *Biochemistry* **1990**, *29*, 8872.

[65] D. J. Gustin, P. Mattei, P. Kast, O. Wiest, L. Lee, W. W. Cleland, D. Hilvert, *J. Am. Chem. Soc.* **1999**, *121*, 1756.

[66] S. T. Cload, D. R. Liu, R. M. Pastor, P. G. Schultz, *J. Am. Chem. Soc.* **1996**, *118*, 1787.

[67] D. R. Liu, S. T. Cload, R. M. Pastor, P. G. Schultz, *J. Am. Chem. Soc.* **1996**, *118*, 1789.

[68] S. Zhang, P. Kongsaeree, J. Clardy, D. B. Wilson, B. Ganem, *Bioorg. Med. Chem.* **1996**, *4*, 1015.

[69] Y. Tang, Ph.D. thesis, The Scripps Research Institute **1996**.

[70] M. Gamper, D. Hilvert, P. Kast, *Biochemistry* **2000**, *46*, 14087.

[71] J. V. Gray, J. R. Knowles, *Biochemistry* **1994**, *33*, 9953.

[72] M. H. J. Cordes, A. R. Davidson, R. T. Sauer, *Curr. Opin. Struct. Biol.* **1996**, *6*, 3.

[73] J. F. Reidhaar-Olson, R. T. Sauer, *Science* **1988**, *241*, 53.

[74] M. Blaber, X. Zhang, B. W. Matthews, *Science* **1993**, *260*, 1637.

[75] H. Xiong, B. L. Buckwalter, H. M. Shieh, M. H. Hecht, *Proc. Natl. Acad. Sci. USA* **1995**, *92*, 6349.

[76] A. P. Brunet, E. S. Huang, M. E. Huffine, J. E. Loeb, R. J. Weltman, M. H. Hecht, *Nature* **1993**, *364*, 355.

[77] L. Castagnoli, C. Vetriani, G. Cesareni, *J. Mol. Biol.* **1994**, *237*, 378.

[78] J. A. Ybe, M. H. Hecht, *Protein Sci.* **1996**, *5*, 814.

[79] W. A. Lim, R. T. Sauer, *Nature* **1989**, *339*, 31.

[80] M. E. Milla, R. T. Sauer, *Biochemistry* **1995**, *34*, 3344.

[81] E. P. Baldwin, O. Hajiseyedjavadi, W. A. Baase, B. W. Matthews, *Science* **1993**, *262*, 1715.

[82] Z. L. Fredericks, G. J. Pielak, *Biochemistry* **1993**, *32*, 929.

[83] P. F. Predki, L. Regan, *Biochemistry* **1995**, *34*, 9834.

[84] P. F. Predki, V. Agrawal, A. T. Brünger, L. Regan, *Nat. Struct. Biol.* **1996**, *3*, 54.

[85] G. MacBeath, P. Kast, D. Hilvert, *Protein Sci.* **1998**, *7*, 325.

[86] S. M. Green, A. G. Gittis, A. K. Meeker, E. E. Lattman, *Nat. Struct. Biol.* **1995**, *2*, 746.

[87] R. R. Dickason, D. P. Huston, *Nature* **1996**, *379*, 652.

[88] R. A. Albright, M. C. Mossing, B. W. Matthews, *Biochemistry* **1996**, *35*, 735.

[89] T. V. Borchert, R. Abagyan, R. Jaenicke, R. K. Wierenga, *Proc. Natl. Acad. Sci. USA* **1994**, *91*, 1515.

[90] M. C. Mossing, R. T. Sauer, *Science* **1990**, *250*, 1712.

[91] W. Schliebs, N. Thanki, R. Jaenicke, R. K. Wierenga, *Biochemistry* **1997**, *36*, 9655.

[92] X. Shao, P. Hensley, C. R. Matthews, *Biochemistry* **1997**, *36*, 9941.

[93] M. J. Bennett, S. Choe, D. Eisenberg, *Proc. Natl. Acad. Sci. USA* **1994**, *91*, 3127.

[94] M. J. Bennett, M. P. Schlunegger, D. Eisenberg, *Protein Sci.* **1995**, *4*, 2455.

[95] G. MacBeath, P. Kast, D. Hilvert, *Protein Sci.* **1998**, *7*, 1757.

[96] L. Lins, R. Brasseur, *FASEB J.* **1995**, *9*, 535.

[97] G. MacBeath, P. Kast, D. Hilvert, *Science* **1998**, *279*, 1958.

[98] M. G. Oakley, P. S. Kim, *Biochemistry* **1997**, *36*, 2544.

[99] K. U. Walter, S. V. Taylor, G. Mäder, P. Kast, D. Hilvert, unpublished.

[100] W. P. C. Stemmer, *Nature* **1994**, *370*, 389.

[101] M. Couturier, F. Bex, W. K. Maas, *Microb. Rev.* **1988**, *52*, 375.

[102] T. A. Y. Ayoubi, W. J. M. VanDeVen, *FASEB J.* **1996**, *10*, 453.

[103] H. A. DeBoer, A. S. Hui, *Methods Enzymol.* **1990**, *185*, 103.

[104] G. R. Dieckmann, D. K. McRorie, D. L. Tierney, L. M. Utschig, C. P. Singer, T. V. O'Halloran, J. E. Penner-Hahn, W. F. DeGrado, V. L. Pecoraro, *J. Am. Chem. Soc.* **1997**, *119*, 6195.

[105] S. F. Betz, W. F. DeGrado, *Biochemistry* **1996**, *35*, 6955.

[106] S. Dalal, S. Balasubramanian, L. Regan, *Nat. Struct. Biol.* **1997**, *4*, 548.

[107] J. E. Rozzelle, A. Tropsha, B. W. Erickson, *Protein Sci.* **1994**, *3*, 345.

[108] K. M. Gernert, M. C. Surles, T. H. Labean, J. S. Richardson, D. C. Richardson, *Protein Sci.* **1995**, *4*, 2252.

[109] P. B. Harbury, J. J. Plecs, B. Tidor, T. Alber, P. S. Kim, *Science* **1998**, *282*, 1462.

[110] X. Jiang, E. J. Bishop, R. S. Farid, *J. Am. Chem. Soc.* **1997**, *119*, 838.

[111] B. I. Dahiyat, S. L. Mayo, *Science* **1997**, *278*, 82.

[112] S. M. Malakauskas, S. L. Mayo, *Nat. Struct. Biol.* **1998**, *5*, 470.

[113] A. R. Davidson, R. T. Sauer, *Proc. Natl. Acad. Sci. USA* **1994**, *91*, 2146.

[114] M. W. West, M. H. Hecht, *Protein Sci.* **1995**, *4*, 2032.

[115] J. U. Bowie, R. T. Sauer, *Proc. Natl. Acad. Sci. USA* **1989**, *86*, 2152.

[116] S. Kamtekar, J. M. Schiffer, H. Xiong, J. M. Babik, M. H. Hecht, *Science* **1993**, *262*, 1680.

[117] S. Roy, G. Ratnaswamy, J. A. Boice, R. Fairman, G. McLendon, M. H. Hecht, *J. Am. Chem. Soc.* **1997**, *119*, 5302.

[118] S. Roy, M. H. Hecht, *Biochemistry* **2000**, *39*, 4603.

[119] S. V. Taylor, K. U. Walter, P. Kast, D. Hilvert, *Proc. Natl. Acad. Sci. USA* **2001**, *98*, 10596.

[120] E. A. Kabat, *Structural Concepts in Immunology and Immunochemistry*, Holt, Rinehart, and Winston, New York, **1976**.

[121] J. T.-F. Wong, *Proc. Natl. Acad. Sci. USA* **1975**, *72*, 1909.

[122] C. Woese, *The Genetic Code*, Harper & Row, New York, **1967**.

[123] F. H. C. Crick, *J. Mol. Biol.* **1968**, *38*, 367.

[124] W. F. DeGrado, C. M. Summa, V. Pavone, F. Nastri, A. Lombardi, *Annu. Rev. Biochem.* **1999**, *68*, 779.

[125] A. D. Keefe, J. W. Szostak, *Nature* **2001**, *410*, 715.

[126] J. C. Moore, H.-M. Jin, O. Kuchner, F. H. Arnold, *J. Mol. Biol.* **1997**, *272*, 336.

[127] U. T. Bornscheuer, J. Altenbuchner, H. H. Meyer, *Biotechnol. Bioeng.* **1998**, *58*, 554.

[128] M. T. Reetz, A. Zonta, K. Schimossek, K. Liebeton, K.-E. Jaeger, *Angew. Chem. Int. Ed. Engl.* **1997**, *36*, 2830.

[129] L. Giver, A. Gershenson, P.-O. Freskgard, F. H. Arnold, *Proc. Natl. Acad. Sci. USA* **1998**, *95*, 12809.

[130] H. Zhao, F. H. Arnold, *Protein Eng.* **1999**, *12*, 47.

[131] H. Joo, Z. Lin, F. H. Arnold, *Nature* **1999**, *399*, 670.

[132] T. Kumamaru, H. Suenaga, M. Mitsuoka, T. Watanabe, K. Furukawa, *Nat. Biotechnol.* **1998**, *16*, 663.

[133] M. J. Olsen, D. Stephens, D. Griffiths, P. Daugherty, G. Georgiou, B. L. Iverson, *Nat. Biotechnol.* **2000**, *18*, 1071.

[134] L. C. Mattheakis, R. R. Bhatt, W. J. Dower, *Proc. Natl. Acad. Sci. USA* **1994**, *91*, 9022.

[135] L. C. Mattheakis, J. M. Dias, W. J. Dower, *Methods Enzymol.* **1996**, *267*, 195.

[136] J. Hanes, A. Plückthun, *Proc. Natl. Acad. Sci. USA* **1997**, *94*, 4937.

[137] C. Schaffitzel, J. Hanes, L. Jermutus, A. Plückthun, *J. Immunol. Meth.* **1999**, *231*, 119.

[138] M. He, M. J. Taussig, *Nucleic Acids Res.* **1997**, *25*, 5132.

[139] R. W. Roberts, J. W. Szostak, *Proc. Natl. Acad. Sci. USA* **1997**, *94*, 12297.

[140] R. Liu, J. E. Barrick, J. W. Szostak, R. W. Roberts, *Methods Enzymol.* **2000**, *318*, 268.

[141] T. Yano, S. Oue, H. Kagamiyama, *Proc. Natl. Acad. Sci. USA* **1998**, *95*, 5511.

[142] A. Crameri, G. Dawes, E. Rodriguez Jr., S. Silver, W. P. C. Stemmer, *Nat. Biotechnol.* **1997**, *15*, 436.

[143] R. W. Roberts, W. J. Ja, *Curr. Opin. Struct. Biol.* **1999**, *9*, 521.

[144] D. S. Tawfik, A. D. Griffiths, *Nat. Biotechnol.* **1998**, *16*, 652.

[145] A. C. Jamieson, S.-H. Kim, J. A. Wells, *Biochemistry* **1994**, *33*, 5689.

[146] A. Crameri, S. A. Raillard, E. Bermudez, W. P. C. Stemmer, *Nature* **1998**, *391*, 288.

[147] Z. Shao, H. Zhao, L. Giver, F. H. Arnold, *Nucleic Acids Res.* **1998**, *26*, 681.

[148] H. Zhao, L. Giver, Z. Shao, J. A. Affholter, F. H. Arnold, *Nat. Biotechnol.* **1998**, *16*, 258.

[149] J. R. Thompson, E. Register, J. Curotto, M. Kurtz, R. Kelly, *Yeast* **1998**, *14*, 565.

[150] C. Yanisch-Perron, J. Vieira, J. Messing, *Gene* **1985**, *33*, 103.

[151] A. Ullmann, F. Jacob, J. Monod, *J. Mol. Biol.* **1967**, *24*, 339.

[152] J. P. Horwitz, J. Chua, R. J. Curby, A. J. Tomson, M. A. Da Rooge, B. E. Fisher, J. Mauricio, I. Klundt, *J. Med. Chem.* **1964**, *7*, 574.

[153] P. Kast, *Gene* **1994**, *138*, 109.

[154] E. T. Kaiser, F. J. Kézdy, *Science* **1984**, *223*, 249.

4

Construction of Environmental Libraries for Functional Screening of Enzyme Activity

Rolf Daniel

Assemblages of environmental microorganisms often encompass a bewildering array of physiological, metabolic, and genetic diversity. Although invisible to the naked eye, microorganisms dominate the biosphere [1]. For example, it has been calculated that 1 g of soil may contain up to 4000 different bacterial species [2]. Current estimates of the total number of microbial species on earth range from 1,000,000 to 100,000,000, but only approximately 5000 microbial species are described in the literature [3, 4]. In addition, the recent research in molecular microbial ecology provides compelling evidence for the existence of many novel types of microorganisms in the environment in numbers and varieties that dwarf those of the comparatively few microorganisms studied to date. These results were obtained by estimation of DNA complexity and discovery of many unique 16S rRNA gene sequences from various environmental sources [5, 6]. Thus, the entirety of the microbial genomes found in nature, termed the metagenome by some authors [7], harbors vastly more genetic information than is contained in the studied and cultured subset. Correspondingly, the potential of microorganisms and their enzymes in the synthesis of new compounds is by no means exhausted. The assessment and exploitation of this enormous microbial diversity offers an almost unlimited pool of new genes, gene products, and biosynthetic pathways for biotechnology and other purposes.

The classical and cumbersome approach to isolate new enzymes from environmental samples is enrichment and screening of a wide variety of microorganisms for the desired activity. The enzymes and the corresponding genes are then recovered from the identified organisms. Traditional cultivation techniques require that the different microorganisms derived from an environmental sample are cultivated on an appropriate growth medium and then separated until individual clones are isolated. In this way, a large fraction of the microbial diversity in an environment is lost as the result of difficulties in enriching and isolating microorganisms in pure culture. It has been estimated that >99 % of microorganisms observable in nature typically cannot be cultivated using standard techniques [8]. To circumvent difficulties and limitations associated with cultivation techniques, several methods for the assessment of the microbial diversity that bypass the cultivation of microorganisms have been developed [9 – 12]. The alternative approach is to use the genetic diversity of the microorganisms in a certain environment as a whole to encounter unknown genes and gene products. One way to exploit the genetic diversity of various environments is the isolation of DNA without culturing the organisms present. Subsequently, the desired genes are amplified by PCR and cloned. The sequences of the primers used are derived from conserved regions of already known genes or protein families [10, 13]. This technique is valuable as a source of 16S rRNA gene sequences in microbial ecology studies [8, 13], but the identification of entirely new genes or gene products for biotechnological purposes by PCR-based methods is limited. The other way, which is described in this Chapter, is to employ the isolated environmental DNA for the construction of

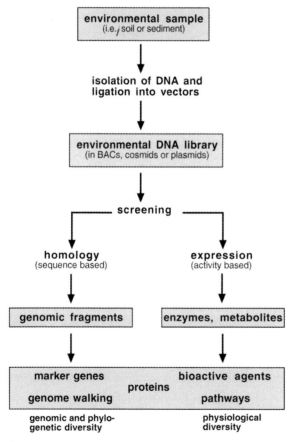

Fig. 4.1. Strategy to access and to exploit the immense genetic diversity of naturally occurring microbial assemblages by construction and screening of environmental libraries.

DNA libraries and to clone directly functional genes from natural samples. The knowledge of sequence information prior to cloning is not required. Another advantage is that existing environmental libraries can be employed for screening of various targets. This approach shows an alternative way to access and exploit the immense pool of genes from microorganisms that have not been cultivated so far. This method has been successfully applied, for example, for the direct cloning of genes encoding 4-hydroxybutyrate dehydrogenases [9], lipases [11, 14], esterases [3, 14], amylases [11], and chitinases [15]. The steps necessary to construct environmental libraries and to apply these libraries for the identification of new genes or gene products are discussed in this Chapter and outlined in Fig. 4.1. The crucial steps comprise isolation and purification of environmental DNA, ligation of the obtained DNA in vectors and subsequently screening of the constructed environmental library for the desired genes.

4.1
Sample Collection and DNA Isolation from Environmental Samples

Crucial for all subsequent molecular analyses is the collection of the sample used for DNA isolation. Sampling is less difficult for terrestrial ecosystems such as soils or microbial mats where the sampling volumes can be kept small and the material can be stored on ice or frozen immediately [16, 17]. Other habitats such as deep marine sediments and hot springs as well as sampling sites requiring intensive sampling effort (i.e. marine environments) may necessitate additional work to collect, store, or process samples at the sampling site. In order to avoid loss of microbial diversity, care must be taken during transport and storage of the samples. This loss is often due to enrichment of specific bacterial groups during storage before freezing and lysis of specimens, which results in loss of nucleic acids. Generally, samples frozen within a period of 2 h maintain the widest spectrum of microorganisms [18].

The cloning of functional genes from natural microbial consortia is dependent on the high quality of the extracted and purified environmental DNA since the enzymatic modifications required during the cloning steps are sensitive to contamination by various biotic and abiotic components that are present in environmental ecosystems. For example, extraction of DNA from soils always results in coextraction of humic substances, which interfere with restriction enzyme digestion and PCR amplification and reduce transformation efficiency and DNA hybridization specificity [19 – 22]. Therefore, extraction methods have been developed to remove or minimize contamination of the purified DNA by humic or other interfering substances. Several protocols for the isolation of bacterial community DNA from various environmental samples have been reported in recent years. These methods are based either on recovery or fractionation of microbial cells and subsequent lysis [23, 24] or on direct lysis of cells in the sample followed by DNA purification [16, 25]. Both approaches are compared, and recommendations for the selection of either method as appropriate for the chosen environmental sample are given.

The direct lysis method was developed for use with high-biomass environmental samples such as surface soils or sludges with total bacterial counts $\geq 10^8$ cells per g of solid material. The DNA extraction procedure includes high temperature, harsh detergent, and mechanical disruption or freezing – thawing for cell lysis. A typical protocol based on the method published by Zhou *et al.* [16] is outlined in Fig. 4.2. This method is currently employed in my laboratory and is suitable for a variety of environmental samples, including soils and sediments [9, 14]. The extraction of DNA from the environmental sample is achieved by lysis of the cells with a high-salt extraction buffer in the presence of sodium dodecyl sulfate (SDS), cetyltrimethy-lammonium bromide (CTAB) and proteinase K. The final step is the purification of the crude DNA and the removal of coextracted substances by chromatography on anion-exchange resins or silica gel (Fig. 4.2). Alternatively, other methods are employed for the final DNA purification, for example, cesium chloride equilibrium density centri-fugation [26], gel filtration [27], and agarose gel electrophoresis [28]. The described procedures recover predominately bacterial DNA from most soil and sediment samples, but non-bacterial and cell-free DNA is also extracted [16, 29]. Direct lysis techniques have been used more often than the fractionation methods because they are suitable for various soils and other environments, they are less labor intensive and yield more DNA [9, 20, 26, 30]. The reported amounts of recovered DNA from different soil samples range from approximately 2.5 to 26.9 μg per g of soil [9, 16, 21]. The DNA isolated by the direct lysis approach presumably represents the microbial diversity of an environmental sample more closely because recovery of microorganisms from the environmental matrix is not required and microorganisms adhering to particles are also lysed. Thus, the entire genetic diversity is used for the preparation of the libraries. In addition, the isolated DNA is ≥ 25 kb and suitable for all molecular procedures such as digestion with restriction enzymes and ligation into vectors [9, 14, 16].

The fractionation method involves the separation of microbial cells from the environmental matrix prior to cell lysis [20, 24]. This approach is advantageous for samples with a low biomass (e.g. water samples) or a high content of substances interfering with DNA isolation (e.g. samples from sewage plants) since recovery and concentration of microorganisms prior to lysis can circumvent some of the difficulties connected with both features. Generally, the separation of the cells from the environmental matrix starts by homogenization of the entire sample in a buffer. Matrix particles and other debris are then removed by a low-speed centrifugation [31]. The unattached microbial cells are subsequently collected by high-speed centrifugation and lysed. For enhancement of the DNA yield multiple rounds of homogenization and centrifugation

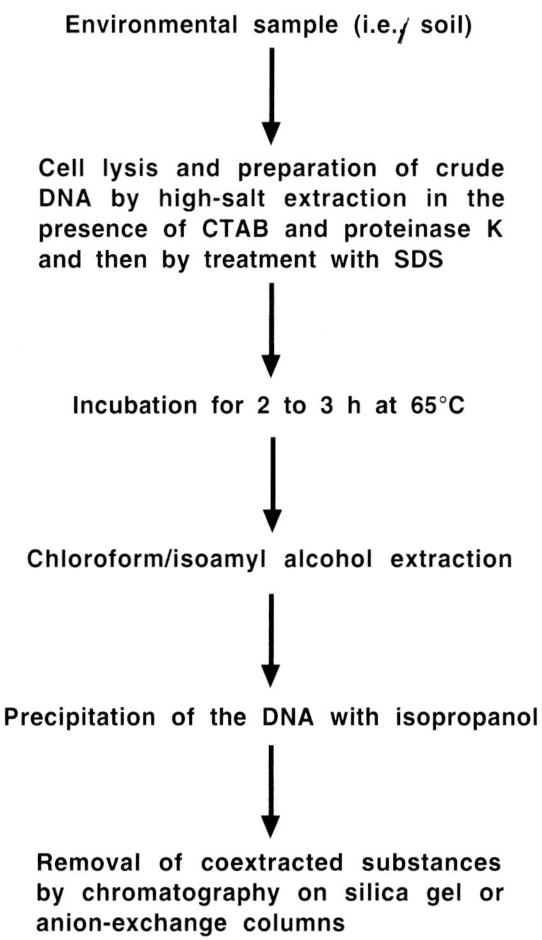

Environmental sample (i.e./ soil)

$\downarrow$

Cell lysis and preparation of crude DNA by high-salt extraction in the presence of CTAB and proteinase K and then by treatment with SDS

$\downarrow$

Incubation for 2 to 3 h at 65°C

$\downarrow$

Chloroform/isoamyl alcohol extraction

$\downarrow$

Precipitation of the DNA with isopropanol

$\downarrow$

Removal of coextracted substances by chromatography on silica gel or anion-exchange columns

Fig. 4.2. Isolation of environmental DNA from soils and sediments by the direct lysis approach. The DNA isolation of soil and sediment samples is based on the direct lysis method of Zhou *et al.* [16]. 50 g of each environmental sample are mixed with 135 ml of DNA extraction buffer (100 mM Tris/HCl, pH 8.0, 100 mM sodium EDTA, 100 mM sodium phosphate, 1.5 M NaCl, 1 % (w/v) CTAB) and 1 ml of proteinase K (10 mg/ml) in GS3 tubes by horizontal shaking for 30 min at 37 °C. Subsequently 15 ml of 20 % (w/v), SDS is added and the samples are incubated in a 65 °C water bath for 2 h with gentle end-over-end inversion every 15 to 20 min. After centrifugation at 6000 × g for 10 min at room temperature the resulting supernatants are transferred into new GS3 tubes. The remaining pellets are extracted two more times by suspending them in 45 ml of extraction buffer and 5 ml of 20 % SDS. The subsequent incubation and centrifugation are done as described above. Supernatants from all extraction steps are combined and mixed with an equal volume of chloroform-isoamyl alcohol (24 : 1 v/v). The aqueous phase is recovered by centrifugation and the DNA is precipitated with 0.6 volume of isopropanol at room temperature for 1 h. A pellet of crude nucleic acids is obtained by centrifugation at 9000 × g for 20 min. After washing with 70 % (v/v) ethanol, the DNA is resuspended in 2 to 4 ml deionized water. The final purification of the DNA and the removal of coextracted substances are performed with the Wizard® Plus Minipreps DNA Purification System" (Promega, Heidelberg, Germany) or the "Qiagen® Midi Kit" (Qiagen, Hilden, Germany).

can be performed [31]. The DNA is then isolated by standard protocols for single organisms [32]. The final DNA purification can be performed as described for the direct lysis approach. The DNA obtained originates almost entirely from bacteria, because the bacteria are separated from the environmental matrix, any cell-free DNA, or larger and heavier eucaryotic cells. Environmental DNA recovered by the fractionation method seems to be less contaminated with matrix compounds or, in the case of soils, with humic substances. In addition, the average size of the isolated DNA is larger than typically obtained by the direct lysis approach [31]. In the case of soil and sediment samples, the fractionation approach may preferentially recover DNA from fast growing bacterial cells that are not adhered or only loosely bound to the environmental matrix. This selectivity results in a loss of genetic diversity in the isolated DNA and, as a consequence, in the constructed environmental libraries.

It is important to note that no single method of cell lysis or DNA purification will be appropriate for all environmental samples. Therefore, researchers should consult protocols that may apply to their systems of study in order to optimize DNA recovery from novel environmental samples.

4.2
Construction of Environmental Libraries

Only a few reports describing the construction of environmental DNA libraries have been published. In all reports *Escherichia coli* is used as the host for the generation and maintenance of the environmental libraries because a variety of genetic tools is available for this organism [9, 11, 12, 14, 28]. The potential of other microorganisms as hosts for environmental libraries will be discussed later. Plasmids [9, 14], cosmids [12] fosmids [28], lambda phages [15], and bacterial artificial chromosomes (BACs) [11, 33] have been used as vectors. The selection of the appropriate vector is dependent on the desired average insert size of the library and the copy number. In addition, the quality of the isolated environmental DNA affects the selection of the vector system. In some cases, environmental DNA still contaminated with humic or matrix substances after purification can only be cloned by using plasmids as vectors. Another advantage of plasmids such as pBluescript or pET is that they exhibit a high copy number in *E. coli*, allowing detection of even weak heterologous expressed genes in activity-based screening programs. This is of great importance since active expression of the environmental genes is often dependent upon the native promoters. This dependence can be reduced by using *cis*-acting vector-based promoters. One drawback of using plasmids for the construction of libraries is that only small DNA molecules can be cloned from the environmental samples. The average insert sizes are $\leq 10\,kb$ (Tab. 4.1). Correspondingly large numbers of clones have to be screened for the desired genes and

Tab. 4.1. Average insert sizes of plasmid-, cosmid- or BAC-based environmental libraries.

DNA source	Cloning vector	Average insert size of the library (kb)	Ref.
Soil	Plasmid (pBluescript SK⁺)	6 to 8	9,14
River sediment	Plasmid (pBluescript SK⁺)	6 to 8	9,14
Enrichment cultures	Cosmid (pWE15)	30 to 40	12
Picoplankton	Fosmid (pFOS1)	40	28, 34
Picoplankton	BAC (pIndigoBAC536)	80	33
Soil	BAC (pBeloBAC11)	27 and 44.5	11

the method is limited to the analysis of single genes or small operons and precludes screening for metabolic activities encoded by large gene clusters. To clone large DNA fragments from environmental DNA, cosmids, fosmids and BACs have been employed [11, 12, 28, 33, 34]. The average insert sizes for environmental cosmid or fosmid libraries range from 20 to 40 kb and those of the BAC libraries from 27 to 80 kb (Tab. 4.1). Recently, the construction of environmental libraries using BAC vectors has received increased attention [7, 11, 33, 35]. BACs are modified plasmids that harbor an origin of replication derived from the *E. coli* F factor. The replication of BACs is strictly controlled, keeping the replicon at one or two copies per cell. These vectors can stably maintain and replicate inserts up to 600 kb [36]. Thus, BACs are excellent vectors for recovering large environmental DNA fragments, which are required to encounter large gene clusters and to characterize the phylogenetic diversity of habitats [33]. The disadvantages of BAC libraries and cosmid libraries are the low copy numbers of the vectors. This may preclude the activity-based detection of genes that are poorly expressed in *E. coli*.

The protocols for the construction of libraries from purified environmental DNA employ standard techniques developed for cloning of genomes from single microorganisms. A typical protocol used routinely in my laboratory for the construction of environmental DNA libraries is depicted in Fig. 4.3. The described method employs plasmids or cosmids such as pBluescript SK⁺ or pWE15, respectively, as cloning vectors. Purified environmental DNA is partially digested with the restriction enzyme *Sau*3AI or sheared, and in order to avoid cloning of very small DNA fragments, size-fractionated by sucrose density centrifugation. Fractions containing DNA fragments of ≥5 kb (plasmid library) or ≥20 kb (cosmid library) are ligated into the appropriate vectors. The resulting products are then transformed or transduced into *E coli*, which is employed for maintenance, amplification, and activity-based screening of the libraries. The quality of the environmental libraries produced is

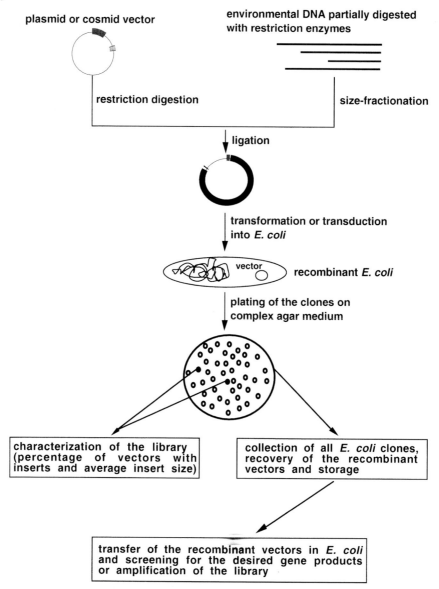

Fig. 4.3. Scheme of the construction of environmental libraries using plasmids or cosmids as vectors. The purified DNA is partially digested and the DNA fragments are size-fractionated, for example, by sucrose density centrifugation. Fractions containing DNA fragments of appropriate sizes are ligated in plasmid or cosmid vectors. The ligation products are then transformed into *E.* *coli*. The quality of the libraries produced is controlled by determination of the average insert size and the percentage of recombinant plasmids with inserts. All obtained recombinant plasmids or cosmids are isolated by using commercial plasmid purification systems and stored. The stored libraries are used as starting material for screening procedures or amplification.

controlled by determination of the average insert size and the percentage of recombinant plasmids containing inserts. For storage of the libraries, the *E. coli* strains are grown on complex agar medium and collected. Subsequently, the recombinant plasmids or cosmids are isolated by standard methods and stored. These recombinant plasmids or cosmids serve as starting material for screening programs.

The construction of BAC libraries from environmental DNA is a difficult challenge because the large size of the DNA has to be maintained during purification. Rondon *et al.* [11] and Béjà *et al.* [33] developed methods in which DNA from soil or planktonic marine microbial assemblages, respectively, can be gently extracted, sized, and ligated into BAC vectors. The initial steps comprise separation of the environmental DNA by pulsed-field gel electrophoresis (PFGE), isolation of high-molecular-weight DNA from the gel and partial digestion with restriction enzymes. Subsequently, the digested DNA is again separated by PFGE; DNA showing an appropriate size range (50 – 300 kb) is excised from the gel, purified, and ligated into the BAC vectors. In contrast to the above-mentioned plasmid or cosmid libraries, the environmental BAC libraries are stored in *E. coli*. Individual clones are transferred to 96-well microtiter plates and stored at – 70 to – 80 °C. In order to perform an activity-based screening, the clones are replicated onto the appropriate indicator medium.

In contrast to the construction of genomic libraries from single organisms, a statistic calculation of the completeness of environmental libraries cannot be done, since the initial number of different microorganisms used for the preparation of the environmental DNA is unknown.

4.3
Screening of Environmental Libraries

The search for and exploitation of natural products and properties have been the mainstay of the biotechnology industry. Taking advantage of microbial diversity by inserting environmental DNA into libraries and then screening for the desired genes and gene clusters is an effective method for the identification of novel biocatalysts, natural products, and new molecular structures. The screening of environmental libraries can be based either on enzyme activity of the recombinant strains or on nucleotide sequence. Clever and simple activity-based strategies are favorable since large numbers of clones have to be tested (Tab. 4.2). For example, the evaluation of 286,000 clones containing DNA from soil samples resulted in the isolation of three novel genes encoding esterases [14].

Activity-based strategies employ tests that are often applied directly to colonies to identify the desired activity. Nontoxic chemical dyes and insoluble or chromophore-bearing derivatives of enzyme substrates, for example, can be incorporated

Tab. 4.2. Examples of activity-based screens of various environmental libraries yielding genes that encode activities relevant for biotechnology.

DNA source	Vector	Number of E. coli clones tested	Screening target	Number of positive E. coli clones	Ref.
Soil	Plasmid	286,000	Esterase	3	14
Soil	Plasmid	930,000	4-Hydroxybutyrate dehy-drogenase	5	9
Estuarine	Lambda	75,000	Chitinase	9	15
Soil	BAC	3,648	DNase	1	11
Soil	BAC	3,648	Antibacterial activity	1	11
Soil	BAC	3,648	Lipase	2	22
Soil	BAC	3,648	Amylase	8	11
Soil	BAC	24,576	Hemolytic activity	29	11

into the growth medium solidified with agar, where they will register specific metabolic capabilities of individual clones. The sensitivity of such screens makes it possible to detect rare clones among large numbers of organisms. Two examples of this strategy are screening for dehydrogenase and lipase/esterase activity. Dehydrogenase activity of E. coli clones can be screened on tetrazolium indicator plates [37] containing an alcohol (i.e. 4-hydroxybutyrate [9]) as test substrate. Tetrazolium in its oxidized state is soluble in water and appears colorless or faintly yellow in solution. Upon oxidation of the test substrate, it is reduced. This yields to the formation of an insoluble deep-red formazan, which is precipitated in the cells. The reduction of tetrazolium is a result of electrons passing from the test substrate, through the enzymatic machinery of central metabolism, the electron transport chain, and ultimately on to tetrazolium [37]. Thus, E. coli colonies capable of catabolizing the test substrate reduce tetrazolium and produce a deep-red formazan, whereas colonies failing to catabolize the test substrate remain uncolored (Fig. 4.4A). For the detection of E. coli clones exhibiting lipase or esterase activity, LB agar plates [32] containing test substrates such as tributyrin or tricaproin can be employed [14]. Positive E. coli clones are detected by zones of clearance around the colonies (Fig. 4.4B).

Activity-based screens can be designed to be highly specific for a specific phenotype of recombinant E. coli strains. A very promising approach is the use of deficient mutants, which require heterologous complementation by genes encoding the affected function for growth under selective conditions. For example, complementation of the Na^+/H^+ antiporter deficient E. coli strain KNabc [38] by recombinant plasmids containing environmental DNA led to the isolation of novel genes encoding $Na^+/$

H⁺ antiporters [39]. In addition to the above-mentioned simple plate assays considerable effort has been and is being expended in the development of high-throughput screening assays, particularly as a response to the need to evaluate large numbers of clones. High-throughput screening involves the robotic handling of very large numbers of clones, the registering of appropriate signals from the assay system, and data management and interpretation. To date, the advent of high-throughput screening of environmental libraries is limited to libraries in which a high percentage of the clones harbor the desired genes. The number of hits in a screen can be increased by using environmental samples for library production which exhibit a natural enrichment of microorganisms containing the desired activity. For example, samples from hot springs are excellent sources for genes encoding thermostable enzymes and the deep and polar seas are environments from which cold-active enzymes can be isolated. The latter find applications in low-temperature operations such as food and leather processing. In addition, some researchers use enrichment cultures for library construction [12]. Activity-based screening programs have been successfully applied for environmental plasmid, cosmid, lambda, and BAC libraries [9, 11, 12, 14, 15, 39]. The phenotypes expressed by the clones include lipase, amylase, chitinase, esterase, 4-hydroxybutyrate dehydrogenase, antibacterial, and hemolytic activities (Tab. 4.2). Sequence analyses of the genes responsible for the observed phenotypes revealed that almost all encode novel gene products. This result confirms that environmental DNA libraries harbor genes from diverse microorganisms, which mostly have not been investigated or even cultivated so far. In all above-mentioned reports, *E. coli*

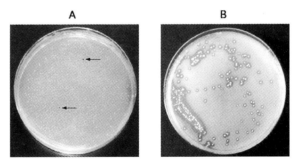

A   **B**

Fig. 4.4: Phenotype of 4–hydroxybutyrate dehydrogenase-positive (A) and lipase/esterase-positive (B) *E. coli* clones. A, 4-Hydroxybutyrate dehydrogenase-positive clones are marked by arrows. Tetrazolium indicator plates [37] containing 4-hydroxybutyrate as test substrate were employed for the screening procedure [9]. Positive clones were identified by formation of a deep-red formazan inside the colonies. B, Lipase/esterase activity of the clones was detected on LB agar [32] containing tributyrin as test substrate [14]. Zones of clearance around the colonies were indicative for lipase/esterase activity.

was used as host for the activity-based screening of environmental libraries. The advantage of working with *E. coli* is that this organism is commonly employed in industrial fermentations. Correspondingly, methods that facilitate batch production, separation, as well as downstream processing are well established. This implies that many of the development stages for the commercial production of valuable products have already been accomplished before the genes are cloned. However, *E. coli* may not be the ideal host in which to express random environmental genes. Nevertheless, the general methodology employing *E. coli* as host has proven successful. In order to enhance the likelihood of detecting any particular gene or gene cluster other hosts such as *Streptomyces* or *Bacillus* species may be also used to express random environmental genes.

Activity-based screening programs proved suitable for plasmid, cosmid, lambda, as well as for BAC libraries. For a nucleotide sequence based screening strategy, the employment of libraries containing large inserts such as cosmid or BAC libraries is favored. Labor-intensive analyses of individual clones or pools of individual clones are often required for this strategy, particularly for the detection of genes encoding interesting enzymes by hybridization using probes derived from known genes. Sequence-based approaches are mostly used in conjunction with replica plating, a technique that is based on the property that colonies of many clones can be imprinted on filter paper or on a piled fabric. Replica plating facilitates transfer of colonies from one surface to another while preserving their spatial arrangement and location. Filter replicas of plates can be probed for the identification of the desired clones in various ways, for example with radiolabeled, chromophore-labeled, or fluorophore-labeled nucleic acid probes or with antibodies directed against specific gene products or cell structures. In contrast to activity-based screens, time-consuming replica plating is a prerequisite for most sequence-based screening strategies. Therefore, a reduction in the number of analyzed clones is advantageous. This can be achieved by using environmental libraries containing very large inserts (e.g. BAC libraries) for sequence-based screens. The majority of sequence-based approaches was performed to retrieve clones containing rRNA genes for phylogenetic surveys of various environments [28, 33, 34]. The desired clones were identified by PCR using domain- or group-specific ribosomal RNA-targeted primers. To retrieve solely rRNA genes from natural habitats, the construction of libraries is not a prerequisite since the application of only the isolated environmental DNA for this purpose is a standard technique [40]. The advantage of the identification of clones harboring rRNA genes on large inserts is that sequencing of the DNA surrounding the rRNA genes is feasible. This can provide access to genomes of uncultivated microorganisms and can yield clues about the physiology of the organisms. One way to gain further information from large-insert libraries is to look for contigs of specific clones, for example by random BAC end sequencing of the arrayed library [41]. Contiguous sequences for specific phylogenetic groups exceeding 200 kb in size have been found [33]. A number of housekeeping genes and unidentified open reading frames will be encountered on the way to gaining information on physiological or biochemical characteristics of uncultivated microorganisms by random sequencing of environmental libraries. Even these genes can provide significant information on evolutionary affinity, gene organization, and codon usage [33]. In addition, sequencing of large-insert environmental libraries allows identification of uncultivated species *in situ* by chromosomal painting techniques [42], reconstruction of metabolic pathways, and preparation of DNA microarrays to monitor distribution, gene expression, and the environmental responses of uncultured microorganisms [33].

However, sequence-based strategies and random sequencing of environmental libraries have not yet been applied to encounter novel genes for biotechnologically important enzymes or other bioactive compounds. Clever and simple activity-based screening strategies are preferred, because they are less labor-intensive, faster, and cheaper than sequence-based screens, and primarily because sequence information prior to screening is not required. Activity-based screening provides access to entirely new types and classes of enzymes or novel pathways catalyzing a certain reaction or the synthesis of bioactive compounds.

Regardless of the strategy used, the screening of environmental libraries allows less troublesome follow-up analysis of the hits in a screen than traditional screening programs. The follow-up analysis includes efforts to isolate and identify active compounds. In traditional natural products screening programs this procedure often requires large amounts of extract and the productive microbe has to be cultured on a large scale. One major drawback of this process is known as the refermentation problem [35]. Microbial cultures, particularly those derived from soil samples, produce an activity of interest the first time when they are grown but they cannot be made to exhibit this activity again when they are refermented. It has been estimated that this difficulty occurs more than 50 % of the time in traditional screening programs [35]. Environmental library screening programs offer some advantages over the traditional approaches with regard to this issue since the DNA encoding the active compounds is already isolated on a plasmid, cosmid, or BAC vector. Correspondingly, the relevant genes can be rapidly identified, sequenced, and analyzed thereby providing valuable information on the compound with the interesting activity. Once characterized, the relevant genes can be engineered to increase production of the desired natural product.

4.4
Conclusions

The discovery of the vast but previously unappreciated microbial diversity inspired many industrial or academic researchers to develop alternative methods to exploit these resources. One of the major bottlenecks of conventional natural product discovery strategies is that the various approaches depend on screening microbial products (e.g. enzymes and metabolites) from cultured organisms since less than 1 % of the microorganisms present in an environment can be cultivated by standard techniques. Therefore, the discovery of novel natural products is restricted to this small proportion. Consequently, molecular methods that do not rely on isolation of microorganisms into pure culture have general utility in harvesting a broader range of microbial diversity for industrial and pharmaceutical purposes. These molecular approaches will not entirely displace traditional natural product screens for certain targets but they provide a complementary way for accessing biomolecules that would otherwise remain undiscovered. One of the exciting new molecular strategies for natural product discovery is to insert environmental DNA containing the genomes of uncultured microorganisms into expression libraries. For retrieval of novel products, the constructed libraries are then screened for genes and gene clusters encoding stable biocatalysts and other microbial products. This strategy provides access to the enormous pool of genes from environmental microbial assemblages. The encountered environmental genes encoding interesting biomolecules can serve as starting material for directed evolution and DNA shuffling processes. The combination of discovery of novel robust activities followed by modification offers great potential for encountering and evolving commercially valuable biomolecules. Nevertheless, some technological hurdles need to be overcome before employment of environmental DNA for natural product discovery can be regarded as a standard method. One difficult challenge is the expression of large multigene clusters of environmental DNA in surrogate host strains. Furthermore, the construction of environmental libraries, in which large numbers of clones harbor the desired genes whose activities can be detected in high-throughput screening procedures, is another important technological challenge. These issues will require the combined efforts of many researchers from different fields. Despite the technological challenges that have to be met, this intriguing new research area of accessing and exploiting the natural biodiversity together with high-throughput screening systems will have a great impact on microbial biotechnology in the future.

References

[1] Whitman, W. B., Coleman, D. C., Wiebe, W. J. *Proc. Natl. Acad. Sci. USA* **1998**, 95, 6578–6583.

[2] Torsvik, V., Goksøyr, J., Daae, F. I. *Appl. Environ. Microbiol.* **1990**, 56, 782–787.

[3] Short, J. M. *Nat. Biotechnol.* **1997**, 15, 1322–1323.

[4] Strickberger, M. W. *Evolution*, Johnes & Bartlett Publishers, Boston, Ma, 1996.

[5] Torsvik, V., Sorheim, R., Goksøyr, J. *J. Ind. Microbiol.* **1996**, 17, 170–178.

[6] Ward, D. M., Weller, R., Bateson, M. M. *Nature* **1990**, 345, 63–65.

[7] Handelsman, J., Rondon, M. R., Brady, S. F., Clardy, J., Goodman, R. M. *Chem. Biol.* **1998**, 5, R245–R249.

[8] Amann, R. I., Ludwig, W., Schleifer, K. H. *Microbiol. Rev.* **1995**, 59, 143–169.

[9] Henne, A., Daniel, R., Schmitz, R. A., Gottschalk G. *Appl. Environ. Microbiol.* **1999**, 65, 3901–3907.

[10] Seow, K.-T., Meurer, G., Gerlitz, M., Wendt-Pienkowski, E., Hutchinson, C. R., Davies, I. *J. Bacteriol.* **1997**, 179, 7360–7368.

[11] Rondon, M. R., August, P. R., Bettermann, A. D., Brady, S. F., Grossman, T. H., Liles, M. R., Loiacono, K. A., Lynch, B. A., MacNeil, I. A., Minor, C., Tiong, C. L., Gilman, M., Osburne, M. S., Clardy, J., Handelsman, J., Goodman, R. M. *Appl. Environ. Microbiol.* **2000**, 66, 2541–2547.

[12] Entcheva, P., Liebl, W., Johann, A., Hartsch, T., Streit, W. *Appl. Environ. Microbiol.* **2001**, 67, 89–99.

[13] Hugenholtz, P., Pace, N. R. *Trends Biotechnol.* **1996**, 190–197.

[14] Henne, A., Schmitz, R. A., Bömeke, M., Gottschalk, G., Daniel, R. *Appl. Environ. Microbiol.* **2000**, 66, 3113–3116.

[15] Cottrell, M. T., Moore, J. A., Kirchman, D. L. *Appl. Environ. Microbiol.* **1999**, 65, 2553–2557.

[16] Zhou, J., Bruns, M. A., Tiedje, J. M. *Appl. Environ. Microbiol.* **1996**, 62, 316–322.

[17] Liesack, W., Stackebrandt, E. *J. Bacteriol.* **1992**, 174, 5072–5078.

[18] Rochelle, P. A., Cragg, B. A., Fry, J. C., Parkes, R. J., Weightman, A. J. *FEMS Microbiol. Ecol.* **1994**, 15, 215–225.

[19] Smalla, K., Cresswell, N., Mendonca-Hagler, L. C., Wolters, A., van Elsas, J. D. *J. Appl. Bacteriol.* **1993**, 74, 78–85.

[20] Steffan, R. J., Atlas, R. M. *Appl. Environ. Microbiol.* **1988**, 54, 2185–2191.

[21] Tebbe, C. C., Vahjen, W. *Appl. Environ. Microbiol.* **1993**, 59, 2657–2665.

[22] Porteous, L. A., Armstrong, J. L. *Curr. Microbiol.* **1993**, 22, 345–348.

[23] Jacobsen, C. S., Rasmussen, O. F. *Appl. Environ. Microbiol.* **1992**, 58, 2458–2462.

[24] Torsvik, V. L. *Soil Biol. Biochem.* **1980**, 12, 15–221.

[25] Tsai, Y.-L., Olson B. H. *Appl. Environ. Microbiol.* **1993**, 57, 1070–1074.

[26] Holben, W. E., Jansson, J. K., Chelm, B. K., Tiedje, J. M. *Appl. Environ. Microbiol.* **1988**, 54, 703–711.

[27] Jackson, C. R., Harper, J. P., Willoughby, D., Roden, E. E., Churchill, P. F. *Appl. Environ. Microbiol.* **1997**, 63, 4993–4995.

[28] Stein, J. L., Marsh, T. L., Wu, K. Y., Shizuya, H., DeLong, E. F. *J. Bacteriol.* **1996**, 178, 591–599.

[29] Holben, W. E., Isolation and purification of bacterial DNA from soil in *Methods of soil analysis, part 2*, R. W. Weaver (Ed.), SSSA Book Series no. 5. Soil Science Society of America, Madison, Wis., 1994.

[30] Leff, L. G., Dana, J. R., McArthur, J. V., Shimkets, L. J. *Appl. Environ. Microbiol.* **1995**, 57, 1141–1143.

[31] Holben, W. E., Isolation and purification of bacterial community DNA from environmental samples in *Manual of environmental microbiology*, C. J. Hurst, G. R. Knudsen, M. J. McInerney, L. D. Stetzenbach, M. V. Walter (Eds.), ASM Press, Washington, D. C., 1996.

[32] Ausubel, F. M., Brent, R., Kingston, R. E., Moore, D. D., Seidman, J. G., Smith, J. A., Struhl, K. (Eds.) *Current protocols in molecular biology*, John Wiley & Sons, NY, 1987.

[33] Béjà, O., Suzuki, M. T., Koonin, E. V., Aravind, L., Hadd, A., Nguyen, L. P., Amjadi, M., Garrigues, C., Jovanovich, S. B., Feldman, R. A., DeLong, E. F. *Environ. Microbiol.* **2000**, 2, 516–529.

[34] Vergin, K. L., Urbach, E., Stein, J. L., DeLong, E. F., Lanoil, B. D., Giovannoni *Appl. Environ. Microbiol.* **1998**, 64, 3075 – 3078.

[35] Osburne, M. S., Grossman, T. H., August, P. R., Macneil, I. A. *ASM News* **2000**, 66, 411 – 417.

[36] Zimmer, R., Gibbins, A. M. V. *Genomics* **1997**, 42, 217 – 226.

[37] Bochner, B. R., Savageau, M. A., *Appl. Environ. Microbiol.* **1977**, 33, 434 – 444.

[38] Nozaki, K., Inaba, K., Kuroda, T., Tsuda, M., Tsuchiya, T. *Biochem. Biophys. Res. Commun.* **1996**, 222, 774 – 779.

[39] Majernik, A., Gottschalk, G., Daniel, R. *J. Bacteriol.* **2001**, 183, 6645 – 6653.

[40] Wintzingerode, F. V., Göbel, U. B., Stackebrandt, E. *FEMS Microbiol. Rev.* **1997**, 21, 213 – 229.

[41] Kelley, J. M., Field, C. E., Craven, M. B., Bocskai, D., Kim, U. J., Rounsley, S. D., Adams M. D. *Nucleic Acids Res.* **1999**, 27, 1539 – 1546.

[42] Giovannoni, S. J., Britschgi, T. B., Moyer, C. L., Field, K. G. *Nature* **1990**, 345, 60 – 63.

5
Investigation of Phage Display for the Directed Evolution of Enzymes

Patrice Soumillion and Jacques Fastrez

5.1
Introduction

Engineering enzymes with new and improved properties or even new activities is becoming a goal within reach of enzymologists. In favorable cases, this can be done by rational design. Frequently, however, our understanding of the structure – function relationship remains too limited for this approach to be fruitful. Consequently, a strategy of directed evolution is frequently being pursued. Libraries of mutants are created by introduction of random mutations. Interesting mutants are then searched either by screening or selection techniques. Screening methods in which each clone of the library is analyzed are probably the most popular but sensitive assays can be difficult to set up, expensive equipment is usually necessary and the maximal size of libraries that can be screened is several orders of magnitude lower than for selection methods. *In vivo* selection is a very powerful and simple method but its application is possible only when the interesting phenotype confers a biological advantage to the host cell [1]. Moreover, *in vivo* conditions are restricted and poorly controllable. Therefore, there is a great interest in developing *in vitro* selection strategies for engineering enzymes endowed with new or improved catalytic properties. Several strategies have been developed with this aim and, in the future, phage display could become a powerful technology for the directed evolution of biocatalysts.

5.2
The Phage Display

Phage display technology enables *in vitro* selection from libraries of proteins or peptides displayed on the surface of filamentous bacteriophages. It was introduced by George P. Smith in 1985 [2] and, since then, several thousands research papers have reported the use of this method (for a review, see [3]). The power of the technology comes from the physical linkage between a piece of genetic information and its expression product *via* a phage particle. This is achieved simply by cloning the genetic information within the phage genome, in fusion with a gene encoding a phage coat protein. On phage morphogenesis, the fusion protein is assembled in the phage particle resulting in a chimeric particle displaying the foreign expression product (peptide or protein) on its surface. This chimeric phage is a small soluble assembly that mimics and simplifies the genotype – phenotype linkage, which is necessary for the evolution of living cells by selection processes. Cloning a library of genes generates a library of phages from which clones of interest can be isolated by *in vitro* selection. Indeed, a protein or a peptide is extracted from the library on the basis of its properties and attached to its genetic information, which can be replicated and amplified by simple infection.

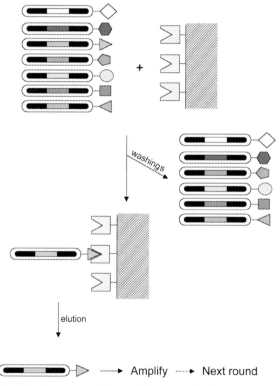

Fig. 5.1. The principle of phage display is based on *in vitro* selection of peptides or proteins expressed at the surface of filamentous phages. In a "biopanning" experiment, specific binders can be captured from a library and their genes can be amplified by infection.

Phage display has been mainly applied to the selection of specific peptide or protein binders because the experimental protocol is quite simple. As shown in Fig. 5.1, the selection consists of an affinity chromatography where the target partner is simply immobilised on a solid support and serves to capture the binding phages from the library. Unbound phages are washed away and bound phages can be eluted in a specific or non-specific manner by competition with soluble target, pH switch, proteolytic or chemical cleavage. Eluted phages are then amplified by infecting *Escherichia coli* and can be analyzed or injected into a new round of selection. The process has been dubbed "biopanning".

Affinity selection is by far the most important application of phage display (for practical reviews, see [4, 5]). Libraries of phage-displayed peptides are now commonly used to select peptides that bind to receptors or enzymes of therapeutic interest. The technology is also very powerful for the discovery of new monoclonal antibodies from collections of phages displayed single chain Fv or Fab fragments.

5.3
Phage Display of Enzymes

5.3.1
The expression vectors

5.3.1.1 Filamentous bacteriophages

The most widespread vectors used in phage display are filamentous bacteriophages such as M13, f1, or fd and related phagemids [6, 7]. These phages infect *E. coli* strains containing the F conjugative plasmid, which encodes the F pilus. The phage particle is composed of a single-stranded circular DNA molecule encapsidated in a long cylinder made of five different proteins. It is resistant to a wide range of pH (2.5 – 11) and temperature (up to 80 °C) conditions. The cylinder length is proportional to the size of the DNA. A wild-type Ff particle (6407 bases) is approximately 930 nm in length and 6.5 nm in diameter. Around 2700 copies of the major g8p coat protein (5.2 kDa) constitute the cylinder. Approximately five copies of g7p (3.6 kDa) and g9p (3.6 kDa) close one extremity of the particle. The other end is made of approximately five copies of g6p (12.3 kDa) and g3p (42.5 kDa). G3p is responsible for phage infection. It is made up of three domains separated by glycine-rich repeats. The amino-terminal domain (residues 1 to 67) and the central domain (residues 87 to 217) are involved in phage infection [8]; their structures have been solved recently [9, 10]. The carboxy-terminal domain (residues 257 to 406) acts as an anchor in the phage particle [11].

Filamentous bacteriophages are not lytic for the infected bacteria but are secreted through a large exit port in the outer membrane. This pore is around 8 nm in diameter and allows passage of one filament [12]. During phage morphogenesis, the single-stranded genome is covered with the coat proteins as it passes through the inner membrane. Before assembly, all the coat proteins reside in the inner membrane with their amino-terminus in the periplasm and their carboxy-terminus in the cytoplasm. G8p and g3p are synthesized with an amino-terminal signal sequence, which is removed when the proteins are inserted into the membrane. For g8p, this insertion is independent of the bacterial Sec system [13] whereas it is required for g3p [14].

Foreign peptides have been displayed as fusion proteins with all five coat proteins of filamentous phages although g8p and g3p are the most widely used. With respect to g7p and g9p, only one example describes the display of a dimeric antibody variable fragment (light and heavy chains) as a double fusion system with both coat protein amino-termini [15]. G6p tolerates fusion with its carboxy-terminus and thus allows the display of cDNA libraries [16]. Despite the interest in cDNA display, few examples have been reported so far and the versatility of this system is hard to estimate. Display on g8p is very popular for expressing short peptides on the phage surface. Beside peptides, several proteins have been successfully displayed as fusions with g8p [17]. This is somewhat surprising, as the display of a protein should lead to a phage particle too large to allow its extrusion through the 8 nm pore. Globular proteins are mostly displayed as fusion with g3p at the tip of the phage; this is probably the recommended choice at present. A wide variety of proteins have been successfully fused to the amino-terminus of g3p by inserting the foreign gene between the sequences encoding the signal peptide and the mature coat protein. Recently, a fusion with the carboxy-terminal of g3p has been reported [18]. This could permit the surface expression of full-length cDNA clones bearing stop codons. An alternative display of cDNAs is a heterodimeric system based on the co-expression of Fos-cDNA and Jun-g3p fusions [19].

Several vectors with various cloning sites and genetic markers have been designed for protein display on g8p and g3p [20]. As filamentous phages are not lytic, a drug-resistance gene has been inserted in the phage genome to allow selection of infected bacteria. The replicative form (RF) of the phage is a circular dsDNA that can be purified as a plasmid DNA and used for cloning. The RFs of most phage vectors are quite large (up to 10 kbp). It is present at a very low copy number (around one per cell). As it is necessary to prepare a large quantity of highly pure RF for the construction of libraries, maxi-preparation and cesium chloride purification are recommended.

In phage vectors, the fusion protein is the only source of coat proteins. If the foreign peptide or protein is tolerated and not degraded *in vivo*, polyvalent display will be obtained. This can lead to the display of several thousands copies of a foreign peptide (g8p fusion) or between three and five copies of a foreign protein (g3p fusion) on a single viral particle. A high level of display is used when searching for weak binders: polyvalency favors binding, an effect called avidity.

A high level of display is, however, not always tolerated and alternative systems have been designed to reduce that level. In these systems, the fusion protein is generally encoded on a phagemid and the production of phagemid particles is triggered by infection with a helper phage, which carries a copy of the wild-type coat protein gene. As this wild-type protein is usually favored in phage assembly, the level of display may be reduced to less than one copy per particle with g3p fusions. Monovalency is required when searching for strong binders. In this case, a single copy is sufficient to bind the phage to an immobilised ligand. A phagemid vector where the fusion protein is under the control of a phage shock promoter has also been designed to facilitate phage display of toxic proteins [21]. This promoter is induced only upon helper phage infection. Compared to phages, phagemid vectors are also easier to handle for DNA manipulation (smaller size, higher copy number). The level of display is, however, difficult to control and, in some cases, much less than one copy is displayed per phage. As the background level of non-specific phages is often high in selections, a low amount of fusion protein expression can decrease significantly the sensitivity of the technique.

5.3.1.2 **Other phages**

Alternative phage display systems have been designed with lambda and T4 phages [22]. These phages accumulate in the cytoplasm and are released into the extracellular medium by cell lysis. Therefore, the fusion protein does not have to move across the inner and outer membranes. This is an advantage over filamentous phages. The system should also facilitate multiple display and surface expression of large or toxic proteins. However, only a few papers report the use of these phages and it is not clear whether they will become as popular as filamentous phages.

Display on lambda phages has been described with two major coat proteins: the gpV tail protein [23, 24] and the gpD head protein [25, 26]. Both proteins accept C-terminal fusions, which offers alternative systems for displaying cDNA libraries. A high level of display is possible as demonstrated by the total modification of gpV [27] and gpD [25] proteins. The structure of gpD has been solved recently [28]; this should help the design of optimal display vectors.

Display on T4 phage has been reported on a minor coat protein called fibritin [29] and on two outer capsid proteins, HOC and SOC [30, 31]. HOC and SOC accept, respectively, N-terminal and C-terminal fusions. High levels of full-length protein display have been obtained with both proteins: between 10 and 40 copies of a protein of 183 amino acid residues were expressed per phage particle [32].

A disadvantage of these systems is probably the lower efficiency and convenience of cloning libraries of DNA molecules. This difficulty has been alleviated by using a Cre-*loxP* recombination strategy: the fusion gene is cloned in a small plasmid and then efficiently recombined *in vivo* into the phage [25].

5.3.2

Phage-enzymes

When displaying an enzyme on a phage surface, evaluation of the level of functional enzymes attached to the phage particle is generally recommended prior to mutagenesis and selection. This level is equal to the ratio between the enzyme and the phage concentrations.

The phage concentration of a solution of a purified phage-enzyme can be estimated by measurement of the optical density at 265 nm. The absorption coefficient can be calculated on the basis of the individual coefficients of all constituents of the phage (amino acids and nucleotides). As g8p is the major contributing protein, the absorption coefficient is proportional to the length of the phage, which itself is proportional to the size of the encapsidated single-stranded DNA. For a 10 kb phage, it is equal to $8.4 \times 10^7 \, M^{-1} cm^{-1}$. The phage concentration can also be evaluated by measuring the protein content [33] or the DNA content [34] with specific dyes and standards. With all these methods, an over-estimate may be possible if the g3p fusion protein is the major source of g3p and the results will not be optimal for correct termination of filamentous phage assembly. Under these conditions, polyphages (phages containing more than one DNA molecule) may be produced. To our knowledge, however, polyphages have been reported for mutants of the g3p protein [35] but not for g3p fusions. It is noteworthy that the concentration of infectious particles is usually between 1 and 5 % of the phage concentration. This is a feature of filamentous phages and the g3p fusion protein does not generally affect phage infectivity, at least at a low level of display.

The concentration of displayed enzyme is evaluated by measuring the activity of the phage-enzyme solution with the assumption that the displayed enzyme has the same activity as the free enzyme in solution. In our experience, this is generally the case.

The presence of the enzyme-g3p fusion protein should also be detected by western blot analysis with an anti-g3P and/or an anti-enzyme antibody. With anti-g3p detection, the relative intensities of the fusion and the free g3p proteins should be in qualitative agreement with the displayed level evaluated by the ratio between the enzyme and the phage concentration.

Obviously, phage display technology requires the protein to be properly folded and stable as a fusion with the coat protein. Successful display may depend on the bacterial host and on growth conditions. General strategies to improve recombinant protein expression in *E.coli* [36] should be transposable to the phage display context.

Inefficient protein folding is potentially a problem, especially when the displayed protein originates from a non-bacterial organism. In particular, eukaryotic proteins are known to be generally difficult to express as active proteins in bacteria. One way to improve folding efficiency is to use engineered hosts that overexpress chaperones. Using this approach, Soderlind *et al.* have described the assisted display of antibody fragments on phages by GroEL and GroES chaperonins [37]. Overexpression of the periplasmic proteins Skp and FkpA has also been shown to improved antibody display [38, 39].

Another possible mechanism that could reduce the level of expression is proteolytic degradation of the fusion protein. Indeed, there is evidence for degradation of fusion proteins by bacterial proteases. The effect of the stability of the enzyme, the structure of the linker connecting it to the phage and culture conditions have been studied in our group with several phage-enzymes. Less stable mutants are poorly displayed on phage. A linker designed to be cleaved by the factor Xa endoprotease (Ile-Glu-Gly-Arg) appears to be more sensitive to proteolysis than a glycine – serine one. Decreasing the temperature of the culture appears to be a general and easy way to increase the level of display by reducing the level of proteolysis. A fivefold increase in the display level of the TEM-1 β-lactamase [40] was observed when decreasing the growth temperature of the infected bacteria from 37 °C to 23 °C. To our knowledge, an increase in display level by producing phages in a protease-deficient strain has never been reported.

Beside proteolysis during phage production, one should keep in mind that purified phages could also lose their displayed proteins because of contaminating proteases. Cocktails of protease inhibitors should be added to prevent proteolysis, especially for long-term storage of phage stocks.

In phagemid vectors, a stop codon is often introduced between the foreign protein and the phage coat protein. Hence, display is only possible when the infected E.coli strain contains an appropriate suppressor. Such a phage display vector behaves as a classical expression vector when a non-suppressor strain is transformed. This system eliminates the need for a recloning step for the production of the free enzyme after selection. Nevertheless, the risk of obtaining a very low level of surface expression is increased as suppression is always partial.

The success of in vitro selection depends on the level of functional display but also principally on the quality and the diversity of the starting library. The attainable diversity can be significantly increased by taking advantage of site-specific in vivo recombination (see Section 5.4). In this case, the bacterial host must express the appropriate site-specific recombinase.

Since 1991, more than 30 different enzymes have been successfully displayed on bacteriophages. Table 5.1 summaries the essential features of these phage-enzymes. As the proteins expressed on filamentous phages have to be exported in the periplasm prior to phage assembly, it is not surprising that the literature reports mainly the display of naturally secreted proteins (proteases, bacterial extracellular or periplasmic enzymes, antibodies...). At present, it is not clear whether cytoplasmic proteins can be generally expressed at the surface of these phages. It should be noted also that the level of display is usually higher when phages are used compared to phagemids that incorporate wild-type coat proteins from the helper phage. Besides this, display on the lytic lambda phage seems to afford a very high level of display.

Tab. 5.1. Essential characteristics of phage displayed enzymes

Displayed enzyme	Size (kD)	Fusion protein	Display format	Display level (enz/phage)	References
Trypsin	24 kD	g3p	phagemid	0.14^b	34
		g8p	phagemid	0.17^b	
Subtilisin	27 kD	g3p	fd	$\sim 0.2^c$	64
Subtilisin mutant	27 kD	g3p	phagemid	nd	66
Subtiligase	27 kD	g3p	na	nd	67
Metallo-protease	34 kD	g3p	fd	$\sim 1.0^c$	I. Ponsard, to be published
Hepatitis C NS3 protease	23 kD	g3p	phagemid	nd	71
Mini-plasminogen	38 kD	g3p	fd	0.1	59
Penicillin acylase	86 kD	g3p	phagemid	$0.06 - 0.12^d$	33
		g8p	phagemid	0.19-0.30^d	33
		g3p	fd	$\sim 1.0^c$	Y.F. Shi, to be published
Prostate specific antigen	27 kD	g3p	phagemid	nd	72
DD-peptidase (PBP4)	50 kD	g3p	fd	$\sim 1.0^c$	S. Lavenne, to be published
DD-peptidase (R61)	37 kD	g3p	fd	1.0-2.0^c	C. Ifi, to be published
β-lactamase	29 kD	g3p	fd	1.0-3.0^c	40
		gpD	λ	nd	26
Metallo-β-lactamase	25 kD	g3p	fd	2.0-5.0^c	69
Alkaline phosphatase	47 kD	g3p	phagemid	nd	73,74
			fd	1.0-2.0	45
Staph. nuclease	14 kD	g3p	fd	nd	75
		g3p	phagemid	$0.05 - 0.5$	48
Ribonuclease A	14 kD	g3p	phagemid	nd	76
Polymerases (Klenow and Stoffel fragments)	68 & 61 kD	g3p	phagemids	$\sim 0.001^a$	68
Glutathione S-transferase	25 kD	g3p	phagemid	nd	49
				$\sim 0.001^a$	66
Hen Lysozyme	14 kD	g3p	phagemid	nd	74
β-galactosidase	464 kD	gpV	λ	~ 0.5	24
		gpD	λ	2.0-34.0	26
Carbonic anhydrase	29 kD	g3p	phagemid	nd	50
GTPase Ras	20 kD	g3p	phagemid & fd	nd	78
Inactive phospholipase A_2	15 kD	g3p	phagemid	nd	73
2,4-dienoyl-CoA reductase	31 kD	g6p	fd	nd	47

Tab. 5.1. Essential characteristics of phage displayed enzymes

Displayed enzyme	Size (kD)	Fusion protein	Display format	Display level (enz/phage)	References
Urate oxidase	35 kD	g6p	fd	nd	47
Thioredoxin	12 kD	g3p	phagemid	nd	77
Superoxide dismutase	16 kD	g3p	phagemid	nd	77
Catalytic antibodies (Fab or scFv)	50 kD or 25 kD	g3p	phagemids	nd	53,54,56, 58, 61,63

na: not available; nd:not determined

a These levels of display are estimated by measuring the ratio between infectious particles before and after trypsin treatment; the helper phage and the phagemid encode respectively a trypsin sensitive and resistant g3p. This could not reflect the amount of foreign protein expressed at the phage surface as the displayed protein could be removed from the phage by bacterial proteases and could also reduce the phage infectivity (see in the text).

b Phage concentration is determined by ethidium bromide visualisation of ssDNA and comparison to a standard of known concentration. Enzyme concentration were estimated by titration of trypsin activity with BPTI.

c,d Phage concentration is determined by the optical density at 265 nm (c) or by measuring the protein content (d). Enzyme concentration is estimated by activity measurment with the assumption that the displayed and soluble enzymes have the same activity. Levels of display are confirmed by western blot analysis (c).

5.4
Creating Libraries of Mutants

The objective of phage display technology is the selection of clones of interest from a library of mutants. The diversity of the library will be the number of individual clones present in the collection whereas its quality will be related to the method used for its creation and to the functional diversity, i.e. the diversity from which all the clones that are unstable, misfolded or degraded have been removed. A good library will feature a high diversity and a high quality.

Different methods are available for generating libraries of mutants. Random point mutations can be introduced within a gene by a mutagenic PCR reaction (error-prone PCR). The PCR reaction is run in the presence of $MnCl_2$ and unbalanced dNTP concentrations, which reduces the fidelity of the polymerase [41]. The average number of mutations introduced per gene in the library can be controlled by the number of PCR cycles. Random mutations can also be created in specific regions of a gene by introducing degenerated oligonucleotides in a site-specific manner, either by direct cloning or by PCR if the oligonucleotides serve as primers. Finally, DNA shuffling and related techniques allows creation of diversity by random combination of fragments generated from a family of homologous genes [42]. The method is also appropriate to create second generation libraries by shuffling a mixture of clones that would have been selected from a first generation library obtained by error-prone PCR. Moreover, DNA shuffling also generates point mutations at a similar frequency to error-prone PCR.

The quality of a library often depends on the protocol used for its creation. For example, if a high number of cycles of error-prone PCR is performed, each mutant generated in the library will have a high number of mutations. Therefore, such a library will contain a high proportion of clones that do not fold properly. In another example, if a degenerated decapeptide is inserted in a surface loop of an enzyme, a large proportion of the library will probably be degraded by proteolysis. Hence, the choice of the method and of the experimental protocol is a crucial step.

A library of phage-displayed mutants is created in two steps. The diversity is initially generated *in vitro* by cloning a library of inserts in the vector and then it is introduced *in vivo* by transformation of the bacteria. In some cases, the final diversity is limited by the number of different DNA molecules present in the DNA mixture prior to transformation. For instance, if a degenerated oligonucleotide encoding a tetrapeptide is inserted in a gene, the maximum diversity will be 1.6×10^5, which is the number of different combinations at the protein level. Nevertheless, taking into account the low frequency of occurrence of some sequences because of the degeneracy of the genetic code, it is necessary to create a library of more than 10^7 individually transformed bacteria to obtain a complete library. In most cases however, the number of different molecules in the DNA mixture is much larger than the maximum number of bacteria that can be reasonably transformed. When a large amount of well-purified and correctly constructed DNA is used, libraries of 10^6 transformants are easily obtained; the upper limit is around 10^9. Alternative strategies that allow the further increase of the size of libraries have been developed [43, 44]. The principle is to generate two partial independent libraries for two parts or subunits of a protein (for instance, the heavy and light chains of an antibody); one library is introduced in bacteria (plasmids or phagemids) and the other is prepared as infectious particles (phagemids or phages); the combinatorial library is then obtained by infection of the first library by the other. The diversity is limited by the infection, which is much more efficient than transformation. In this way, between 10^{11} and 10^{12} different clones could be generated quite easily. Moreover, an *in vivo* site-specific recombination step allowing the fusion of the two vectors can be added.

5.5
Selection of Phage-enzymes

5.5.1
Selection for binding

In the initial characterization of a phage-enzyme, it may be useful to check that the phages bind specific inhibitors or monoclonal antibodies (mAb) specific for the folded enzyme. This is easily done by "biopanning" (Fig. 5.1). Phages are incubated in the presence of an immobilised inhibitor (or mAb) and eluted by addition of soluble inhibitor or by changing the conditions (for instance the pH) to reduce the affinity. The ratio between the numbers of eluted versus adsorbed phages is compared to the background observed with unrelated phages. It is inferred that the phages display a properly folded enzyme. This ratio is also related to the level of display. This method was used initially to characterize phage-enzymes displaying the *E. coli* alkaline phosphatase [45] or rat trypsin [34]. The first one binds to an immobilised arsenate inhibitor, the second to ecotin, a natural *E. coli* protease inhibitor.

It is of interest to engineer additional binding sites within an enzyme with the purpose of generating enzyme variants whose activity could be modulated by binding of allosteric effectors. These hybrid enzymes would find application as diagnostic tools as the presence of the allosteric effector in a sample could be detected by measuring the specific activity versus that observed in a reference solution. As a first step to create hybrid enzymes endowed with affinity for other proteins, libraries of insertion mutants of the class A *β*-lactamase TEM-1 were created in which surface loops were extended by replacement of 1 – 3 residues of the wild-type sequence by random peptide sequences of 3 – 7 residues. *In vivo* selection was applied to the individual libraries to obtain bacterial clones producing phage-enzymes active enough to provide resistance against *β*-lactam antibiotics. Selected clones were recombined into a large library in which two loops have accepted random insertions. The original and recombined libraries of active phage-enzymes were then selected for binding to monoclonal antibodies raised against a totally unrelated protein, prostate specific antigen (PSA). Application of two orthogonal selections allows the use of huge high-quality libraries. Several selected clones were characterized. It was observed that the affinities for the mAbs were in the nanomolar to micromolar range and that the activity was modulated mainly downwards by factors up to 10. PSA could be detected in a competitive immunoassay at concentrations in the nanomolar range [46].

Biopanning on antibodies raised against proteins of a specific organelle has also been used with the purpose of discovering new proteins or enzymes. A rat liver cDNA library was cloned in fusion with phage gene g6p and the phages displaying fusion proteins were submitted to an immunoaffinity selection process with antibodies raised against peroxysomal subfractions. Beside two previously identified peroxysomal enzymes, a catalase and an urate oxydase, this protocol has led to the discovery of a new peroxysomal enzyme, a 2,4-dienoyl-CoA reductase, whose activity is required for the degradation of unsaturated fatty acids [47].

5.5.2
Selection for catalytic activity

Selection for catalytic activity is obviously more tricky than selection for binding. Indeed, it is necessary to find a way to couple the ability of active phage-enzymes to catalyze the transformation of a substrate to their specific extraction from a mixture. Several strategies have been developed. The first strategies were relatively simple, they were applied in projects aimed at changing the specificity of an enzyme and used affinity chromatography on a support on which analogues of substrates or products were immobilised. In a more sophisticated approach taking into account the fact that enzymes have to be complementary to the transition state instead of the ground state form of the substrate, attempts have been made to select phage-enzymes on immobilised transition-state analogues. Affinity labeling has also been used to select active phage-enzymes on the rationale that enzymatic activities rely frequently on the presence in the active site of residues whose reactivity is exacerbated by the local environment. In parallel, suicide substrates have been used on the rationale that they specifically use the enzymatic activity to become activated before irreversibly blocking the active site. Finally, new methods are emerging in which the direct interaction with substrates is used to select active mutants.

5.5.2.1   Selection with substrate or product analogues

Selection on inhibitors designed as substrate or product analogues has been applied to two enzymes: staphylococcal nuclease and a glutathione transferase. Staphylococcal nuclease is a Ca^{2+}-dependent phosphodiesterase that catalyses the hydrolysis of DNA with a slight specificity for thymidine on the 5' side of the cleavage point. Libraries of mutants of phage-displayed staphylococcal nuclease were generated by error-prone PCR (epPCR). Biopanning on immobilised substrate analogues, thymidine- or guanosine-containing phosphorothioate oligonucleotides (Fig. 5.2), enriched the libraries in binders. The best mutants selected on the thymidine substrate analogue (**1**) were nearly as active as the wild-type enzyme; those selected on the guanosine substrate analogue (**2**) were ten times less active but showed the expected change in specificity [48].

Glutathione transferases (GST) are enzymes involved in detoxification processes. They catalyze the conjugation between toxic compounds and glutathione (GSH, **3** in Fig. 5.3A) as a preliminary step to biological elimination. Several classes of

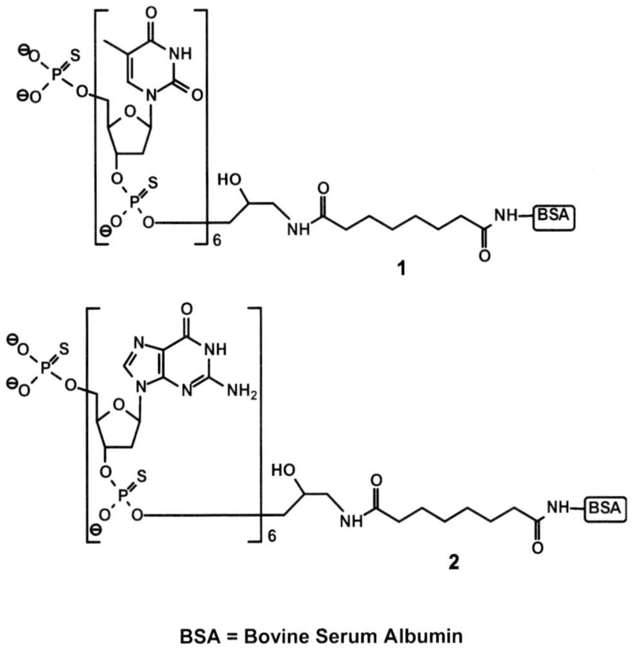

BSA = Bovine Serum Albumin

Fig. 5.2. Substrate analogues were used to select variants of the staphylococcal nuclease [48]. The phosphothioate functions are resistant to the nuclease activity.

GSTs are known; they are homologous but differ widely in their specific activities. GST A1-1 is active in GSH conjugation to aromatic substrates activated for nucleophilic substitution like chloro-dinitrobenzene (**4**). In an effort to change the specificity towards substrates bearing a negative charge on their aromatic ring, Widersten and Mannervik [49] created libraries of phage-displayed GSTs: 10 amino acid residues in the aromatic electrophile binding site were randomly mutated. Biopanning on product-like affinity ligands (e.g. **7** in Fig. 5.3B) allowed novel GSTs with altered substrate specificity to be extracted from these libraries. Characterization of isolated mutants indicated that good binders were selected but the specific activity of these enzymes was reduced one thousand-fold compared to the wild-type enzyme.

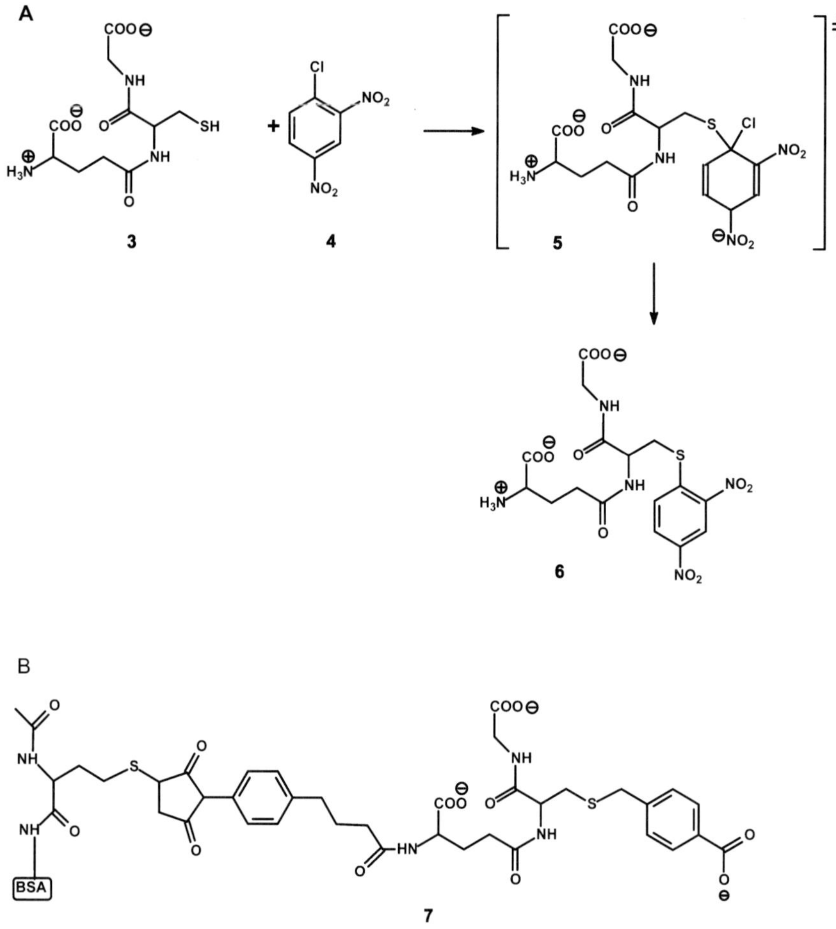

Fig. 5.3. (A) The glutathione-S-transferase (GST) catalyses the coupling between glutathione (**3**) and 1–chloro-2,4-dinitrobenzene (**4**). (B) A product analogue was used to select GSTs with altered specificities [49].

5.5.2.2 Selection with transition-state analogues

The theory of enzymatic catalysis, as first enunciated by Pauling, states that enzymes must be complementary to the transition state rather than to the ground-state form of the substrate. According to this theory, the ratio k_{cat}/k_{uncat} is equal to the ratio of affinity constants of the enzyme for the substrate and the transition state, K_S/K_S (Scheme 5.1).

Recognition of this fundamental feature of enzymes has led to the design of transition-state analogues (TSA) as potent inhibitors of enzymes and to their use as haptens to induce the immune system into generating antibodies endowed with catalytic ac-

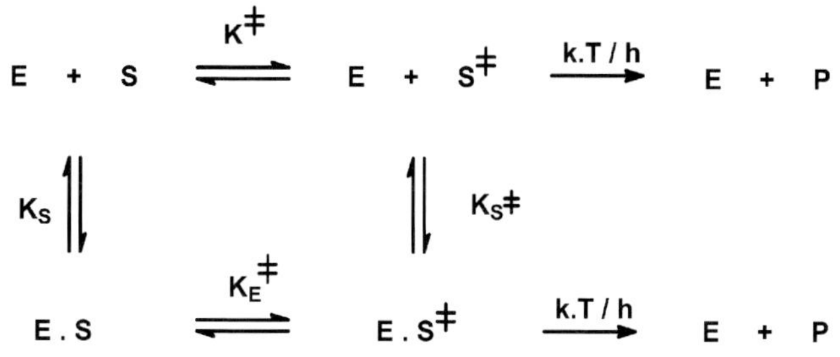

$$E + S \xrightleftharpoons[\;]{K^{\ddagger}} E + S^{\ddagger} \xrightarrow{k.T/h} E + P$$

$$K_S \updownarrow \qquad\qquad K_S{}^{\ddagger} \updownarrow$$

$$E.S \xrightleftharpoons[\;]{K_E{}^{\ddagger}} E.S^{\ddagger} \xrightarrow{k.T/h} E + P$$

$$K_S = \frac{[E]\,[S]}{[E.S]} \qquad\qquad K_S{}^{\ddagger} = \frac{[E]\,[S^{\ddagger}]}{[E.S^{\ddagger}]}$$

$$K^{\ddagger} = \frac{[S^{\ddagger}]}{[S]} \qquad\qquad K_E{}^{\ddagger} = \frac{[E.S^{\ddagger}]}{[E.S]}$$

$$\frac{k_{cat}}{k_{uncat}} = \frac{K_E{}^{\ddagger}}{K^{\ddagger}} = \frac{K_S}{K_S{}^{\ddagger}} \approx \frac{K_S}{K_{TSA}}$$

Scheme 5.1. Thermodynamic cycle demonstrating the theory of enzymatic catalysis. The acceleration rate is correlated to the affinity of the enzyme for the transition state versus the substrate.

tivity, sometimes called abzymes. For a perfect TSA, the ratio K_S/K_{TSA} would be equal to k_{cat}/k_{uncat}.

Carbonic anhydrase is a metallo-enzyme that catalyses very efficiently the hydration of CO_2. It has also been shown to catalyze the hydrolysis of activated carboxylic esters like p-nitrophenylacetate. It is inhibited by sulfonamides, which are considered to be TSAs. In a project aimed at understanding the structural determinants of metal-ion affinity in proteins, three of the amino acid residues interacting directly with the histidine metal ligands were replaced by a library of 1728 variants in the phage-displayed carbonic anhydrase II. The library was then selected for zinc-ion affinity by chromatography on an immobilised sulfonamide that forms a coordination bond to the zinc through its nitrogen. Selected mutants were shown to bind zinc ion with a range of affinities, from equal affinity down to 100–fold lower than the wild-type enzyme. Eighty percent of the variants, however, had a CO_2 hydrase activity close to that of the wild-type enzyme (within a factor 3) [50].

Selection by affinity chromatography on a transition-state analogue has also been used to extract active enzymes from a library of phage-displayed GST A1-1 mutants. Four residues in the electrophilic substrate binding site were randomly mu-

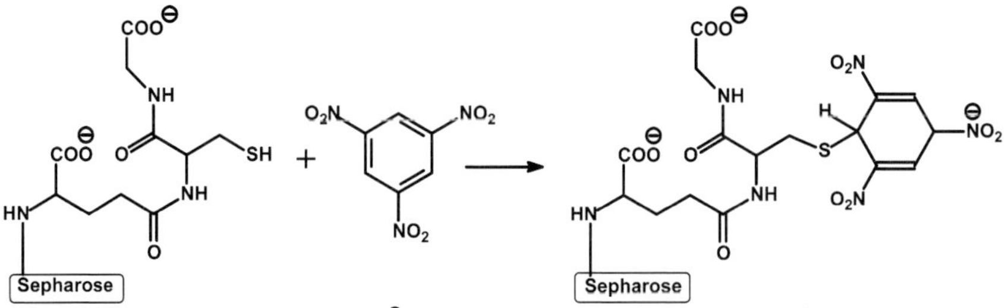

Fig. 5.4. Phage-displayed glutathione-S-transferase mutants were selected with an immobilised transition-state analogue [51].

tated. The σ-complex (**9**) formed by reaction between GSH and 1,3,5-trinitrobenzene (TNB, **8**) was used as an affinity ligand (Fig. 5.4). It mimics the transition state of a nucleophilic aromatic substitution by GSH on halogeno-nitrobenzene derivatives but does not lead to a complete reaction because there is no leaving group. Several mutants were characterized after four rounds of biopanning. The catalytic efficiency of one of them was determined on several substrates: it was 20 to 90-fold lower than the wild-type enzyme on chloronitrobenzene substrates. This mutant stabilizes the σ-complex formed between GSH and TNB and also catalyses its formation. The ratio between the affinities of the mutant and the wild-type enzyme for the σ-complex (K_{TSA}) has been measured, it is the same as the ratio between the k_{cat}/K_m values determined with 1-chloro-2,4-dinitrobenzene, as predicted by theory. One surprising observation is that catalysis of GSH conjugation with ethacrynic acid (a Michael addition reaction) was 13-fold more efficient with this mutant than with the wild-type enzyme [51].

In these two examples, the principle of transition-state stabilization as a source of enzymatic catalysis is successfully exploited for the selection of active mutants. Selection with a TSA appears to be able to enrich libraries in variants with high catalytic activity.

Selection with transition-state analogues has also been used to extract catalytic antibodies from libraries of phage-antibodies. McCafferty *et al.* [52] have shown that antibodies with nanomolar affinities for a phosphonate transition-state analogue could easily be extracted from a library derived from immunized mice.

Baca *et al.* [53] have used random mutagenesis followed by biopanning on a phosphonate in order to try to increase the catalytic activity of first-generation catalytic antibodies. Starting from a mAb that catalyses the hydrolysis of different amino acid esters (e.g. **10** to **11** in Fig. 5.5A), they have "humanized" the original antibody to improve its expression in *E. coli* and adapt it to phage-display. Combinatorial libraries of mutants were then created in the hapten binding site and subjected to multiple rounds of selection by biopanning on phosphonate **12** (Fig. 5.5B). Individual clones were then picked for sequencing and characterization. Affinity for the phosphonate

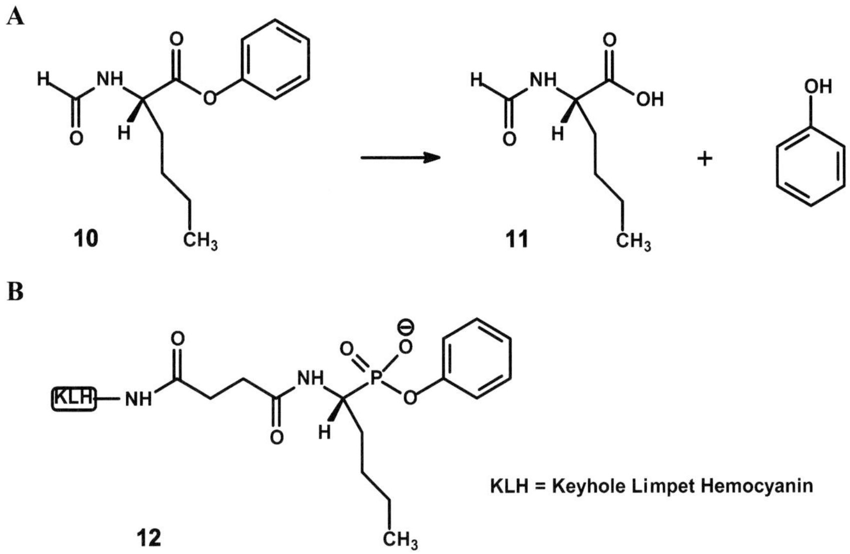

A

10 CH_3 **11** CH_3

B

KLH = Keyhole Limpet Hemocyanin

12 CH_3

Fig. 5.5. (A) A reaction catalyzed by an esterase antibody and (B) the transition-state analogue derivative used for the selection [53].

transition-state analogue increased 2 to 8-fold but none of the affinity-matured mutants showed an improved catalytic efficiency *versus* the original mAb. Surprisingly, a weaker binding mutant was found to have a 2-fold higher activity. In this example, biopanning on TSA was not really successful for the selection of mutants with increased activity.

Conversely, Fujii *et al.* [54] were successful in using affinity maturation for a TSA to improve catalysis. A mAb was generated by immunization with a transition-state analogue. It catalyses the regioselective deprotection of an acylated carbohydrate (13 to 14 in Fig. 5.6A) with a significant rate enhancement. Model building suggested the mutagenization of six amino acid residues in the heavy chain CDR3. A library of phage-displayed antibodies in which these residues were randomly mutated was created. Biopanning on bovine serum albumin (BSA)-conjugated transition-state analogue 15 (Fig. 5.6B) and screening of 124 randomly picked clones identified six variants whose Fab fragments were produced in soluble form. The most active mutant had a k_{cat} 12–fold higher than the wild-type mAb but a 9-fold increased K_m. Interestingly, in going from the starting mAb to the mutant, the ratio k_{cat}/k_{uncat} increased by a factor 12.2 (from 184 to 2248), a value comparable to the change in the K_m/K_{TSA} ratio (14.4) in agreement with theory. The origin of the lower catalytic activity of the TSA-affinity matured mAb reported by Baca *et al.* [53] was suggested by these authors to result from the fact that optimization of TSA binding may lead to some loss of the quality of the alignment of a Ser-His catalytic dyad that had been shown earlier [55] to be involved in the mechanism.

A

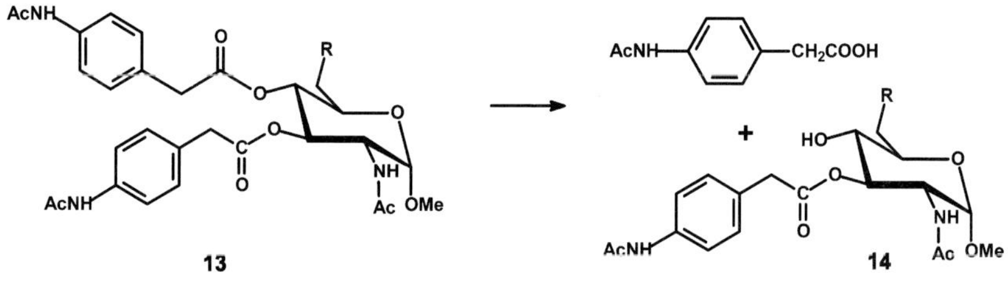

R = F or p-AcNH-C6H4-CH2-COO

B

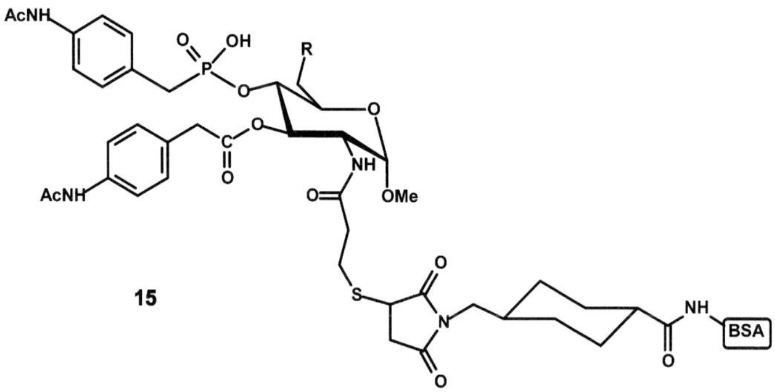

Fig. 5.6. (A) A regioselective ester hydrolysis catalyzed by an antibody and (B) the derivatized transition-state analogue used for the selection [54].

5.5.2.3 Selection of reactive active site residues by affinity labeling

The active sites of enzymes frequently contain residues whose side chains play an important role in catalysis particularly because their reactivity is exacerbated by the local environment. Affinity labeling by reagents designed to dock in the active site and that able to form a stable covalent bond with a reactive side chain has been used to extract active enzymes or abzymes from libraries.

The group of Lerner has designed a reagent able to capture phage-antibodies containing a reactive cystein within their hapten binding site. BSA-conjugated α-phenethyl pyridyl disulfide (**16** in Fig. 5.7A) was immobilised on wells of microtiter plates and used to capture phage-antibodies containing a reactive cystein in their binding site. Two of the ten phages randomly picked from the product of a single round of selection contained an unpaired cystein. One of them was studied in detail and shown to catalyze the hydrolysis of a thioester whose electrophilic carbonyl occupies the position of the reactive sulfur during selection (**17** in Fig. 5.7B). The reaction operates with formation of a covalent acyl-intermediate. A 30-fold rate enhancement over background is measured; it is modest on the acylation step but hydrolysis of the acylated cystein

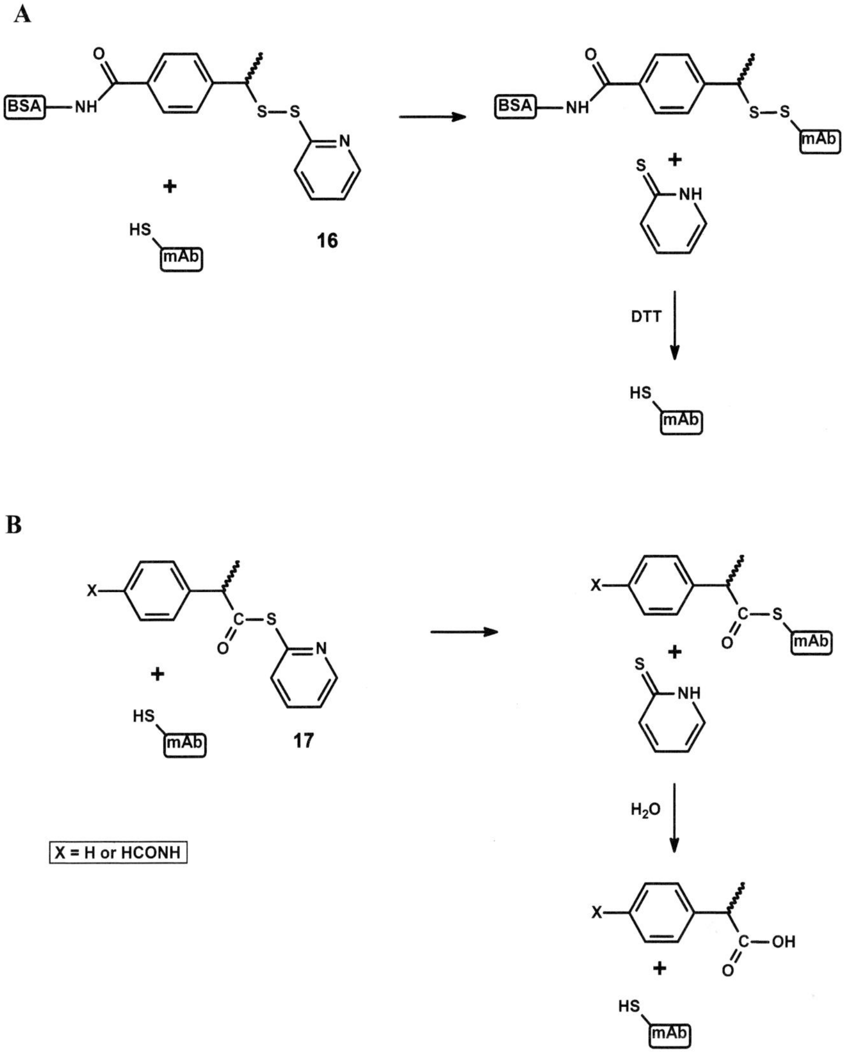

A

B

X = H or HCONH

Fig. 5.7. (A) An active site affinity labeling reagent used to select a reactive cystein in an antibody binding site. (B) One of the selected antibodies catalyses the hydrolysis of a thioester [56].

intermediate appears to be catalyzed more efficiently (10^4-fold versus the spontaneous hydrolysis of a thioester) [56].

Aldol condensation reactions are catalyzed by amines and the active sites of many aldolases contain an essential lysine residue. Using a strategy of reactive immunization with a 1,3-diketone (**18** in Fig. 5.8), Wagner et al. were able to generate antibodies with aldolase activity. These were shown to possess a highly reactive lysine residue in

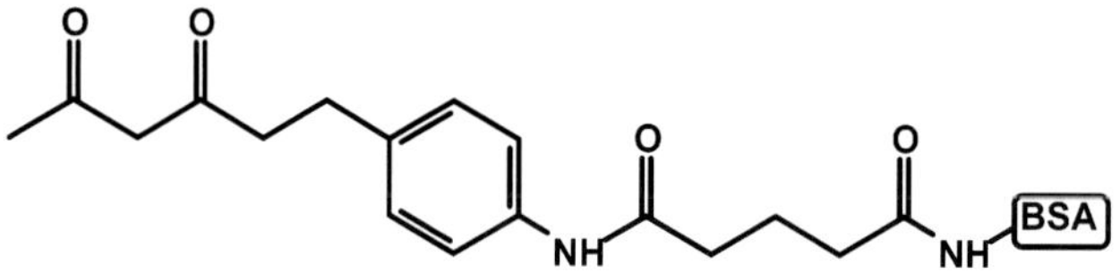

18

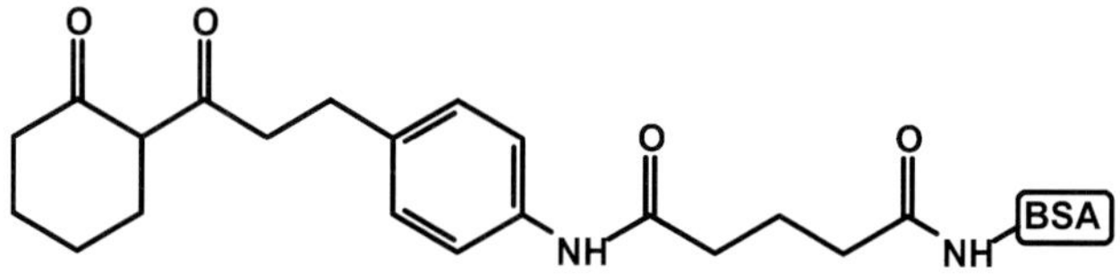

19

Fig. 5.8. 1,3-diketone derivatives allow selection of aldolase antibodies by reaction with an essential lysine residue in their binding sites [57].

their active site [57]. In an effort to reconstruct the active site of one aldolase antibody in order to alter its substrate specificity, Tanaka *et al.* constructed libraries of phage-displayed antibodies by fusing a library containing the CDR1 and CDR2 of naive heavy chain variable domains with the CRD3 containing the essential lysine of the isolated aldolase antibody. Selection on BSA-conjugated **18** and **19** allowed the isolation of new enantioselective antibodies of the targeted specificity; one of them had a ratio k_{cat}/k_{uncat} of 2×10^4, 3 to 10–fold higher than the parent antibody [58].

Chloromethylketones have been used extensively to inhibit serine and cystein proteases. They react specifically with the active site histidine that activates the essential serine for nucleophilic attack on ester or amides substrates. A biotinylated chloromethylketone (biotin-Phe-Pro-Arg-CH$_2$Cl) has been used by Lasters *et al.* to show that micro-plasminogen, a proenzyme constituted from fragments of human plasminogen, could be activated efficiently not only by the physiological activator urokinase but also by streptokinase and staphylokinase. Phages displaying the proenzyme were incubated with the activating proteins, trapped with the active site labeling reagent and adsorbed on streptavidin-coated support; detection of immobilised phages with horseradish-peroxidase-conjugated anti-M13 antibodies was possible, following proenzyme activation [59].

5.5.2.4 Selection with suicide substrates

The irreversible inhibition of enzymes by suicide substrates occurs as a consequence of activation steps in which the target enzymes transform these substrates into inhibitors using their normal mechanism.

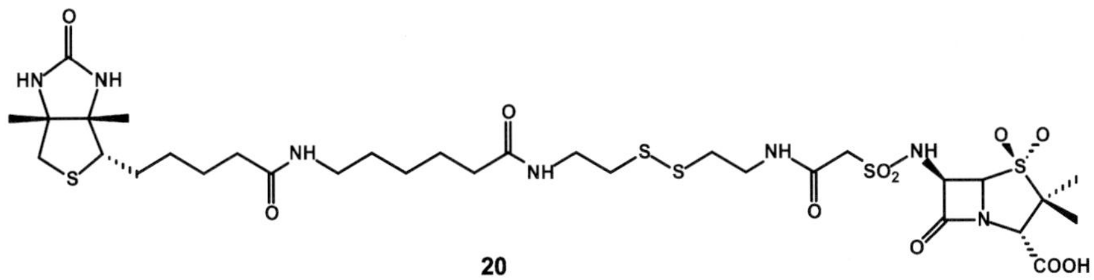

Fig. 5.9. A biotinylated suicide inhibitor allows selection of phagedisplayed β-lactamases [40, 60].

Soumillion *et al.* have designed a selection strategy based on a suicide substrate to extract active phage-enzymes from mixtures containing inactive mutants. The phage-displayed TEM-1 β-lactamase and an inactive mutant were incubated under kinetic control with a biotinylated penicillinsulfone (**20**, Fig. 5.9). The active phage-enzymes were preferentially labeled and extracted from the mixtures by binding to streptavidin-coated beads. Cleavage of a chemical bond within the biotinylated label or between the displayed enzyme and g3p allowed the phages to be recovered. The enrichment factor was 50 in favor of the active phage-enzyme [40]. The method has been shown to be suitable for selecting the most active enzymes from a mixture of mutants [60].

This strategy has been applied to select catalytic antibodies from phage-displayed libraries. Two catalytic single-chain antibodies catalyzing the hydrolysis of ampicillin with rate accelerations k_{cat}/k_{uncat} of 5200 and 320 ($k_{cat} = 0.29$ and $0.018\,\text{min}^{-1}$) were isolated from combinatorial libraries prepared from mice immunized with penam sulfone conjugates and selected with a biotinylated penam sulfone [61].

This strategy has also been applied for the selection of active β-lactamases from a library of mutants also containing penicillin-binding proteins. For this purpose, the protocol had to be modified to circumvent a difficulty of selections with suicide substrates in mechanisms involving a covalent intermediate. If inhibition arises from a covalent intermediate (**Y** in Scheme 5.2, an acyl-enzyme in the case of serine β-lactamases), enzymes whose rate of release of this intermediate (hydrolysis of the acyl-enzyme) is slow will be efficiently selected as the efficiency of inhibition depends on the ratio of rate constants k_4/k_3 (Scheme 5.2). To prevent the selection of enzymes with inadequate turnover, a counter-selection step was included in the protocol: the library of mutants was incubated with substrate in order to block them as covalent intermediates before adding the biotinylated inhibitor. The library could be enriched from 6 ppm to 25 % active β-lactamases in four rounds of selection [62].

Antibodies possessing glycosidase activity have been induced by immunization of mice with conjugates of the transition-state analogue **21** (Fig. 5.10A). A combinatorial phage-Fab library was then constructed by PCR amplification of the heavy and light chains. Selection with a BSA-conjugated suicide inhibitor **22** (Fig. 5.10B) was then

$$E + S \underset{k_{-1}}{\overset{k_1}{\rightleftarrows}} E.S \xrightarrow{k_2} Y \xrightarrow{k_3} E + P$$

$$Y \xrightarrow{k_4} Ei$$

Scheme 5.2. Kinetic scheme for a suicide inhibition arising from a covalent intermediate (Y).

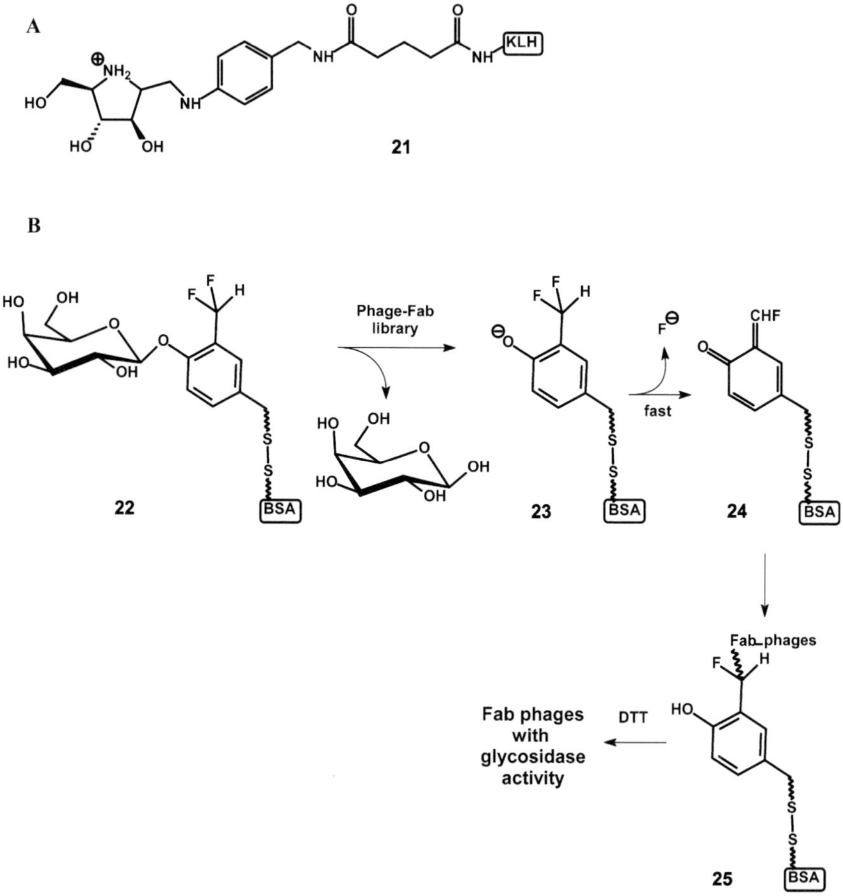

Fig. 5.10. (A) Transition-state analogue used to generate glycosidase antibodies by immunization. (B) A suicide inhibitor derivative was used to select improved antibodies (Fab) displayed on phages [63].

applied: cleavage of the glycosidic bond releases an *o*-difluoromethyl-phenol **23**, fast elimination of fluoride generates a very reactive quinone methide **24**, which reacts with the proteins of the phage. Only clones that catalyze the hydrolysis of the glyco-sidic bond are susceptible to capture on immobilised BSA-conjugate. The phages are released by reduction with dithiothreitol (DTT) of the disulfide bridge connecting the inhibitor to BSA in **25**. After four rounds of panning, the selected genes were recloned to express soluble Fab fragments in a *β*-galactosidase deficient strain. Plating on chro-mogenic substrate allowed clones producing active Fabs to be screened. One protein was purified and shown to catalyze the hydrolysis of *p*-nitrophenyl-*β*-galactopyranoside with a k_{cat} of 0.007 min^{-1} and a K_m of 0.53 mM, corresponding to a rate enhancement k_{cat}/k_{uncat} of 7×10^4. For comparison, the best catalyst obtained out of 22 clones using the classical hybridoma technology was 700 times less active [63].

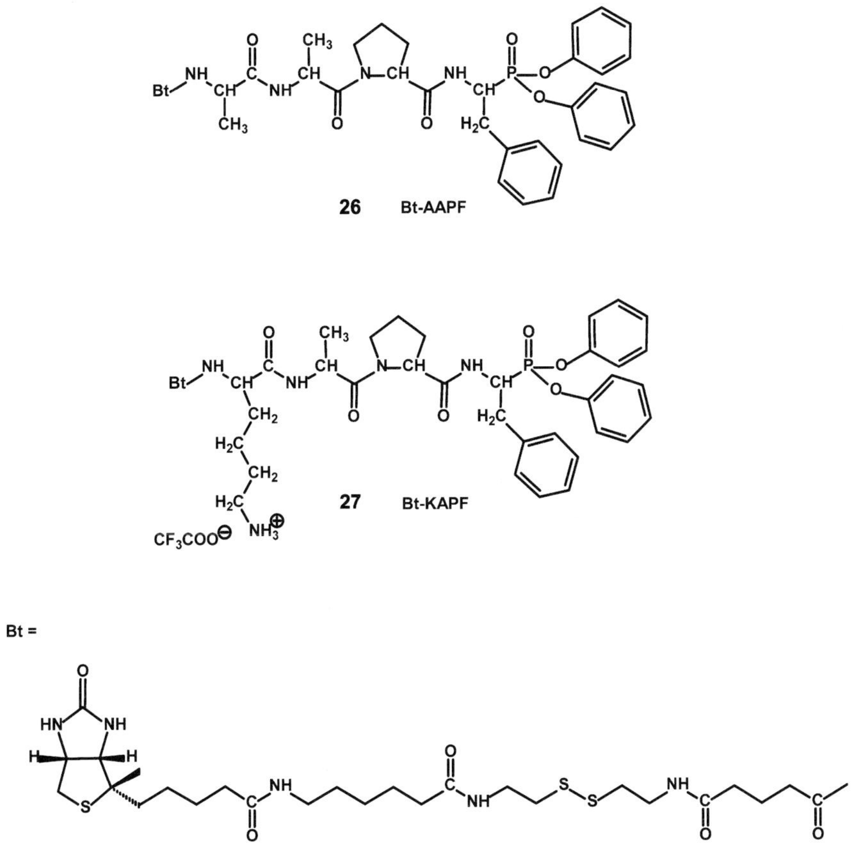

26   Bt-AAPF

27   Bt-KAPF

Bt =

Fig. 5.11. Phosphonylating inhibitor derivatives used to select subtilisin variants [64].

Subtilisin, a bacterial serine protease, has been displayed in fusion with g3p. Subtilisins are produced and secreted as pre-proenzymes that mature into active enzymes autocatalytically. Maturation occurred normally on phage but the phage-enzyme had to be protected from autoproteolysis by addition or periplasmic co-expression of the proteic inhibitor cI2. The free enzyme could be released from the complex upon addition of an anionic detergent. A library of variants in which residues 104 and 107, forming part of the S4 binding pocket, were randomized was constructed. Reaction with biotinylated phosphonylating agents AAPF (**26**) or KAPF (**27**) (Fig. 5.11), whose status is intermediate between covalent transition-state analogues and suicide inhibitors, enriched the mixture either in phage-enzymes with wild-type-like specificity or with activity on substrates possessing a lysine in P4 position, respectively. Parameters influencing the selection and efficiency of maturation were investigated [64].

5.5.2.5 Selections based directly on substrate transformations

The selection techniques described above are indirect as they rely on inhibitors. They have important limitations. Transition-state analogues are frequently unable to mimic adequately the essential features of true transition states. For many enzymatic activities, there are simply no transition-state analogues or suicide substrates.

In order to avoid these limitations, efforts have been made to find ways of coupling substrate turnover directly to a selection process. Two groups have devised a method that involves the attachment of a substrate to a phage-enzyme in a way that allows its "intraphage" interaction with the enzyme. Phage-displaying active enzymes are able to convert the substrate into product. They can be captured from mixtures with product specific reagents or antibodies.

The first implementation of this strategy was described by Pedersen *et al.* [65] (Fig. 5.12). The gene encoding staphylococcal ribonuclease (SNase), a nuclease requiring Ca^{2+} for activity, was cloned into a phagemid. A helper-phage was constructed to display, on the g3p coat protein, an acid peptide of 39 amino acid residues containing a free cystein. On addition of this helper-phage to bacteria hosting the phagemid, particles displaying both the acid peptide and SNase are produced. A conjugate of the substrate was prepared that connects it through flexible linkers, on the one hand, to a basic peptide sequence containing two free cysteins, and, on the other hand, to biotin. Incubation of the phagemid particles with the substrate conjugate allowed absorption by interaction with the acid peptide; formation of a disulfide bridge between the cysteins of the phagemid and of the conjugate was thought to stabilize the association although this was not demonstrated. This construct was shown to bind efficiently to streptavidin-coated beads under conditions where SNase is inactivated by EDTA capture of Ca^{2+} ions. Addition of soluble Dnase I releases all the bound phages. Addition of Ca^{2+} releases 20 % bound phages. A control phagemid was con-

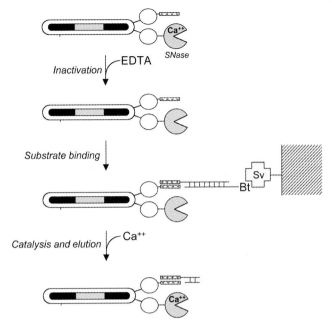

Fig. 5.12. Strategy for the selection of phage-displayed staphylococcal nuclease (SNase) with a DNA substrate derivative featuring a biotin (Bt) on one
end and a basic peptide on the other [65]. The substrate is attached to the
phage *via* an acidic peptide and the phage is captured with immobilised
streptavidin (Sv).

structed in which SNase was replaced by an Fab fragment. Selection from mixtures of
SNase- and Fab-displaying phagemid allowed selection of the former with an enrichment factor of 100.

The second implementation was described by Demartis *et al.* [66] (Fig. 5.13). Phage-
enzymes displaying enzyme-calmodulin chimeras fused to g3p were created. Two
phage-enzymes were constructed and studied in more detail. The submut-CaM-phage
displays a mutant of subtilisin in which the essential histidine has been mutated to
alanine; it is active on peptides supplying the essential histidine in the AAHY sequence. The GST-CaM-phage displays the glutathione transferase of *Schistosoma ja-
ponicum*. Substrates were conjugated to a biotinylated peptide that forms stable complexes with calmodulin in the presence of calcium ions and dissociates on addition of
calcium chelators. Submut-CaM-phages were incubated with three peptides containing respectively, (1) the product of the submut proteolytic cleavage whose N-terminal
sequence is DYKDE, (2) the substrate featuring a GAAHY-DYKDE N-terminal (- indicates the cleavage site) and (3) a negative control. The C-terminal of all these peptides
contains the calmodulin-binding motif. The phages were rescued using a product-specific mAb recognizing an N-terminal DYKDE sequence. The number of phages recov-

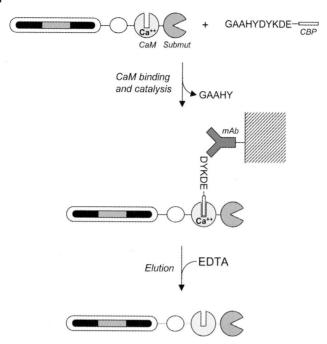

Fig. 5.13. Strategy for the selection of a subtilisin mutant (Submut) displayed on phage [66]. Calmodulin (CaM) is also displayed and allows the capture of one product of the reaction *via* the calmodulin binding peptide (CBD).

ered was 3300 and 110 times larger with peptides 1 and 2 respectively. The fact that more phages were recovered in (1) versus (2) suggests that the enzyme was not completely active on phage. When a mixture of submut-CaM- and GST-CaM-phages were incubated in the presence of peptide (2) application of the selection protocol enriched the mixture 54-fold in submut-CaM-phage.

A similar experiment was run with GST-CaM-phages (Fig. 5.14): the substrates were a calmodulin-binding peptide capped with glutathione on its N-terminus and a biotinylated derivative of 4-chloro-3,5-dinitrobenzoic acid. After reaction, the phages could be captured on streptavidin-coated beads and released by calcium complexation. The ratio of phages recovered from reactions with GST-CaM-phage *versus* submut-CaM-phage was rather low (factor 6). This was interpreted as resulting from the presence of a spontaneous conjugation between the two substrates during the course of the experiment.

A product-capture strategy was also used to select improved subtiligases. Subtiligase is a double mutant of subtilisine that catalyses the ligation of peptides. A library of $>10^9$ variants involving 25 residues of the active site was constructed on the phage-displayed enzyme. Variants that ligated a biotinylated peptide on their own ex-

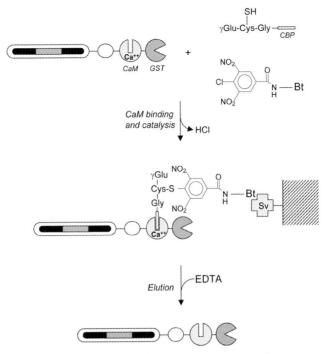

Fig. 5.14. Strategy for the selection of glutathione-S-transferase (GST) [66]. The two substrates are derivatized respectively with the calmodulin binding peptide (CBD) and a biotin (Bt). The product comprising both modules binds to the phage-displayed calmodulin (CaM) and the phage is then captured with immobilised streptavidin (Sv).

tended N-terminus (a 15-residue optimized extension) were selected by capture on streptavidin-coated beads. Several mutants with increase ligase activity were isolated. The selection protocol also afforded ligases with improved stability [67].

A possible limitation of the selection protocols described above arises from the intramolecular nature of the process. A single catalytic event, transforming the phage-bound substrate into product during the time required to complete the experiment is sufficient to lead to selection of the phage-enzyme. Consequently, poorly active enzymes are likely to be selected.

To try to circumvent this limitation, Jestin *et al.* [68] have designed and tested a selection scheme involving two chemically independent reactions: the reaction catalyzed by the enzyme and a chemical reaction leading to phage labeling by the substrate or product of the enzymatic reaction. Selection of labeled phages with a product-specific binder (e.g. a monoclonal antibody) should allow recovery of the product-labeled phages only (Fig. 5.15).

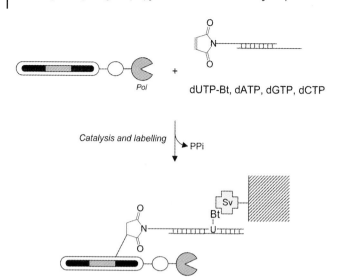

Fig. 5.15. Strategy for the selection of a phage-displayed polymerase (Pol) [68]. The primer template substrate is derivatized with a maleimide reagent and one nucleotide substrate is derivatized with biotin (Bt). The phage is labeled by the product *via* the maleimide and subsequently captured with immobilised streptavidin (Sv).

The Klenow and the Stoffel fragments of DNA polymerase I from *E. coli* and *Thermus aquaticus* were displayed on g3p. Both fragments lack the 5'→3' exonuclease domain, the Stoffel fragment also lack 3'→5' exonuclease activity. The substrate was an oligonucleotide labeled on its 5'-end with a maleimyl group. Phages displaying the Stoffel fragment were incubated with substrate in the presence of dNTPs, biotinylated dUTP and a template. After reaction the phages could be captured on streptravidin-coated magnetic beads. This shows that the product of polymerization can label the phages, presumably on the major coat protein g8p, by reaction through the maleimyl group. A oligonucleotide featuring a maleimyl group on its 5'-end and a biotin group on its 3'-end was used as a positive control. The process was used to select active *versus* denatured polymerase: a mixture of phages displaying the *E. coli* Klenow and *T. aquaticus* Stoffel fragments was heated to 60 °C to denature the *E. coli* polymerase and submitted to selection. A 14 to 26-fold enrichment in the stable enzyme was observed depending on the pH. Finally, the wild-type *T. aquaticus* enzyme was enriched 35 to 125-fold from a mixture with a low-activity mutant using the selection protocol. It is not entirely clear whether the reaction with the maleimyl group occurs after complexation with the substrate of product or before complexation and reaction, in which case this strategy becomes equivalent to the two previous ones. Furthermore, the high affinity of the polymerase for its substrate may contribute to the immobilization of the biotinylated product without involvement of the maleilation reaction.

The most recent selection protocol based on substrate transformation uses catalytic elution [69] (Fig. 5.16). It is applicable to enzymes that require a cofactor for activity, particularly metallo-enzymes provided that they are still able to bind their substrate after cofactor removal. The phage-displayed enzymes are first inactivated by metal extraction; then they are adsorbed on streptavidin-coated beads on which biotinylated substrate has been immobilised at high density. Addition of the cofactor restores activity and allows phage elution since affinity for the product is lower than for the substrate. The strategy was demonstrated with a phage-displayed metallo-β-lactamase. In model selections between wild-type enzyme and low-activity mutants, enrichment factors amounting to 20 – 40 were observed. Mutants with wild-type-like activity could be extracted from a library whose mean activity was 1.6 % of that of the wild-type enzyme. The efficiency of selection appears to increase with the level of display of the enzyme on phage, suggesting that the avidity phenomenon can play a role in the adsorption to the beads.

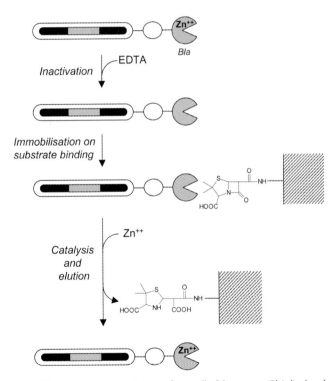

Fig. 5.16. Strategy for the selection of a metallo-β-lactamase (Bla) displayed on phage [69]. The phage-enzyme is inactivated by extracting the metallic cofactor and captured with an immobilised penicillin substrate. Addition of the metallic cofactor results in the catalytic elution of the phage-enzyme.

5.6
Conclusions

When looking at the enzymes that have been displayed on filamentous phage, it appears that most of them are extracellular enzymes, designed by nature to be secreted. A few cytoplasmic enzymes have been displayed but the level of display is apparently rather low in these cases. The largest enzymes that have been efficiently displayed so far are penicillin acylase and the dimeric alkaline phosphatase with molecular weights of 86 and 94 kD respectively. Other phages like λ and T4 could accept larger enzymes like β-galactosidase (MW = 464 kD). However, no selection for catalytic activity has been reported with enzymes displayed on these phages.

Several selection schemes have been designed and tested. However, it remains difficult to draw more than tentative conclusions about the relative efficiency of these schemes as no phage-enzyme library has been submitted to a comparative test. From results obtained with GST A1-1, it appears that selection on an immobilised TSA is more efficient than on an immobilised product analogue [70]. This is not too surprising. A broad comparison of results obtained with TSAs and suicide substrates does not allow firm conclusions to be drawn about the superiority of one system versus the other. In both cases, it is essential to pay full attention to the choice and the design of the inhibitor. TSAs have to mimic as closely as possible the main features of the true transition state and reactions with suicide substrates have to recruit the crucial characteristics of the mechanism of the catalyst. Both highly active enzymes and medium activity abzymes have been selected from libraries using the two kinds of inhibitors, however, to our knowledge, no example has been reported so far in which selection with these inhibitors has been used to evolve a first generation poorly active enzyme into a very active one. Selections directly based on substrate transformations have the potential to avoid limitations of the use of inhibitors but they are still at an early stage of development and only one example has been reported in which selection from a library was relatively successful.

In terms of achievements, selection from libraries of phage-displayed enzymes has afforded several enzymes with modified specificities and several abzymes. A combination of *in vivo* selection and *in vitro* selection for binding to ligands unrelated to the substrates has also opened the way to the engineering of the regulation of enzymatic activity.

References

[1] J. Fastrez, *Mol. Biotechnol.* **1997**, *7*, 37–55.

[2] G.P. Smith, *Science* **1985**, *228*, 1315–1317.

[3] G.P. Smith, V.A. Petrenko, *Chem. Rev.* **1997**, *97*, 391–410.

[4] S.S. Sidhu, H.B. Lowman, B.C. Cunningham, J.A. Wells, *Methods Enzymol.* **2000**, *328*, 333–363.

[5] A. Pluckthun, C. Schaffitzel, J. Hanes, L. Jermutus, *Adv. Protein Chem.* **2000**, *55*, 367–403.

[6] P. Model, M. Russel, Filamentous bacteriophage. *In* "The Bacteriophages" (R. Calendar Ed.), Vol. 2, pp. 375–456. Plenum, New York, **1988.**

[7] R.E. Webster, J. Lopez, Structure and assembly of the class 1 filamentous bacteriophage. *In* "Virus Structure and Assembly" (S. Casjens Ed.), pp 235–268, Jones & Bartlett, Boston, **1985.**

[8] L. Riechmann, P. Holliger, *Cell* **1997**, *90*, 351–360.

[9] P. Holliger, L. Riechmann, R.L. Williams, *J. Mol. Biol.* **1999**, *288*, 649–657.

[10] J. Lubkowski, F. Hennecke, A. Pluckthun, A. Wlodawer, *Nat. Struct. Biol.* **1998**, *5*, 140–147.

[11] J. Rakonjac, J.N. Feng, P. Model, *J. Mol. Biol.* **1999**, *289*, 1253–1265.

[12] N.A. Linderoth, M.N. Simon, M. Russel, *Science* **1997**, *278*, 1635–1638.

[13] A. Kuhn, D. Troschel. Distinct steps in the insertion pathway of bacteriophage coat proteins. *In* "Membrane Biogenesis and Protein Targeting" (W. Newport & R. Lill, eds), pp 33–47. Elsevier, New York, **1992.**

[14] M.P. Rapoza, R.E.Webster, *J. Bacteriol.* **1993**, *175*, 1856–1859.

[15] C. Gao, S. Mao, C.H. Lo, P. Wirsching, R.A. Lerner, K.D. Janda, *Proc. Natl Acad. Sci. U.S.A.* **1999**, *96*, 6025–6030.

[16] L.S. Jespers, J.H. Messens, A. De Keyser, D. Eeckhout, I. Van den Brande, Y.G. Gansemans, M.J. Lauwereys, G.P. Vlasuk, P.E. Stanssens, *Biotechnology (N Y)* **1995**, *13*, 378–382.

[17] S.S. Sidhu, G.A. Weiss, J.A. Wells, *J. Mol. Biol.* **2000**, *296*, 487–495.

[18] G. Fuh, S.S. Sidhu, *FEBS Lett.* **2000**, *480*, 231–234.

[19] R. Crameri, R. Jaussi, G. Menz, K. Blazer, *Eur. J. Biochem.* **1994**, *226*, 53–58.

[20] N. Armstrong, N.B. Adey, S.J. McConnell, B.K. Kay. Vectors for phage display. *In* "Phage display of peptides and proteins. A laboratory manual." (B.K. Kay, J. Winter & J. McCafferty, eds), pp 35–53. Academic Press, San Diego, **1996.**

[21] J. Beekwilder, J. Rakonjac, M. Jongsma, D. Bosch, *Gene* **1999**, *228*, 23–31.

[22] L. Castagnoli, A. Zucconi, M. Quondam, M. Rossi, P. Vaccaro, S. Panni, S. Paoluzi, E. Santonico, L. Dente, G. Cesareni, *Comb. Chem. High Throughput Screen.* **2001**, *4*, 121–133.

[23] I.S. Dunn, *J. Mol. Biol.* **1995**, *248*, 497–506.

[24] I.N. Maruyama, H.I. Maruyama, S. Brenner, *Proc. Natl Acad. Sci.U.S.A.* **1994**, *91*, 8273–8277.

[25] N. Sternberg, R.H. Hoess, *Proc. Natl Acad. Sci.U.S.A.* **1995**, *92*, 1609–1613.

[26] Y.G. Mikawa, I.N. Maruyama, S. Brenner, *J Mol Biol* **1996**, *262*, 21–30.

[27] I.S. Dunn, *Gene* **1996** *183*, 15–21.

[28] F. Yang, P. Forrer, Z. Dauter, J.F. Conway, N.Q. Cheng, M.E. Cerritelli, A.C. Steven, A. Pluckthun, A. Wlodawer, *Nat. Struct. Biol.* **2000**, *7*, 230–237.

[29] V.P. Efimov, I.V. Nepluev, V.V. Mesyanzhinov, *Virus Genes* **1995**, *10*, 173–177.

[30] Z.J. Ren, G.K. Lewis, P.T. Wingfield, E.G. Locke, A.C. Steven, L.W. Black, *Protein Sci.* **1996**, *5*, 1833–1843.

[31] J. Jiang, L. Abu-Shilbayeh, V.B. Rao, *Infect. Immun.* **1997**, *65*, 4770–4777.

[32] Z.J. Ren, L.W. Black, *Gene* **1998**, *215*, 439–444.

[33] R.M.D. Verhaert, J. van Duin, W.J. Quax, *Biochem. J.* **1999**, *342*, 415–422.

[34] D.R. Corey, A.K. Shiau, Q. Yang, B.A. Janowski, C.S. Craik, *Gene* **1993**, *128*, 129–134.

[35] J.W. Crissman, G.P. Smith, *Virology* **1984**, *132*, 445–455.

[36] S.C. Makrides, *Microbiol. Rev.* **1996**, *60*, 512–538.

[37] E. Soderlind, A.C.S. Lagerkvist, M. Duenas, A.C. Malmborg, M. Ayala, L. Danielsson, C.A.K Borrebaeck, *Bio-Technology* **1993**, *11*, 503–507.

[38] H. Bothmann, A. Pluckthun, *Nat. Biotechnol.* **1998**, *16*, 376–380.

[39] H. Bothmann, A. Pluckthun, *J. Biol. Chem.* **2000**, *275*, 17100–17105.

[40] P. Soumillion, L. Jespers, M. Bouchet, J. Marchand-Brynaert, G. Winter, J. Fastrez, *J. Mol. Biol.* **1994**, *237*, 415–422.

[41] R.C. Cadwell, G.F. Joyce, *PCR Methods Appl.* **1994**, *3*, S136–140.

[42] A.A. Volkov, F.H. Arnold, *Methods Enzymol.* **2000**, *328*, 447–456.

[43] P. Waterhouse, A.D. Griffiths, K.S. Johnson, G. Winter, *Nucleic Acids Res.* **1993**, *21*, 2265–2266.

[44] F. Geoffroy, R. Sodoyer, L. Aujame, *Gene* **1994**, *151*, 109–113.

[45] J. McCafferty, R.H. Jackson, D.J. Chiswell, *Protein Eng.* **1991**, *4*, 955–961.

[46] D. Legendre, P. Soumillion, J. Fastrez, *Nat. Biotechnol.* **1999**, *17*, 67–72.

[47] M. Fransen, P.P. Van Veldhoven, S. Subramani, *Biochem. J.* **1999**, *340*, 561–568.

[48] J. Light, R.A. Lerner, *Bioorg. Med. Chem.* **1995**, *3*, 955–967.

[49] M. Widersten, B. Mannervik, *J. Mol. Biol.* **1995**, *250*, 115–122.

[50] J.A. Hunt, C.A. Fierke, *J. Biol. Chem.* **1997**, *272*, 20364–20372.

[51] L.O. Hansson, M. Widersten, B. Mannervik, *Biochemistry* **1997**, *36*, 11252–11260.

[52] J. McCafferty, K.J. Fitzgerald, J. Earnshaw, D.J. Chiswell, R. Link, R. Smith, J. Kenten, *Appl. Biochem. Biotechnol.* **1994**, *47*, 157–173.

[53] M. Baca, T.S. Scanlan, R.C. Stephenson, J.A. Wells, *Proc. Natl. Acad. Sci. U.S.A.* **1997**, *94*, 10063–10068.

[54] I.Fujii, S. Fukuyama, Y. Iwabuchi, R. Tanimura, *Nat. Biotechnol.* **1998**, *16*, 463–467.

[55] G.W. Zhou, J. Guo, W. Huang, R.J. Fletterick, T.S. Scanlan, *Science* **1994**, *265*, 1059–1064.

[56] J.D. Janda, C.H.L. Lo, T. Li, C.F. Barbas III, P. Wirsching, R.A. Lerner, *Proc. Natl. Acad. Sci. U. S. A.* **1994**, *91*, 2532–2536.

[57] J. Wagner, R.A. Lerner, C.F. Barbas III, *Science* **1995**, *270*, 1797–1800.

[58] F. Tanaka, R.A. Lerner, C.F. Barbas III, *J. Am. Chem. Soc.* **2000**, *122*, 4835–4836.

[59] I. Lasters, N. Van Herzeele, H.R. Lijnen, D. Collen, L. Jespers, *Eur. J. Biochem.* **1997**, *244*, 946–952.

[60] S. Vanwetswinkel, J. Marchand-Brynaert, J. Fastrez, *Bioorg. Med. Chem. Lett.* **1996**, *6*, 789–792.

[61] F. Tanaka, H. Almer, R.A. Lerner, C.F. Barbas III, *Tetrahedron Lett.* **1999**, *40*, 8063–8066.

[62] S. Vanwetswinkel, B. Avalle, J. Fastrez, *J. Mol. Biol.* **2000**, *295*, 527–540.

[63] K.D. Janda, L.-C. Lo, C.-H.L. Lo, M.-M. Sim, R. Wang, C.-H. Wong, R.A. Lerner, *Science* **1997**, *275*, 945–948.

[64] D. Legendre, N. Laraki, T. Graslund, M.E. Bjornvad, M. Bouchet, P.A. Nygren, T.V. Borchert, J. Fastrez, *J. Mol. Biol.* **2000**, *296*, 87–102.

[65] H. Pedersen, S. Holder, D.P. Sutherlin, U. Schwitter, D.S. King, P.G. Schultz, *Proc. Natl. Acad. Sci. U.S.A.* **1998**, *95*, 10523–10528.

[66] S. Demartis, A. Huber, F. Viti, L. Lozzi, L. Giovannoni, P. Neri, G. Winter, D. Neri, *J. Mol. Biol.* **1999**, *286*, 617–633.

[67] S. Atwell, J.A. Wells, *Proc. Natl. Acad. Sci. U.S.A.* **1999**, *96*, 9497–9502.

[68] J.-L. Jestin, P. Kristensen, G. Winter, *Angew. Chem. Int. Ed.* **1999**, *38*, 1124–1127.

[69] I. Ponsard, M. Galleni, P. Soumillion, J. Fastrez, *ChemBioChem* **2001**, *2*, 253–259.

[70] M. Widersten, L.O. Hansson, L. Tronstad, B. Mannervik, *Methods Enzymol.* **2000**, *328*, 389–404.

[71] N. Dimasi, A. Pasquo, F. Martin, S. Di Marco, C. Steinkuhler, R. Cortese, M. Sollazzo, *Protein Eng.* **1998**, *11*, 1257–1265.

[72] R. Eerola, P. Saviranta, H. Lilja, K. Pettersson, M. Lovgren, T. Karp, *Biochem. Biophys. Res. Commun.* **1994**, *200*, 1346–1352.

[73] R. Crameri, M. Suter, *Gene* **1993**, *137*, 69–75.

[74] K. Maenaka, M. Furuta, K. Tsumoto, K. Watanabe, Y. Ueda, I. Kumagai, *Biochem. Biophys. Res. Commun.* **1996**, *218*, 682–687.

[75] J. Ku, P.G. Schultz, *Bioorg. Med. Chem.* **1994**, *2*, 1413–1415.

[76] K. Korn, H.H. Foerster, U. Hahn, *Biol. Chem.* **2000**, *381*, 179–181.

[77] J.A. Watson, M.G. Rumsby, R.G. Wolowacz, *Biochem. J.* **1999**, *343*, 301–305.

[78] T. Wind, S. Kjaer, B.F. Clark, *Biochimie* **1999**, *81*, 1079–1087.

6
Directed Evolution of Binding Proteins by Cell Surface Display: Analysis of the Screening Process

K. Dane Wittrup

"Even a blind chicken finds a few grains of corn now and then." *Lyle Lovett.*

6.1
Introduction

The recent rapid accumulation of success stories attests to the power of directed evolution for the purposeful alteration of protein properties [1 – 3]. Whether for engineering catalytic activity, thermal stability, or binding affinity, it is clear that given our present rudimentary understanding of protein structure – function relationships the most rapid route to an improved protein is via Darwinian mutation and selection. At the same time, evolutionary experiments can help to build our understanding of how proteins work, for example with respect to antibody affinity maturation, hormone – receptor recognition, protein binding hotspots, thermal stability, and protein folding [4 – 9]. In this way, evolutionary approaches may someday contribute to their own obsolescence in favor of silico rational design. That day is far off, however.

   At this point in time the great majority of protein engineering efforts utilize phage display, first described in 1985 by George Smith [10]. Alternatively, a growing list of interesting protein mutants have been attained via cell surface display methodology, providing motivation to explore this avenue further [11 – 18]. A particular advantage of cell-displayed mutant libraries is that the screening machinery becomes transparent, in a sense – one can directly observe the underlying kinetics, equilibria, and statistics of the screening process, and thereby design improved search strategies. In general, such analysis is not possible with the submicroscopic particles that are utilized in phage, ribosome, or RNA display. With these methods, artifacts such as protein stickiness, genetic instability, or the stochastic statistics inherent with small numbers of displayed molecules are undetected until the end of the screen, when the final products are examined.

In this Chapter I will analyze the process of protein-directed evolution by cell surface display, with particular emphasis on the kinetics, thermodynamics, and statistics of library screening. Although these ideas were originally formulated for the yeast display system, they should be generally applicable to display on other cell types. The discussion will be restricted to binding proteins, with antibodies as the benchmark for this class. Enzyme engineering introduces additional complexity to the system, both scientifically (in terms of diverse catalytic mechanisms) and technically (in terms of physically linking diffusible reaction products to their molecular source [19]). Nevertheless, some principles discussed here will also be relevant to engineering enzymes by directed evolution.

The first law of directed evolution is that you get what you screen for [20]. Anecdotes abound in the unpublished literature of negative results concerning unintended outcomes of screening experiments that were, in retrospect, fully consistent with the ground rules established for the evolutionary process. This being the case, careful attention to the physical chemistry and statistics of screening can be rewarded with markedly improved results.

Tab. 6.1.

I. Library Construction	II. Mutant Isolation
Mutagenesis	**Differential labeling**
Error-prone PCR	**Definition of objective function**
mutagenesis rate?	immobilization?
Site specific	fluorescence?
saturation or spiking?	**Stringency**
which sites?	pH, chaotropes, T, shear?
Shuffling	**Equilibrium**
source of diversity?	ligand concentration?
crossover frequency?	**Kinetic competition**
	competition time?
Expression	**Screening**
Expression Systems	**Oversampling**
Molecular conjugates	# mutants examined
ribosomes	relative to genetic diversity?
puromycin linkage	**Instrumentation**
Viruses	by hand ($10^3 - 10^4$)
phage	robotic ($10^5 - 10^6$)
baculovirus	flow cytometry ($10^7 - 10^9$)
Cells	**Retention criterion**
E. coli	expression normalization?
yeast	yield vs. purity?
insect	purification passes?
Stability maturation	
protease sensitivity	
ER quality control	

Directed evolution is always comprised of two components: I) generation of diversity; and II) isolation of improved variants. Each of these aspects are in turn divisible, as shown in Tab. 6.1, into the subclassifications of mutagenesis, expression, differential labeling, and screening. The discussion will follow this outline.

6.2
Library Construction

6.2.1
Mutagenesis

A variety of methods of mutagenesis are available to generate library diversity at the nucleic acid level. Perhaps the most commonly practiced approach is the use of error-prone PCR, which removes any biases potentially introduced into the experimental design by incorrect hypotheses regarding the best sites for alteration. Despite this "hypothesis-free" aspect of error-prone PCR mutagenesis, a design choice is still necessary regarding the error rate. At low mutagenesis rates, single changes will predominate and subsequent interpretation will be easier; however, the probability of finding a given single change will be lower, and beneficially interacting double mutations will not be found. At higher mutagenesis rates, a more comprehensive sampling of single changes is performed, and beneficial double mutations are sampled; however, deleterious mutations will accumulate to a greater extent. The existence of these counteracting tendencies ensures the existence of an optimal mutagenesis rate, and there is empirical evidence that it may lie towards higher error rates than often practiced [13, 21, 22]. Since the optimal error rate is unknown, it may be wise to sample a spread of mutation frequencies in each library to be screened.

Deep searches of sequence space (i.e. exhaustive sampling of all possible mutations) can be performed by saturation mutagenesis at particular sites in a protein. Selection of the right site to mutagenize can provide large functional gains in affinity, stability, or activity in one round of mutagenesis and screening (e.g. [15]). Conversely, deep searches in the wrong place can yield negligible improvements [23 – 25]. Several algorithms have been proposed for identifying the most favorable sites for exploration. Pastan and coworkers suggest that the "hotspot" sites in antibody genes that are mutated at high frequency by B cells during somatic hypermutation and affinity maturation *in vivo* represent intrinsically favorable locations, and have experimentally used this tactic with success [26]. Arnold & coworkers have proposed that consensus mutations that occur frequently following a first, random, round of mutagenesis, might represent good spots for further investigation [27]. And finally, Voigt *et al.* have developed an algorithm for computational identification of sites that are minimally coupled to the overall protein structure, which are thereby more "labile" to change and less likely to incur deleterious consequences upon mutation [28]. At this point no general recommendation is possible for choosing sites for saturation mutagenesis.

DNA shuffling has proven a powerful search strategy in sequence space, with many large phenotypic improvements attained [2]. The design choices for shuffling include the various mechanisms for the shuffling reaction itself, recently reviewed [29]; and the source of genetic diversity for the shuffling reaction, whether from an ensemble of improved point mutants or a less-related family of homologues [30].

6.2.2
Expression

The primary criterion for selection of any display format, whether cell surface display or *in vitro*, must be the biosynthetic requirements of the protein to be engineered. The phylogeny of available display formats has recently been reviewed [31]. The synthetic generation of diversity at the nucleic acid level is fairly straightforward, but translation of this underlying diversity into a collection of folded, functional proteins is always subject to a biological filter, or expression bias. Proteins co-evolve with a particular biosynthetic milieu that is responsible for a number of posttranslational events necessary for assumption of their functional form. For example, extracellular and secreted eucaryotic proteins rely on the machinery of the endoplasmic reticulum to introduce and shuffle disulfide bonds in a controlled redox environment, covalently attach polysaccharides at Asn-X-Ser/Thr sites, transiently interact with chaperones such as BiP and GRP94, and finally pass a quality control check with calnexin before packaging in vesicles for transport towards the cell surface. Perhaps not surprisingly, such secretory domains do not always fold efficiently in the bacterial periplasm or *in vitro* transcription/translation systems. These shortcomings can be at least partially overcome with clever molecular genetic engineering of non-eucaryotic expression systems [32 – 35].

As an alternative to engineering expression systems to handle a given protein domain efficiently, a number of methods have arisen for stability maturation to improve the intrinsic expression properties of the protein itself. In one approach, vulnerability to proteolysis is used as a proxy indicator for instability, and mutant proteins with greater protease resistance are selected by phage display [36, 37]. Solubility and proteolytic stability of cytoplasmically expressed GFP fusions can be used as screening criteria for stabilized proteins [38]. In an analogous approach, the quality control apparatus of the eucaryotic secretory pathway is utilized to retain and degrade unstable proteins targeted for display on the yeast cell wall [39]. Efficiency of transit through the eucaryotic secretory pathway is closely related to protein stability [40, 41]. An advantage of the yeast display method is the continuously variable nature of surface expression levels, allowing gradual improvements to be selected across a span of at least two orders of magnitude [17].

6.3
Mutant Isolation

6.3.1
Differential labeling

Precise definition of the screening parameter is essential – if a particular equilibrium binding constant or dissociation rate are desired, these quantities should be linked as tightly as possible to the measured parameter. In the language of computational optimization, this screening parameter is the "objective function" that the search process will evaluate. In evolutionary language, this parameter is termed "fitness." Occasionally a "stringency" criterion such as low pH, high chaotrope concentration, or high temperature has been utilized as a proxy variable for screening for the actual variable of interest, but the connection between these environmental conditions and a particular K_d or k_{off} is tenuous at best, and could steer the screening process in unintended and undesirable directions.

Fluorescent labeling presents compelling advantages for quantitative analysis. Fluorescence intensity is proportional to label concentrations across several orders of magnitude, and high-throughput instrumentation is available for precise simultaneous quantitation of labeling intensity at multiple wavelengths. The use of soluble ligands enables labeling governed by well-defined monovalent kinetics and equilibria, without avidity artifacts from multivalency or surface immobilization artifacts. A plethora of fluorescent secondary labeling reagents are commercially available, providing flexibility in experimental design (e.g. http://www.probes.com/handbook/).

The combinatorial library should be labeled so as to maximize the relative signal difference between the generally small number of improved mutants and the vast excess of unimproved mutants. For affinity maturation, two general methods have been used: kinetic or "off-rate" screening by competition with excess unlabeled ligand, and equilibrium screening by lowered ligand concentration. Although the term "optimize" is often loosely used synonymously with "improve", there is in fact one best way to differentially label cells for either equilibrium or kinetic screening. Detailed derivations and discussion of optimal labeling protocols are provided elsewhere [42], and are outlined briefly here.

The principle underlying equilibrium screening is illustrated in Fig. 6.1. The key design choice is the concentration of labeled ligand to use. As ligand concentration is decreased, labeling intensity decreases according to a simple monovalent binding isotherm. However, all detection systems possess a noise floor below which the signal will not drop, hence wild-type and mutant cells are identically labeled at both very high or very low intensity. The ratio of mutant to wild-type fluorescence is optimal at some

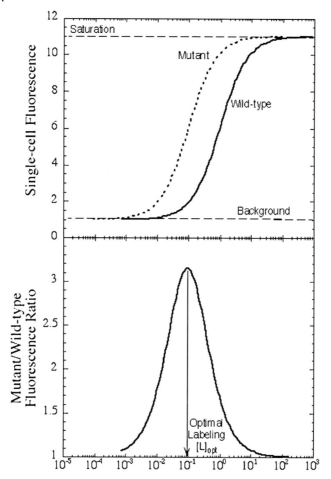

Fig. 6.1. Equilibrium labeling for optimal discrimination of mutant clones. At high ligand concentrations, both mutant and wild-type cells are labeled to saturation; at low ligand concentrations, both exhibit background labeling. At an intermediate ligand concentration, the ratio of mutant to wild-type fluorescence is maximal.

intermediate value. The optimal ligand concentration is a function of wild-type and mutant K_d, and the signal-to-noise ratio of the measurement. This relationship is given in the following equation:

$$\frac{[L]_{opt}}{K_d^{wt}} = \frac{1}{\sqrt{S_r K_r}}$$

where $[L]_{opt}$ is the concentration of ligand yielding the maximum ratio of mutant to wild-type fluorescence, S_r is the maximum signal-to-background ratio for cells saturated with fluorescent ligand, and K_r is the minimum affinity improvement desired (e.g. $K_r = 5$ if mutants improved = 5-fold in affinity are desired). S_r is the ratio of fluorescence of cells saturated with fluorescent ligand over autofluorescence of unlabeled cells, and is dependent on the particular flow cytometer and efficiency of protein expression. S_r can be quickly estimated from the relative positions of the positively stained peak and the peak for nondisplaying cells in a histogram [43]. A general rule of thumb to identify several-fold improved mutants by flow cytometric screening of yeast-displayed libraries is that the optimal ligand concentration is approximately 10 to 20-fold lower than the wild-type K_d.

For excess ligand, the time to reach equilibrated labeling is approximately $5.0/(k_{on}[L]+k_{off})$. For example, consider an antibody/antigen pair with typical binding parameters: $k_{on} = 10^5\,\text{M}^{-1}\text{s}^{-1}$, $k_{off} = 10^{-4}\,\text{s}^{-1}$, and $[L] = 0.1\,\text{nM}$. If the antibody and antigen are mixed together at time zero, then at 1 min the complex concentration has reached 0.7 % of its final value; at 1 h, the complex is at 33 % of its final concentration; and finally at 12 h 99 % of the equilibrium antibody/antigen complex concentration is attained. The slow pace of equilibration of antibody/antigen interactions at low ligand concentrations is often not appreciated.

If ligand concentration is in excess over the summed concentration of displayed binding protein, the analysis just described is correct. In practice this can be accomplished by decreasing the number of displaying cells per assay volume. If, however, bulk ligand is depleted from the labeling solution nevertheless, the analysis would have be adjusted to account for the change in final ligand concentration. Longer equilibration times would be necessary in situations of ligand depletion, complicating the analysis somewhat.

Kinetic screening is illustrated in Fig. 6.2. The library is first labeled to saturation with ligand, and then at time zero an excess of unlabeled ligand competitor is added. Label intensity then falls exponentially with rate constant k_{off} to the background level. The key design choice is the competition time prior to screening. As with equilibrium labeling, there is an optimum value for this parameter that strikes a balance between intrinsic kinetic differences of the mutants and signal detection limits. The optimal competition time is given by the following relationship:

$$k_{off}, wt^t_{opt} = 0.293 + 2.05\,log\,k_r + \left(2.30 - 0.759\frac{1}{k_r}\right)log\,S_r$$

where t_{opt} is the optimal duration of competition, S_r is the signal-to-background ratio of flow cytometrically analyzed cells, and k_r is the minimum fold-improvement desired in k_{off}. S_r is best calculated as the ratio of fluorescence of displaying cells saturated with

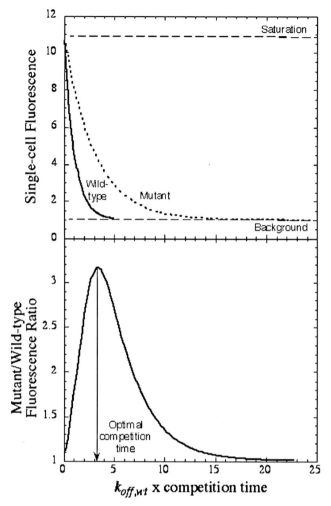

Fig. 6.2. Kinetic competition labeling for optimal discrimination of mutant clones. Both wild-type and mutant clones are labeled to saturation, and at time zero excess unlabeled ligand competitor is added. Single cell fluorescence drops exponentially to background levels, which are the same for both mutant and wild-type cells. Optimal discrimination is obtained at an intermediate competition time.

fluorescent ligand over that of displaying cells following competition to complete dissociation. In practice, for flow cytometric screening of yeast displayed libraries, a competition time of approximately $5/k_{off,wt}$ is optimal. The optimal labeling strategy described above could in principle also be applied to discriminate on the basis of the association rate constant k_{on}.

6.3.2
Screening

The first issue to resolve regarding library screening is how many mutants to analyze. Setting aside practical considerations for the moment, one wishes to screen a sufficient excess of clones over the intrinsic library diversity so as to minimize random losses (i.e. false negatives). The required level of oversampling can be defined in two alternative ways: to attain a given probability of loss for a *given* clone, or to attain a given probability of loss for *any* clone. The latter criterion requires a much greater oversampling than the former, as illustrated in the following simple calculation. Consider a library with diversity 10^6, and an individual mutant present at a frequency of 1 in 10^6. If a number of total clones N is analyzed, then the number of representatives of that particular mutant in the sample is a Poisson random number. The probability that there are zero representatives of the mutant in the sample is $e^{-10^{-6}N}$. For 95 % statistical confidence that there is at least one representative clone for this given mutant, the sample size must be 3×10^6. Hence, a three-fold oversampling of the under-

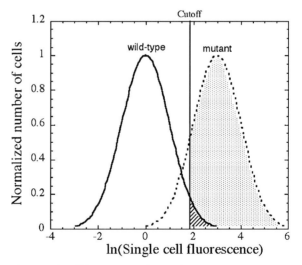

Fig. 6.3. Cutoff fluorescence selection for screening. Instrumentation, labeling, and biological noise introduce spreading into a fluorescence measurement, such that the fluorescence probability distributions for wild-type and mutant cells overlap. The logarithm of single-cell fluorescence as measured by flow cytometry is generally well-approximated by a symmetrical Gaussian curve. A cutoff fluorescence value is selected for screening, with all cells above that value sorted out. The enrichment factor for the mutants is the ratio of (dotted + striped areas)/(striped area), and the probability of retention of a given mutant clone at a single pass is the (striped + dotted area)/(all area under mutant curve).

lying diversity is a reasonable rule of thumb by this criterion. Now consider repeating this analysis for *all* possible mutants. If the probability of loss for each individual clone is 5 %, then the probability that every possible mutant is represented in the sample is 0.95^{10^6}. Hence, for 95 % statistical confidence that all possible mutants have been represented, the probability of loss for any given clone must be less than 5×10^{-8}. To achieve this low probability of loss, a sample size of 17×10^6 must be used. In general, this degree of oversampling (i.e. approximately 20–fold) is not experimentally unreasonable.

The practical limitation on library oversampling depends on screening instrumentation, for which flow cytometry must be considered the gold standard. Commercially available flow cytometers analyze multicolor fluorescence signals and sort desired cells within user-defined gates at a rate of 50,000 cells per second [44]. A salient comparison would be to microplate-based robotic screening. Considering each cell in a combinatorial library to be functionally equivalent to a microplate well, a flow cytometer can screen the equivalent of several *million* 1536-well microplates per day.

With all separations there must be a tradeoff between yield and purity – high loss probabilities inevitably accompany high enrichment ratios. The underlying cause for this phenomenon is illustrated in Fig. 6.3. Although the mean mutant signal intensity can be maximally differentiated from wild type as discussed above, the expression, labeling and measurement processes introduce a point spread function that can cause

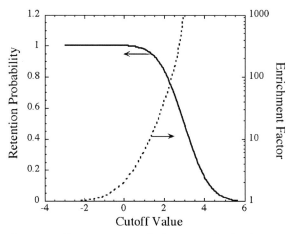

Fig. 6.4. Yield vs. purity in cutoff selection. For the distributions represented in Fig. 6.3, retention probability (solid line) and enrichment factor (dotted line) were calculated as a function of the cutoff fluorescence value. For example, a cutoff value of 1.4 gives 95 % probability of retention of a given mutant cell, but a fairly modest enrichment factor of 11.8–fold. A cutoff of 3.0 increases the enrichment factor two orders of magnitude to 1000 × , at the cost of an increase in probability of clone loss to 50 %.

the signals to overlap, sometimes substantially. The key design choice is where to set the cutoff criterion for rejection of a clone; high values result in good enrichment ratios but high probability of loss, while low values provide poor enrichment ratios but low loss probabilities. These counteracting trends are illustrated in Fig. 6.4. Needless to say, one nevertheless desires both high enrichment ratios and low loss probabilities.

Fortunately, this catch-22 situation only applies for a single-pass through the screening system. As with many classic chemical engineering unit operations such as distillation, multistaged separation can propagate small single-pass enrichment ratios to larger values. Hence, the use of multiple separation passes can accomplish both high enrichment ratios and low probabilities of clonal loss.

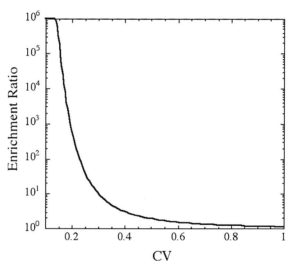

Fig. 6.5. Peak spreading strongly affects enrichment ratio at fixed probability of retention. The coefficient of variance CV is equal to the ratio of the standard deviation to the mean, and is a measure of peak breadth. For example, in both curves shown in Fig. 6.3 the CV is 1.0. The enrichment ratio was calculated for a situation in which mutant fluorescence intensity was double wild-type fluorescence intensity, the mutant was initially present at 1 in 10^6 cells, and the probability of retention was fixed at 95 %. The logarithmic fluorescence intensity was assumed to follow a Gaussian distribution. Fixing the probability of retention defines the cutoff fluorescence value for screening at a given CV. Enrichment ratio drops precipitously with increasing CV, as the mutant and wild-type fluorescence distributions begin to overlap. At a CV of 0.2, the enrichment factor is 600. However, at a CV of 0.4, the enrichment factor has dropped to 3! Clearly, every effort should be expended to minimize peak spreading and subsequent overlap of the mutant and wild-type fluorescence distributions.

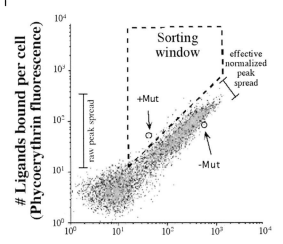

Protein fusions displayed per cell (Fluorescein fluorescence)

Fig. 6.6. Normalization by expression level effectively decreases peak spreading by a significant factor. The dot plot shown here is actual flow cytometric data for a homogenous population of yeast cells displaying a particular single chain antibody. On the X axis, the number of antibody molecules present on the cell surface has been quantified by immunofluorescent labeling of an epitope tag at the C terminus of the antibody fragment. On the Y axis, the number of bound antigen molecules is quantified with a different colored fluorophore. In such experiments, there is generally a nondisplaying population corresponding to 20 – 50 % of the cells. In this data set, the nondisplaying population is the peak centered at an X and Y intensity of $\sim$10^1. It is apparent that the levels of surface display are broadly distributed, with fluorescein fluorescence intensity values ranging from $\sim$30 – 1000. This leads to substantial overlap of mutant and wild-type ligand binding curves, as illustrated for the individual cells marked with arrows as +Mut and – Mut on the figure. Although the +Mut cell exhibits higher *specific* ligand binding than the – Mut cell for its given surface expression level, the raw ligand binding signal is *below* the ligand binding intensity for – Mut. It would therefore be *impossible* to isolate this +Mut cell from the – Mut cell on the basis of ligand binding intensity alone, due to the raw peak spread. However, a diagonal sorting window can be drawn as indicated, that normalizes for varying surface expression levels. Use of a diagonal sorting window effectively lowers the peak spread substantially, as indicated in the upper right corner.

The predominant concern should be the loss of rare clones in the screening process, a risk that is greatest in the first screening pass since desirable mutants are at their lowest frequency in the initial library. In subsequent passes, desirable mutants have been enriched and therefore the loss probability drops substantially. As a general rule, one should use a very conservative first pass screen, for example isolating the top 5 % of the labeled library. The modest 20-fold enrichment factor attained in this first pass nevertheless can diminish the importance of the purity vs. yield tradeoff for subsequent screening rounds, since the larger number of enriched mutant clones substantially lowers the overall probability of loss. For example, consider a stringent cutoff value for which the probability of loss for a given mutant is 90 %. If there is one cell expressing the desired mutant in the initial library, there is a 90 % chance that it will be lost. However, after a 20-fold enrichment under more conservative conditions that ensure the mutant is not lost, it will be present at 20 cells in the population. The probability that all 20 cells in the population will be lost under the stringent conditions would be $(0.90)^{20}$, or only a 12 % chance that the mutant will be completely lost from the library.

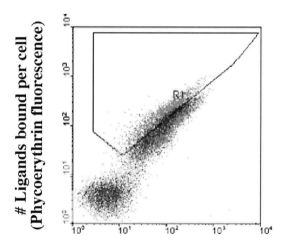

Protein fusions displayed per cell (Fluorescein fluorescence)

Fig. 6.7. Fine discrimination of a mutant population by expression normalization. Data are shown for an actual affinity maturation experiment, using off-rate screening. The mutant population (red) was enriched from the "wild-type" population (blue) using the R1 polygon window drawn on the dot plot. The mutant was determined to have a dissociation rate constant 5–fold slower than wild-type. Note the large extent of overlap in the ligand binding signal, such that it would be extremely difficult to isolate the mutant from the wild-type on the basis of the absolute level of ligand binding alone.

The difficulty of separation is highly dependent on peak spreading, as shown in Fig. 6.5. It is therefore critical to minimize the peak width as far as possible. This would be difficult for cell display methods if only single color fluorescent labeling were used, because the primary source of variability is biological. Flow cytometry instrumentation point spread functions generally contribute below 2 % to the overall coefficient of variance (CV = standard deviation/mean), but typical overall CVs for yeast display are approximately 50 – 100 % for the logarithmic fluorescence intensity.

Fortunately, two-color fluorescent labeling can be used to normalize ligand binding levels vs. expression levels, as shown in Fig. 6.6. The net effect of such normalization is to dramatically reduce the effective CV, by implicitly calculating the ratio of bound ligand levels to the level of cell-surface protein display. An example of the discrimination achievable by such normalization is illustrated for the actual case of an enriched mutant population from a yeast display affinity maturation project, in Fig. 6.7.

6.4
Summary

Navigating the protein functional landscape via directed evolution carries risks and rewards: the risk of finding idiosyncratic mutants that meet the screening criterion in unexpected, potentially trivial ways; and the reward of rapidly improving the properties of interest once the screening process has been systematically analyzed and optimized. The perspectives outlined here should provide a roadmap for designing successful protein engineering projects using cell surface display.

Acknowledgments
Thanks to Jennifer Cochran, Katarina Midelfort, and Andrew Girvin for critical comments on this chapter. Thanks to Christilyn Graff for providing the data for Figs 6.6 and 6.7 prior to publication. The yeast display work reviewed here was supported by grants from the National Institutes of Health and the Whitaker Foundation.

References

[1] Arnold, F. H.. "Combinatorial and computational challenges for biocatalyst design." Nature (2001) 409(6817): 253–7.

[2] Ness, J. E., S. B. Del Cardayre, J. Minshull and W. P. Stemmer. "Molecular breeding: the natural approach to protein design." Adv Protein Chem (2000) 55: 261–92.

[3] Pluckthun, A., C. Schaffitzel, J. Hanes and L. Jermutus. "In vitro selection and evolution of proteins." Adv Protein Chem (2000) 55: 367–403.

[4] Atwell, S., M. Ultsch, A. M. De Vos and J. A. Wells. "Structural plasticity in a remodeled protein-protein interface." Science (1997) 278(5340): 1125–8.

[5] DeLano, W. L., M. H. Ultsch, A. M. de Vos and J. A. Wells. "Convergent solutions to binding at a protein-protein interface." Science (2000) 287(5456): 1279–83.

[6] Sauer, R. T.. "Protein folding from a combinatorial perspective." Fold Des (1996) 1(2): R27–30.

[7] Silverman, J. A., R. Balakrishnan and P. B. Harbury. "From the Cover: Reverse engineering the (beta/alpha)8 barrel fold." Proc Natl Acad Sci U S A (2001) 98(6): 3092–7.

[8] Spiller, B., A. Gershenson, F. H. Arnold and R. C. Stevens. "A structural view of evolutionary divergence." Proc Natl Acad Sci U S A (1999) 96(22): 12305–10.

[9] Wedemayer, G. J., P. A. Patten, L. H. Wang, P. G. Schultz and R. C. Stevens. "Structural insights into the evolution of an antibody combining site." Science (1997) 276(5319): 1665–9.

[10] Smith, G. P. "Filamentous fusion phage: novel expression vectors that display cloned antigens on the virion surface." Science (1985) 228(4705): 1315–7.

[11] Boder, E. T., K. S. Midelfort and K. D. Wittrup. "Directed evolution of antibody fragments with monovalent femtomolar antigen-binding affinity." Proc Natl Acad Sci U S A (2000) 97(20): 10701–5.

[12] Boder, E. T. and K. D. Wittrup. "Yeast surface display for screening combinatorial polypeptide libraries." Nat Biotechnol (1997) 15(6): 553–7.

[13] Daugherty, P. S., G. Chen, B. L. Iverson and G. Georgiou. "Quantitative analysis of the effect of the mutation frequency on the affinity maturation of single chain Fv antibodies." Proc Natl Acad Sci U S A (2000) 97(5): 2029–34.

[14] Georgiou, G. "Analysis of large libraries of protein mutants using flow cytometry." Adv Protein Chem (2000) 55: 293–315.

[15] Holler, P. D., P. O. Holman, E. V. Shusta, S. O'Herrin, K. D. Wittrup and D. M. Kranz. "In vitro evolution of a T cell receptor with high affinity for peptide/MHC." Proc Natl Acad Sci U S A (2000) 97(10): 5387–92.

[16] Olsen, M. J., D. Stephens, D. Griffiths, P. Daugherty, G. Georgiou and B. L. Iverson. "Function-based isolation of novel enzymes from a large library." Nat Biotechnol (2000) 18(10): 1071–4.

[17] Shusta, E. V., P. D. Holler, M. C. Kieke, D. M. Kranz and K. D. Wittrup. "Directed evolution of a stable scaffold for T-cell receptor engineering." Nat Biotechnol (2000) 18(7): 754–9.

[18] Wittrup, K. D.. "Protein engineering by cell-surface display." Current Opinion in Biotechnology (2001) 12: 395–399.

[19] Wittrup, K. D. "The single cell as a microplate well." Nat Biotechnol (2000) 18(10): 1039–40.

[20] Schmidt-Dannert, C. and F. H. Arnold. "Directed evolution of industrial enzymes." Trends Biotechnol (1999) 17(4): 135–6.

[21] Zaccolo, M. and E. Gherardi. "The effect of high-frequency random mutagenesis on in vitro protein evolution: a study on TEM-1 beta-lactamase." J Mol Biol (1999) 285(2): 775–83.

[22] Zaccolo, M., D. M. Williams, D. M. Brown and E. Gherardi. "An approach to random mutagenesis of DNA using mixtures of triphosphate derivatives of nucleoside analogues." J Mol Biol (1996) 255(4): 589–603.

[23] Burks, E. A., G. Chen, G. Georgiou and B. L. Iverson. "In vitro scanning saturation mutagenesis of an antibody binding pocket." Proc Natl Acad Sci U S A (1997) 94(2): 412–7.

[24] Chen, G., I. Dubrawsky, P. Mendez, G. Georgiou and B. L. Iverson. "In vitro

scanning saturation mutagenesis of all the specificity determining residues in an antibody binding site." Protein Eng (1999) **12**(4): 349–56.

[25] Daugherty, P. S., G. Chen, M. J. Olsen, B. L. Iverson and G. Georgiou. "Antibody affinity maturation using bacterial surface display." Protein Eng (1998) **11**(9): 825–32.

[26] Chowdhury, P. S. and I. Pastan. "Improving antibody affinity by mimicking somatic hypermutation in vitro." Nat Biotechnol (1999) **17**(6): 568–72.

[27] Miyazaki, K. and F. H. Arnold. "Exploring nonnatural evolutionary pathways by saturation mutagenesis: rapid improvement of protein function." J Mol Evol (1999) **49**(6): 716–20.

[28] Voigt, C. A., S. L. Mayo, F. H. Arnold and Z. G. Wang. "Computational method to reduce the search space for directed protein evolution." Proc Natl Acad Sci U S A (2001) **98**(7): 3778–83.

[29] Volkov, A. A. and F. H. Arnold. "Methods for in vitro DNA recombination and random chimeragenesis." Methods Enzymol (2000) **328**: 447–56.

[30] Chang, C. C., T. T. Chen, B. W. Cox, G. N. Dawes, W. P. Stemmer, J. Punnonen and P. A. Patten. "Evolution of a cytokine using DNA family shuffling." Nat Biotechnol (1999) **17**(8): 793–7.

[31] Shusta, E. V., J. VanAntwerp and K. D. Wittrup. "Biosynthetic polypeptide libraries." Curr Opin Biotechnol (1999) **10**(2): 117–22.

[32] Bessette, P. H., F. Aslund, J. Beckwith and G. Georgiou. "Efficient folding of proteins with multiple disulfide bonds in the Escherichia coli cytoplasm." Proc Natl Acad Sci U S A (1999) **96**(24): 13703–8.

[33] Bothmann, H. and A. Pluckthun. "Selection for a periplasmic factor improving phage display and functional periplasmic expression." Nat Biotechnol (1998) **16**(4): 376–80.

[34] Hanes, J. and A. Pluckthun. "In vitro selection and evolution of functional proteins by using ribosome display." Proc Natl Acad Sci U S A (1997) **94**(10): 4937–42.

[35] Kim, D. M. and J. R. Swartz. "Prolonging cell-free protein synthesis with a novel ATP regeneration system." Biotechnol Bioeng (1999) **66**(3): 180–8.

[36] Kristensen, P. and G. Winter. "Proteolytic selection for protein folding using filamentous bacteriophages." Fold Des (1998) **3**(5): 321–8.

[37] Sieber, V., A. Pluckthun and F. X. Schmid. "Selecting proteins with improved stability by a phage-based method." Nat Biotechnol (1998) **16**(10): 955–60.

[38] Waldo, G. S., B. M. Standish, J. Berendzen and T. C. Terwilliger. "Rapid protein-folding assay using green fluorescent protein." Nat Biotechnol (1999) **17**(7): 691–5.

[39] Shusta, E. V., M. C. Kieke, E. Parke, D. M. Kranz and K. D. Wittrup. "Yeast polypeptide fusion surface display levels predict thermal stability and soluble secretion efficiency." J Mol Biol (1999) **292**(5): 949–56.

[40] Kowalski, J. M., R. N. Parekh, J. Mao and K. D. Wittrup. "Protein folding stability can determine the efficiency of escape from endoplasmic reticulum quality control." J Biol Chem (1998) **273**(31): 19453–8.

[41] Kowalski, J. M., R. N. Parekh and K. D. Wittrup. "Secretion efficiency in Saccharomyces cerevisiae of bovine pancreatic trypsin inhibitor mutants lacking disulfide bonds is correlated with thermodynamic stability." Biochemistry (1998) **37**(5): 1264–73.

[42] Boder, E. T. and K. D. Wittrup. "Optimal screening of surface-displayed polypeptide libraries." Biotechnol Prog (1998) **14**(1): 55–62.

[43] Boder, E. T. and K. D. Wittrup. "Yeast surface display for directed evolution of protein expression, affinity, and stability." Methods Enzymol (2000) **328**: 430–44.

[44] Ashcroft, R. G. and P. A. Lopez. "Commercial high speed machines open new opportunities in high throughput flow cytometry (HTFC)." J Immunol Methods (2000) **243**(1-2): 13–24.

7
Yeast n-Hybrid Systems for Molecular Evolution

Brian T. Carter, Hening Lin, and Virginia W. Cornish

7.1
Introduction

In 1989 Fields and Song introduced the "Yeast Two-Hybrid Assay", which provides a straightforward method for detecting protein-protein interactions *in vivo* [1]. There are three significant advantages of this *in vivo* assay that led almost immediately to its widespread use: first, it is technically straightforward and can be carried out rapidly; second, the sequence of the two interacting proteins can be read off directly from the DNA sequence of the plasmids encoding them; and third, it does not depend on the identity of the interacting proteins and so is generally applicable. The yeast two-hybrid (Y2H) assay was based on the observations that eukaryotic transcriptional activators can be dissected into two functionally independent domains, a DNA-binding domain (DBD) and a transcription activation domain (AD), and that hybrid transcriptional activators can be generated by mixing and matching these two domains [2, 3]. It seems that the DNA-binding domain only needs to bring the activation domain into the proximity of the transcription start site, suggesting that the linkage between the DBD and the AD can be manipulated without disrupting activity [4]. Thus, the linkage in the Y2H assay is the noncovalent bond between the two interacting proteins.

As outlined in Fig. 7.1, the Y2H system consists of two protein chimeras, and a reporter gene downstream from the binding site for the transcriptional activator. If the two proteins of interest (X and Y) interact, they effectively dimerize the DNA-binding protein chimera (DBD-X) and the transcription activation protein chimera (AD-Y). Dimerization of the DBD and the transcription AD helps to recruit the transcription machinery to a promoter adjacent to the binding site for the transcriptional activator, thereby activating transcription of the reporter gene.

The assay was demonstrated initially using two yeast proteins known to be physically associated *in vivo* [1]. The yeast SNF1 protein, a serine-threonine protein kinase, was fused to the Gal4 DBD, and the SNF1 activator protein SNF4 was fused to the Gal4 AD.

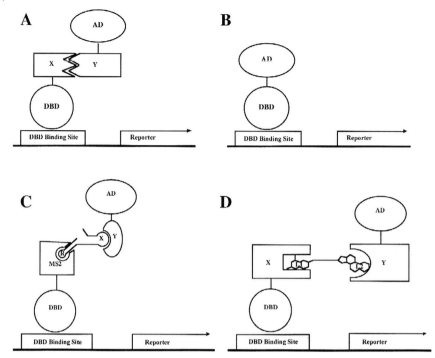

Fig. 7.1. Different yeast n-hybrid systems that have been developed to study protein-protein, protein-DNA, protein-RNA, and protein – small molecule interactions. **A.** In the original version of the Y2H system, transcriptional activation of the reporter gene is reconstituted by recruitment of the activation domain (AD) to the promoter region through direct interaction of protein X and Y, since protein X is fused to a DNA-binding domain (DBD) and protein Y is fused to the AD [1]. **B.** In the Y1H assay, the AD is fused directly to the DBD [109]. This assay can be used to screen either DBDs that can bind to a specific DNA sequence or the *in vivo* binding site for a given DBD. **C.**

Compared with the Y2H system, the Y3H system used to detect RNA-protein interactions has one additional component, a hybrid RNA molecule [90]. One half of the hybrid RNA is a known RNA (R) that can binds the MS2 coat protein (MS2) with high affinity and serves as an anchor. The other half is RNA X, whose interaction with protein Y is being tested. **D.** Another version of the Y3H system can be used to detect small molecule-protein interactions [96]. Ligand L1 which interacts with protein X can be linked to ligand L2, and thus, if L2 interacts with Y, transcriptional activation of the reporter gene will be reconstituted.

A Gal4 binding sequence was placed upstream of a β-galactosidase reporter gene and was integrated at the *URA3* locus of the yeast chromosome. Plasmids encoding the protein fusions were introduced into this yeast strain, and β-galactosidase synthesis levels were quantified using standard biochemical assays. Control experiments established that neither the DBD and AD domains on their own nor the individual protein chimeras induced β-galactosidase synthesis above background levels. β-galactosidase synthesis levels were increased 200 – fold when the DBD-SNF1 and SNF4-AD fusion proteins were introduced together. By comparison, the direct DBD-AD fusion protein activated β-galactosidase synthesis levels 4000-fold.

Since the initial paper by Fields and Song, there have been significant technical improvements in the method. DNA-binding domains and transcription activation domains have been optimized to reduce false positives and increase the transcription read-out. A variety of reporter plasmids have been engineered to detect a broad range of protein-protein interactions. Much more is understood about the nature of "false positives" and how to rout them out. Moreover, in response to the utility of this approach, several laboratories have begun to develop transcription-based assays that can be carried out in bacteria, or protein-protein interaction assays based on alternate readouts such as enzyme complementation or fluorescence resonance energy transfer (FRET).

While envisioned initially as a method for verifying positive protein-protein interactions, it was quickly realized that the method was well suited to screening libraries of proteins to identify new interactions [5]. Now in widespread use, the Y2H assay has

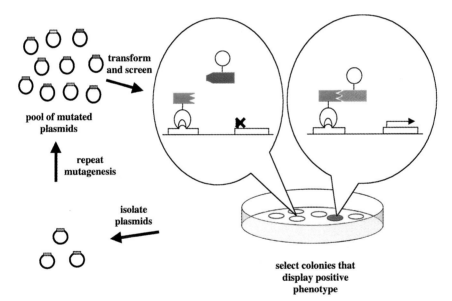

Fig. 7.2. The Y2H assay provides a convenient general method for evolving proteins with new properties. As shown, in the first step the plasmid borne DNA for the protein fused to the AD is mutated over a specific region to yield a library of protein variants. These plasmids are transformed into a host organism that also contains the plasmid encoding a protein-DBD fusion as well as a reporter strain, which will have a defined phenotype depending on whether or not the two proteins interact. In this illustration, cells that contain a pair of interacting proteins turn blue, though the assay can also be run as a growth selection. The plasmids encoding the AD fusion proteins are isolated from these blue cells. These plasmids can then either be sequenced or carried on for further rounds of mutation and selection to increase the affinity of the protein-protein interaction.

been used to identify thousands of new protein-protein interactions [6]. Surprisingly, while the same features that make it an attractive method for screening cDNA libraries should apply to directed evolution (Fig. 7.2), applications of the method to protein engineering were slower to follow. The introduction of yeast one-hybrid (Y1H) assays for detecting DNA-protein interactions and yeast three-hybrid (Y3H) assays for detecting RNA-protein and small molecule-protein interactions suggests a wide array of possible applications (Fig. 7.1). Here we look at the impressive successes n-hybrid assays have seen recently in the evolution of proteins with new properties.

7.2
Technical Considerations

7.2.1
Yeast two-hybrid assay

Since the introduction of the Y2H system in 1989, there have been a number of improvements to the basic system. Different DNA-binding and activation domains have been introduced. Convenient vectors with different bacterial drug-resistance markers, yeast origin of replication (ori), and yeast auxotrophic markers are now even commercially available. An array of common yeast genetic markers has been introduced into the Y2H system, making it possible to test large pools of protein variants (ca. 10^6) using growth selections. In addition to the basic activator system, reverse and split-hybrid systems have been developed to detect the disruption of protein-protein interactions and most recently a transcriptional repressor-based system was reported. Finally, we are beginning to understand the nature of false positives common with Y2H selections and how to minimize them. As biochemical and structural studies begin to unveil the mechanism of transcription in eukaryotes, we can even think about how the nature of a given protein-protein interaction determines the levels of transcription activation in the Y2H system.

DNA-binding and transcription activation domains

Since the first Y2H system was reported by Fields and Song [1], several laboratories have focused on the optimization of the methodology and have generated a number of variations on the basic system. While most improvements have focused on making the system better suited for detection of direct protein-protein interactions from cDNA libraries, these variations should also prove useful for applications in molecular evolution. Several of the Y2H systems described below are commercially available from companies including Stratagene, which markets the Gal4 system, Origene and Invitrogen, for the LexA system, and Clontech, which offers versions of both systems.

While finding the system that works best for a particular experiment is still largely a matter of trial and error, the availability of several different systems makes it likely that a suitable system for any given experiment has already been developed.

Fields and Song's original Y2H system was built from a Gal4 DNA-binding domain (DBD) and a Gal4 activation domain (AD) (Fig. 7.3) [1, 5]. Both domains were placed under the control of a truncated version of the alcohol dehydrogenase promoter *ADH1**, which is an intermediate strength, constitutive promoter. The plasmids en-

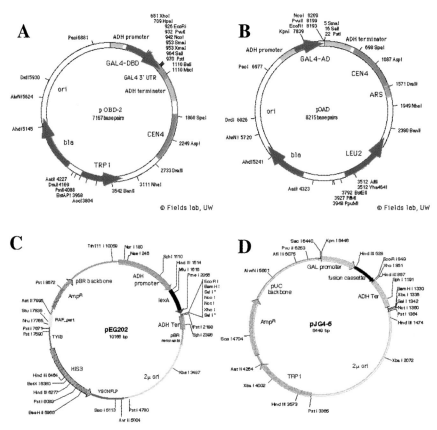

Fig. 7.3. Plasmid maps for the DBD and AD fusion proteins in the Fields and Song Gal4 and the Brent LexA/B42 Y2H systems. **A.** Plasmid pOBD-2 encodes the Gal4 DNA-binding domain fusion [110]. **B.** Plasmid pOAD encodes the Gal4 activation domain fusion [110]. **C.** Plasmid pEG202 encodes for the LexA$_{202}$ DNA-binding domain [13]. **D.** Plasmid pJG4 – 5 encodes for the B42 activation domain [20]. All maps show several key features of the plasmids. The CEN4 ori and *TRP1*, *LEU2*, or *HIS3* auxotrophic markers allow propagation and selection in yeast. The *amp*^{R} gene allow propagation in *E. coli*. All of the plasmids have a multiple cloning region C-terminal to the DBD and AD, facilitating construction of the C-terminal fusion proteins. Both the Gal4 DBD and Gal4 AD are under control of the constitutive truncated *ADH1** promoter while the LexA DBD is under control of the constitutive full length *ADH1* promoter, and the B42 AD is under control of the inducible *Gal1* promoter.

coding these domains are shuttle vectors and thus allow propagation in both yeast and *E. coli*. Both plasmids contain a pBr bacterial ori and an ampicillin antibiotic resistance marker (amp^R), as well as the high-copy yeast 2 μ ori and the auxotrophic markers *TRP1* and *LEU2* for selective maintenance of the plasmids in yeast.

Since the origin of the Gal4 system over a decade ago, a number of modifications to this system have been made to improve its compatibility with a broad range of proteins. Since some proteins prove toxic to yeast cells when expressed at high concentrations, plasmids with a centromere (CEN) ori instead of a 2 μ ori have been developed [7]. The CEN ori maintains the plasmid copy number at one to five per cell, as opposed to ca. 60 per cell in the case of the 2μ ori. Another improvement, designed for proteins that do not express well, was the development of plasmids that use the full length *ADH1* promoter, which directs higher protein expression levels than the truncated version [8]. To facilitate subcloning of different proteins and cDNA libraries, multiple cloning sites have been introduced at both the N- and C-termini of the Gal4 DBD and AD genes, allowing both N- and C-terminal fusions to be made in all three reading frames in both the forward and reverse directions [9]. Thus, if a protein exhibits protein stability or folding problems as a C-terminal fusion to the Gal4 DBD, this problem can often be corrected by moving the protein to the AD or making an N-terminal fusion.

The final step of any Y2H screen is to isolate the plasmid encoding the protein of interest, either for sequencing or for further rounds of mutation and selection. In a typical Y2H experiment, the yeast contain plasmids encoding both the DBD and AD protein fusions and often a third or fourth plasmid encoding a reporter gene or a third protein partner. Thus, a technical problem arises of how to isolate selectively the plasmid encoding the protein being evolved. One approach is to incorporate counter-selectable markers, such as *URA3*, into all of the plasmids except the plasmid encoding the selected protein variant. *URA3* encodes orotidine-5'-phosphate decarboxylase [10]. *URA3* not only is an essential enzyme in the uracil biosynthetic pathway, but also converts non-toxic 5-fluoroorotic acid (5-FOA) into toxic 5-fluorouracil (5-FU). Thus, plasmids containing the *URA3* marker can be selected against in the presence of 5-FOA. Alternatively, the plasmids can be isolated from yeast and transformed directly into *E. coli*. The desired plasmid can then be isolated using the yeast auxotrophic markers *TRP1* or *LEU2* in *E. coli* strain KC8, that has been made *trp1*⁻ and *leu2*⁻, and is commercially available from Clontech. Alternatively, the plasmid encoding the evolved protein variant can be engineered to contain a unique antibiotic marker, such as amp^R, a kanamycin resistance gene (kan^R), a chlormaphenicol resistance gene (cam^R) or a zeocin resistance gene (zeo^R). Mumberg *et al.* have designed plasmids for the Gal4 system that allow the ori (2 μ or CEN), promoter (*ADH1, CYC1, TEF2,* or *GPD*), and auxotrophic marker (*HIS3, LEU2, TRP1,* or *URA3*) to all be readily exchanged [11].

About four years after the original Y2H system was developed, two new systems were developed that moved away from Gal4 [12, 13]. A major advantage of these systems is that they can employ the *GAL* promoter for protein expression. The *GAL* promoter is the most tightly regulated promoter in yeast and can be induced to high levels with galactose, completely repressed by glucose, and tuned with varying ratios of galactose and glucose [14]. Of course this promoter could not be used in the Gal4 system because Gal4 is part of the *GAL* promoter system. Controllable expression of the protein fusions offers a number of advantages, particularly for molecular evolution. A protein can be expressed at high levels during early rounds of selection. The protein expression levels can then be decreased during later rounds of selection to discriminate the most active protein variants. Inducible systems also offer a rapid means to check for false positives, by rescreening for the ability to activate reporter genes in the absence of inducer.

In place of the Gal4 DBD, the LexA$_{202}$ DBD from *E. coli* was used [15], and the *GAL* upstream activating sequence (UAS) was replaced with multiple copies of the LexA operator. In place of the Gal4 AD, Vojtek's version of this system uses the AD from the herpes simplex virus VP16 protein [12]. This LexA based system is extremely similar to the original Gal4 Y2H, and thus many of the components can be interchanged. Vojtek's system uses the same auxotrophic markers for the DBD and AD plasmids, *TRP1* and *LEU2*, respectively, as the Gal4 system. It uses the *ADH1** promoter to control expression of the LexA$_{202}$ fusion protein and the *ADH1* promoter to control expression of the VP16 fusion constructs, and these plasmids also have a 2μ ori. Some major differences of this system are the use of the extremely strong VP16 activation domain, and the fact that while the VP16 AD contains the same simian virus 40 (SV40) large T antigen nuclear localization sequence (NLS) as the Gal4 AD, the LexA$_{202}$ DBD does not contain a NLS [16]. This design was specifically intended to make the concentration of the AD fusion proteins in the nucleus greater than the LexA$_{202}$ fusion proteins, increasing the chances that a LexA$_{202}$-AD complex would be bound upstream of the reporter gene. Variants of the LexA/VP16 system have been developed including vectors that allow N-terminal fusions to LexA$_{202}$ [17] and vectors which use the yeast CEN ori and have the VP16 AD fusion under control of the inducible *GAL* promoter [18]. Both of these latter adaptations have proven helpful as the strong VP16 AD often proves toxic to the yeast cells when maintained at high concentrations.

Brent's lab also developed a Y2H system based on the LexA$_{202}$ DBD (Fig. 7.3) [13]. In Brent's Y2H system, however, the Gal4 AD is replaced with the weaker B42 AD. B42 is an artificial AD, a small, acidic protein isolated from a random library of *E. coli* genomic DNA using a one-hybrid selection [19]. While Brent's system still uses the yeast 2μ ori, the auxotrophic markers on the plasmids encoding the DBD and AD fusions, *HIS3* and *TRP1*, respectively, are different than those in the original Gal4 system.

The LexA$_{202}$ protein fusions are under the control of the *ADH1* promoter and again contain no NLS, while the B42 fusions are controlled by the strong, inducible *GAL1* promoter and include the SV40 large T antigen NLS. As with the original Gal4 system, a large number of variants to Brent's system have been developed. The *amp^R* gene has been replaced with both the *kan^R* and *cam^R* genes to facilitate the eventual isolation of the library plasmid [20]. Plasmids encoding both N- and C-terminal DBD and AD fusions have been constructed [21]. The plasmid encoding LexA has been altered in a number of ways, including incorporation of a NLS [22] and use of a CEN ori with replacement of the ADH1 promoter with the *GAL1* promoter [23]. In addition, it is worth noting that LexA$_{202}$ is a homodimer with an N-terminal DNA-binding domain and a C-terminal dimerization domain. While the Brent system typically uses full-length LexA$_{202}$, there are also versions that use LexA$_{87}$, which lacks the C-terminal dimerization domain and so is monomeric [24]. Also, while the LexA/B42 based system was designed as a plasmid-based system, plasmids are available that are specifically designed to facilitate integration of these plasmids into the yeast chromosome. Chromosomal integration effectively lowers the protein expression levels and also eliminates cell-to-cell variations in plasmid copy number and hence protein expression.

More recently additional systems have been developed that rely on other DBDs. Golemis and co-workers developed a system that uses the DBD from the bacteriophage λ repressor cI protein [25]. Interestingly, the plasmid encoding the cI fusion protein carries the *zeo^R* gene, which makes both yeast and *E. coli* containing the plasmids resistant to zeocin. Also, a system has been developed that uses the DBD from the human estrogen receptor [26].

Reporters

The Y2H system should be particularly well suited to molecular evolution experiments because it can be run as a growth selection and because the stringency of the selection is tunable (Tab. 7.1). In addition, a secondary screen using a chromogenic reporter gene such as *lacZ* can be used to rule out many false positives. In early rounds of selection where proteins with low levels of activity need to be detected, sensitive reporter genes can be employed. In later rounds where the most active variants need to be distinguished, more selective reporter genes can be used. Y2H reporter plasmids provide three levels of control – the upstream activating sequences (UAS), the promoter, and the reporter gene itself. The choice of reporter plasmid is a balance between optimal sensitivity and acceptable background.

The UAS provides the first level of control. The UAS is the region of DNA recognized by the DBD, and binding of a DBD-AD complex to the UAS is responsible for transcription activation. It has been established that the levels of transcription activation correlate with the number of DBD binding sites [27]. Thus, it is common to pre-

Tab. 7.1. Common strains used in yeast two-hybrid systems

Strain	Reporter	Genotype	Ref
Gal4 Two-Hybrid Strains			
PJ69–4A	*ADE2, HIS3, lacZ, MEL1*	*MATa gal4Δ gal80Δ his3-200 leu2-3 trp1-901 ura3-52 GAL2-ADE2 LYS2::GAL1-HIS3 met2::GAL7-lacZ*	38
Y187	*lacZ, MEL1*	*MATα ade2-101 gal4 gal80 his3-200 leu2-3 trp1-901 URA3::GAL1-lacZ*	113
Y190	*HIS3, lacZ, MEL1*	*MATa ade2-101 cyh2^R gal4Δ gal80Δ his3-200 leu2-3 trp1-901 LYS2::GAL1-HIS3 URA3::GAL1-lacZ*	113
Reverse Gal4 Two-Hybrid Strains			
MaV95 (MaV96, MaV97, MaV99)	*HIS3, URA3, lacZ, MEL1*	*MATa ade2-101 can1^R cyh2^R gal4Δ gal80Δ his3-200 leu2-3 trp1-901 ura3-52 LYS2::GAL1-HIS3 GAL1-lacZ SPAL5::URA3 (SPAL7::URA3, SPAL8::URA3, SPAL10::URA3)*	28
Repressed Transactivator Gal4 Strains			
YJMH1	*URA3*	*MATα ade1 ade2-101 ara1 gal4-542 gal80-538 his3-200 leu2-3,112 lys2-80 met trp1-901 ura3-52 LEU2::GAL1-URA3*	33
MaV108	*URA3*	*MATa gal4Δ gal80Δ his3-200 leu2-3 trp1-901 SPAL10::URA3*	33
W303::131	*lacZ*	*MATa ade2 can1 his3 leu2 trp1 ura3 URA3::GAL1-lacZ*	33
H617::131	*lacZ*	*MATα ade2 can1 his3 leu2 trp1 ura3 srb10::HIS3 URA3::GAL1-lacZ*	33
LexA/VP16 Two-Hybrid Strain			
L40	*HIS3, lacZ*	*MATa ade2 his3 leu2 trp1 LYS2::lexAop(4×)-HIS3 URA3::lexAop(8x)-lacZ*	12
LexA/B42 Two-Hybrid Strains			
EGY48 (EGY191, EGY195)	*LEU2*	*MATα his3 trp1 ura3 lexAop(6x)-LEU2 (lexAop(2x)-LEU2, lexAop(4x)-LEU2*	13, 40
Dual Bait Lex and cI Strains			
SKY48 (SKY191, SKY473)	*LEU2, LYS2*	*MATα his3 trp1 ura3 lexAop(6x)-LEU2 cIop(3x)-LYS2 (lexAop(2x)-LEU2, lexAop(4x)-LEU2)*	25
Split-Hybrid LexA/B42 Strain			
YI584	*HIS3*	*MATa/MATα ade2 his3Δ200 leu2–3, 112 trp1-901 URA3::lexAop(8x)-TetR LYS2::tetop(2x)-HIS3*	32

pare a series of reporter genes with 1 – 4 tandem binding sites. Recall that $LexA_{202}$ binds DNA as a homodimer such that the most sensitive LexA reporter genes have 8 LexA operators. The more sensitive reporters, of course, have a higher background of false positives.

The promoter provides the second level of control. It should be emphasized that, in addition to activated transcription, there is basal transcription dictated by the promoter sequence. The sensitivity of a reporter can be enhanced by increasing the basal levels of transcription. Again, it is a balance between optimal sensitivity and acceptable background. For example, when Vidal and co-workers were developing the reverse Y2H system, they needed a reporter that would have very low levels of basal transcription. All of the available Gal4-inducible promoters developed for use with the Fields and Song Y2H system gave basal levels of transcription that were too high. To solve this problem, they used the *SPO13* promoter which contains the *cis*-acting upstream repressing sequence *URS1*, which substantially reduced basal transcription [28].

The final component of the reporter plasmid is the reporter gene itself. An advantage of working in yeast is that geneticists have developed a number of clever reporter systems over the years. The Y2H assay can be run as a colorimetric screen or a positive or negative growth selection. The original Y2H system used the *lacZ* gene to provide a colorimetric readout for protein-protein interaction [1]. The *lacZ* gene encodes the *E. coli* β-galactosidase enzyme, which has broad substrate specificity and can cleave a number of chromogenic substrates, including 5-bromo-4-chloro-3-indolyl-β-D-galacto-pyranoside (X-gal) and o-nitrophenyl-β-D-galactopyranoside (ONPG), that can be detected readily on plates or in liquid culture. In addition to *lacZ*, several other colorimetric reporter genes are available. *MEL1* and *gusA* encode α-galactosidase and β-glucurnidase, respectively, and can be detected with the chromogenic substrates 5 – bromo-4-chloro-3-indolyl-α-D-galactoside (X-α-gal) and 5-bromo-4-chloro-3-indolyl-β-D-glucuronic acid (X-gluc) [25]. The gene encoding green fluorescent protein (GFP) has been used as a colorimetric reporter in the Y2H system and can be assayed without the addition of any exogenous substrate [29].

A large advance in the Y2H field was made when the original colorimetric reporter gene was replaced by a yeast auxotrophic marker, making it possible to test large pools of proteins (ca. 10^6) using growth selections [8]. The most common auxotrophic reporter genes used in yeast are *HIS3*, *LEU2, ADE2*, and *URA3*, all essential enzymes in the biosynthesis of amino acids or nucleotides. Since these genes are also used as the selective markers for maintaining the DBD and AD plasmids in yeast, care must be taken in choosing a reporter that is compatible with the other Y2H components. The *HIS3* gene provides a sensitive reporter with a high level of basal transcription [8]. The sensitivity of the reporter can be regulated with 3 – amino-1,2,4-triazole (3-AT), which competitively inhibits the *HIS3* gene product. The *LEU2* and *ADE2* genes also provide robust reporter genes. The stringency of these reporters can be controlled by supplementing the growth media with varying concentrations of leucine and adenine, respectively. The advantage to *URA3*, as mentioned previously, is that it provides both a positive selection for growth in the absence of uracil and a negative selection for conversion of 5-FOA to 5-FU, which is toxic [10]. In addition the *CYH2* gene has been used for negative selection, since yeast expressing Cyh2 are sensitive to cycloheximide, which blocks peptide elongation during translation [30].

Variations on the basic two-hybrid system

Several clever variations on the basic Y2H system have been developed that make it possible to assay for disruption of a protein-protein interaction. Vidal and co-workers introduced a reverse Y2H system, which detects disruption of a protein-protein interaction using a counter-selectable marker (Fig. 7.4A) [28]. The reverse Y2H system uses a Gal4 DBD and AD and a *URA3* reporter gene. First, the protein-protein interaction is confirmed based on growth in media lacking uracil. Then, molecules that disrupt the interaction can be detected based on disruption of *URA3* transcription and growth in the presence of 5-FOA. Schreiber and co-workers have reported a variant of this system using the LexA$_{202}$ DBD and B42 AD [31].

In the split-hybrid system disruption of a protein-protein interaction is detected using a dual reporter system (Fig. 7.4B) [32]. The protein-protein interaction directs transcription of the *E. coli tet* repressor protein (TetR), a DNA-binding protein. A *HIS3* gene in turn is under control of two tandem TetR operators. This system is based on the Vojtek systems with a LexA$_{202}$ DBD and a VP16 AD [12]. The initial protein-protein interaction activates transcription of TetR, and TetR then suppresses production of *HIS3* simply by physically blocking transcription by the basal transcription machinery. Thus, a selection for growth in the absence of histidine can be used to evolve a protein that disrupts the protein-protein interaction.

Most recently, Hirst and co-workers developed a repressor, as opposed to an activator, system (Fig. 7.5) [33]. While intended simply for proteins that activate transcription when fused to the DBD, this system opens many new possibilities using competing activation and repression domains. The so-called repressed transactivator system (RTS) is based on the Gal4 Y2H system, only the Gal4 AD is replaced with the *N*-terminal repression domain (RD) of the yeast repressor protein TUP1 [34]. The system was designed for an initial negative growth selection using a *URA3* reporter followed by a *lacZ* screen. The *URA3* gene is under control of the *GAL1* promoter. Thus, if no protein-protein interaction occurs, the *URA3* gene is activated, and the yeast are sensitive to 5–FOA. However, if the Gal4 DBD and TUP1 RD fusion proteins interact, then TUP1 RD is recruited to the *GAL1* promoter and represses transcription of *URA3*, allowing survival in the presence of 5-FOA.

A

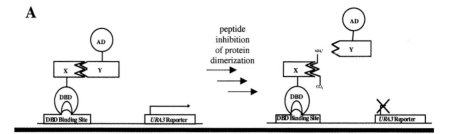

B

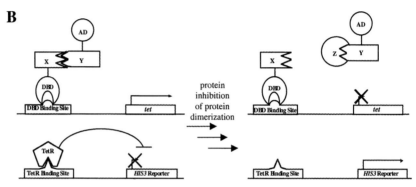

Fig. 7.4. Variations of the original Y2H assay have been designed to detect the disruption of protein-protein interactions. **A.** In the reverse Y2H system, a direct protein-protein interaction between X and Y activates the transcription of *URA3*, which converts 5-FOA to a toxic compound (left); however, if the interaction between X and Y is disrupted, then transcription of *URA3* is disrupted and the cells can grow in media containing 5-FOA (right) [28]. **B.** In the split-hybrid system, a direct protein-protein interaction between X and Y ac-

tivates the transcription of *tet*, which encodes the bacterial DNA-binding protein TetR [32]. TetR in turn binds upstream of the *HIS3* reporter gene and prevents transcription of this gene, which is essential for growth in media lacking histidine (left). Disruption of this interaction between X and Y, shown in this case by a third protein Z, turns off transcription of the *tet* gene. In the absence of TetR, the *HIS3* gene is no longer repressed, allowing the cells to grow in media lacking histidine (right).

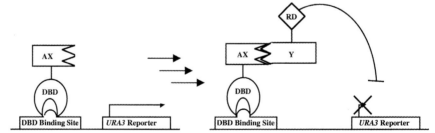

Fig. 7.5. The repressed transactivator system can be used with proteins that activate transcription independently [33]. Since the protein being screened, AX (activating protein X), has the intrinsic ability to activate transcription of the *URA3* reporter gene, expression of AX makes the

cells susceptible to 5-FOA (left). Recruitment of the repression domain (RD) of TUP1, through a direct interaction between the activation protein AX and protein Y, represses transcription of the *URA3* gene, allowing the cells to grow in media containing 5-FOA (right).

False positives and false negatives

The major drawback to all *in vivo* selections is false positives. False positives are particularly likely in early rounds of selection, where proteins with low levels of activity need to be detected and so more sensitive reporters with higher basal levels of transcription are being employed. Often cells that survive such a selection do not contain a productive protein-protein interaction, but rather have an unrelated mutation that generates the desired phenotype. Since the Y2H assay is now quite common, researchers are beginning to have a sense of the sources of false positives in this assay. There is an excellent discussion of false positives commonly observed in the Y2H assay by Golemis in a recent publication [35]. Also, several groups maintain compendiums of common false positives [6, 36]. In addition, there are now standard protocols for choosing a system to begin with that will minimize false positives, for minimizing false positives during the selection process, and for routing out false positives at the end of a selection experiment (Fig. 7.6) [37].

One key to minimizing false positives is to choose the optimal Y2H format before beginning a selection. For example, if one wanted to evolve a peptide aptamer that

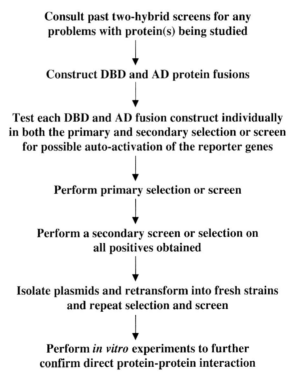

Fig. 7.6. Basic outline of a typical Y2H experiment, emphasizing steps taken at each stage to eliminate false positives.

bound a cyclin-dependent kinase (Cdk) with high affinity, the first step would be to find a system that was reasonably sensitive, but where the Cdk target and peptide aptamer scaffold gave acceptable background levels of transcription activation. The experiment might begin with the Brent Y2H system with a LexA$_{202}$-Cdk fusion protein and a B42-peptide fusion protein and a sensitive *LEU2* reporter gene under control of six tandem LexA operators. If with this format, yeast colonies were observed in leu$^-$ selective media after only three days, several permutations could be tested to minimize the background. The orientation of the interaction could be reversed; Cdk could be fused to B42 and the peptide to LexA$_{202}$. A less sensitive reporter, such as the *LEU2* gene under control of two tandem LexA operators, could be substituted. Alternatively, a protein complementation assay that is not transcription-based could be employed.

In the course of the selection experiment, the most common way to rule out false positives is to carry out several experiments in tandem using different reporter plasmids. Only "hits" identified with multiple reporters are considered to be valid. This can be done by using different DBD/UAS combinations, for example, a LexA$_{202}$- and a cI-system [25]. Alternatively, the promoter sequence can be varied to eliminate AD-fusions that interact with DNA in the promoter region. A Gal4 system has been developed with reporters that are identical, except the promoter region is varied among the *GAL1*, *GAL2*, and *GAL7* promoters [38]. Finally, different auxotrophic markers can be employed, such that a *HIS3* and *URA3* selection are run in parallel and only peptide aptamer sequences that activate both reporters are assumed to really interact with Cdk.

As with other *in vivo* selections, it is critical to validate hits at the end of the selection experiment with a secondary screen. The most common secondary screen used with the Y2H assay is a *lacZ* screen. If either the DBD or AD fusion is under control of an inducible promoter, this screen can be carried out both under inducing and non-inducing conditions to ensure that transcription activation is protein dependent. Y3H systems, where transcription activation depends on a bridging RNA or small molecule, and reverse Y2H systems, where a third protein is disrupting the interaction, provide a built-in control. One can simply check that transcription activation is in fact dependent on the third component. As with any *in vivo* selection, the evolved plasmid should be isolated and retransformed into a fresh yeast selection strain to ensure that the phenotype is plasmid-dependent. Ultimately, the interaction will be confirmed with co-immunoprecipitation experiments or other *in vitro* binding assays [39].

A phenomenon that is less well understood is that of false negatives, proteins that are known to interact but do not activate transcription in the Y2H system. While false negatives may not seem like much of a problem for molecular evolution experiments where the goal is just to evolve a protein with the desired property, it should be recalled that typically the assumption is made at intermediate rounds of selection that consensus sequences are emerging. There are many anecdotal reports of false negatives. The most clear-cut example was provided in a systematic study of the Y2H system by Go-lemis and co-workers [40]. In this study, they found that Max and Myc, two eukaryotic helix-loop-helix proteins known to form heterodimers *in vitro* with a 5 nM K_D, activated transcription as LexA-Max/B42–Myc, but not when the orientation was reversed to LexA-Myc/B42-Max. We have observed a similar phenomenon with our dexametha-sone-methotrexate (Dex-Mtx) Y3H system [41]. In this system a Dex-Mtx heterodimer bridges a LexA$_{202}$-dihydrofolate reductase (DHFR) chimera and a B42-glucocorticoid receptor (GR) chimera to activate transcription [42]. Mtx is known to bind DHFR with low picomolar affinity, and Dex binds GR with low nanomolar affinity. Dex-Mtx-in-duced transcription is much more pronounced when DHFR is fused to LexA and GR to B42, than when the orientation is reversed. Even more striking, Dex-Mtx-in-duced transcription is receptor dependent. The bacterial DHFR from *E. coli* leads to high levels of transcription activation, while there is no detectable transcription activation with the mammalian DHFR from the mouse. This difference is seen de-spite the fact that the *E. coli* and murine DHFRs have similar three-dimensional folds, conserved active-site residues, and their overall affinities for Mtx are indistin-guishable [43, 44].

A related concern is that the levels of transcription activation may not be strictly correlated with the strength of a protein-protein interaction. Again, there is anecdotal evidence to this effect, but the clearest evidence comes from the study by Golemis and Brent on the Y2H assay [40]. They showed that for a given reporter gene, the levels of transcription activation were roughly correlated with the strength of the protein-pro-tein interaction, but the response was not linear. Furthermore, they suggested that with the most sensitive reporters for the Brent system, the limit of detection was a K_D around 1 μM. Other than the work by Golemis and Brent and from our lab, little is known about how the nature of an interaction correlates with the levels of transcrip-tion activation. It should be emphasized that even these studies have considered only a handful of examples. Furthermore, while the phenomenon has begun to be documen-ted, variations in the levels of transcription activation are still not understood at the mechanistic level. A more precise mechanistic picture of transcription activation in the Y2H system should go hand in hand with current efforts to understand transcription activation and repression in eukaryotes at natural promoters [45 – 48].

7.2.2
Alternative assays

While similarities in transcription among eukaryotic organisms allow the Y2H assay to be transferred readily to mammalian cell lines, new transcription-based assays must be developed for bacteria. There are several potential advantages to working in bacteria. The primary advantage for molecular evolution experiments is that the transformation efficiency of *E. coli* is several orders of magnitude greater than that of yeast, allowing on the order of 10^8 protein variants, as opposed to 10^6 in yeast, to be tested at once. In addition, molecular biology techniques have been optimized in *E. coli*, and the rapid doubling time of *E. coli* decreases the time required for the selection experiments. Surprisingly, it has only been in the last few years that bacterial protein-protein interaction assays have gained popularity [49].

One approach to developing transcription-based assays in bacteria has taken advantage of the fact that many bacterial repressors and activators are dimeric proteins with structurally distinct DNA-binding and dimerization domains (Fig. 7.7A). Hu *et al.* demonstrated that the C-terminal dimerization domain of λ-repressor could be replaced with the leucine zipper dimerization domain from the yeast transcriptional activator GCN4 [50]. The chimeric protein is stable and can functionally replace λ-repressor *in vivo*, providing immunity to superinfection by λ bacteriophage or efficient repression of an artificial λ-promoter *lacZ* reporter gene fusion. Conceptually similar systems have been developed around both LexA and AraC and can detect both homo-dimers and heterodimers [51 – 53]. The main drawback to these systems seems to be a limited dynamic range. Full-length dimeric λ-repressor is only able to repress *lacZ* transcription levels 10–fold, and the background levels of X-gal hydrolysis for the "fully repressed" *lacZ* reporter are still quite high.

Kornacker and co-workers have been able to improve the dynamic range of *lacZ* repression using an AraC-LexA system based on repression of transcription via DNA bending (Fig. 7.7B) [54]. In their system, three tandem LexA operators have been engineered between the transcriptional and translational start sites of a *lacZ* reporter gene under control of the *araBAD* promoter. Expression of AraC or an AraC-fusion protein activates transcription of this promoter. If a LexA-fusion protein is expressed that can interact with the AraC-fusion protein, then the interaction produces a bend in the promoter DNA. This bend impedes transcription of the *lacZ* gene, presumably disrupting RNA polymerase (Pol) binding or activity. In this system, the levels of *lacZ* transcription can vary up to 100-fold depending on the strength of the protein-protein interaction. The next step is to engineer a toxic reporter gene for this system so that large numbers of protein variants can be tested using a growth selection.

Based on their biochemical and genetic studies of the mechanism of transcription activation by λ-repressor, Hochschild and Dove realized that they could construct a bacterial activator system (Fig. 7.7C) [55, 56]. They simply placed the λ operator upstream of a transcription start site. Protein-protein interactions could then be detected based on the ability of a λ-repressor-fusion protein to stabilize the interaction of RNA Pol with the promoter when the wild-type (wt) α-subunit of RNA Pol was replaced with an interacting α-subunit-fusion protein. In this system, the interaction of Gal4 with Gal11, which has a K_D of 10^{-7} M, increased *lacZ* transcription 45-fold. Recently, a

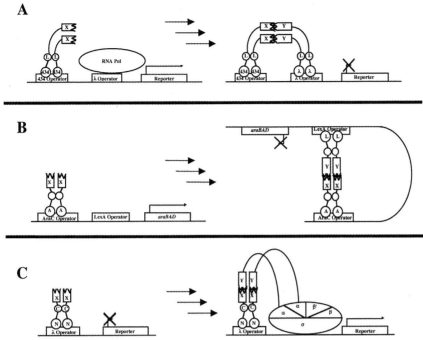

Fig. 7.7. Bacterial two-hybrid (B2H) systems increase the number of protein variants that can be tested in each selection. **A.** The first B2H systems were based on natural bacterial repressor proteins [50]. In the example shown, protein X is fused to the C-terminal end of a 434 repressor-leucine (L) zipper fusion, which binds to the 434 operator. Occupation of the 434 operator does not effect binding of RNA Pol or transcription of the reporter gene (left). However, if protein X interacts with protein Y, then the λ-repressor fusion is recruited to the low-affinity λ operator, preventing RNA Pol from binding to the promoter region and thus, repressing transcription of the reporter gene (right). **B.** In the AraC-LexA system, repression is based on DNA bending [54]. The AraC-protein X chimeras activate transcription of the *araBAD* operon (left). Tandem LexA operators have been engineered between the transcription and translation start sites of the araBAD operon. The interaction of the AraC-X and LexA-Y proteins causes substantial DNA bending and repression of transcription (right). **C.** A bacterial activator system developed by Hochschild [55]. λ-Repressor was found to activate transcription simply by stabilizing the RNA Pol-promoter complex. Thus, a protein X-λ-repressor fusion and a protein Y fusion to the N-terminal domain of the α-subunit of RNA Pol activate transcription of a reporter gene if protein X and Y interact.

HIS3 reporter has been engineered for this system such that it can be run as a growth selection [57]. Interestingly, a conceptually similar system has been developed in yeast based on Pol III-dependent transcription [58].

Bacterial two-hybrid (B2H) systems have only just begun to be developed. These systems show promise both for transcription repression and activation assays. For these methods to become competitive with the Y2H system, the fusion-protein and reporter plasmids will have to be optimized, just as the original Y2H system was re-engineered. A technical complication in working with bacteria, as opposed to yeast, is that it is difficult to express multiple plasmids in the cell, and chromosomal integration is not as trivial as in yeast. Thus, it will be particularly important for bacterial systems to engineer strains containing a variety of robust reporter genes for different applications.

Alternatively, there is interest in developing assays for detecting protein-protein interactions that are not based on transcriptional activators or repressors. As our understanding of the mechanism of transcription activation and repression in eukaryotes improves, there is increasing evidence that these processes may require precise protein-protein interactions, making them less malleable than originally presumed [45 – 48]. In addition, transcription assays require nuclear localization of the interacting proteins and other constraints. Thus, there may be advantages to other types of assays. Several clever approaches have been devised. Generally, these approaches rely on the induced interaction of two complementary fragments of a protein that results in either reconstitution of enzymatic activity or FRET (Tab. 7.2).

Several laboratories have designed high-throughput assays based on reconstitution of an essential enzyme from two protein fragments. Here the question seems to be whether or not reconstitution of an enzyme active site will be too sensitive to conformation to detect a range of protein-protein interactions. Clearly, the choice of enzyme is critical. An enzyme is needed that can be split into two fragments. The fragments must not be able to associate on their own, even when overexpressed. Dimerization of the fragments must restore proper folding and function. The enzyme must provide a clean phenotype for selection and screening – ideally one that is compatible with bacteria, yeast and higher eukaryotes.

One such assay is based on the synthesis of adenosine 3',5'-cyclic monophosphate (cAMP) by adenylate cyclase [59]. The adenylate cyclase from *B. pertussis* can be split into two functionally complementary fragments, T18 and T25, allowing protein dimerization to be assayed based on dimerization of T18 and T25 and reconstitution of

Tab. 7.2. Protein complementation assays used to perform two hybrid assays

Enzyme	Assay	Cell Type	Ref.
ubiquitin	screen and selection	*S. cerevisiae*	72, 73, 74
β-galactosidase	screen	C2C12 myoblast	60
adenylate cyclase	screen and selection	*E. coli*	59
dihydrofolate reductase	selection	*E. coli*	61
dihydrofolate reductase	screen and selection	CHO	62
dihydrofolate reductase	screen	tobacco and potato protoplasts	65
green fluorescent protein	screen	HeLa, *E. coli*	66, 68

adenylate cyclase activity. Since the cAMP/CAP complex induces transcription of several genes, including the *lac* operon, it is possible to screen for adenylate cyclase activity using β-galactosidase plate assays or selections for growth on lactose. Here the issue is a finicky phenotype. The adenylate cyclase knock-out strain required for this assay grows poorly and is unstable, tending to revert to a cya$^+$ phenotype. Similarly, Blau and co-workers have designed a system around dimerization of the a- and ω-subunits of β-galactosidase [60].

More recently Michnick and co-workers have introduced a dihydrofolate reductase complementation system, which seems to be particularly robust [61 – 65]. They attribute the success of this system to the fact that the N-terminal (1 – 105) and C-terminal (106 – 186) DHFR fragments do not fold until they are dimerized. In addition to the obvious selection for essential metabolites dependent on the reduction of dihydrofolate to tetrahydrofolate, protein-protein interactions are detected based on the retention of a fluorescein-methotrexate conjugate. Several other enzymes have been employed for the design of complementation assays, including green fluorescent protein, which allows screens based on fluorescence or FRET [66 – 68]. As with the bacterial transcription assays, these complementation systems are new. It will be interesting to see if, as the selections are optimized, these systems prove competitive with the Y2H assay.

Alternatively, protein-protein interactions can be detected simply based on protein localization. Building from their observation that the guanyl nucleotide exchange factor son of sevenless (hSos) can complement yeast with a temperature-sensitive mutation in the yeast exchange factor Cdc25, Aronheim and Karin developed an assay based on recruitment of Ras or Ras nucleotide exchange factors to the plasma membrane (Fig. 7.8A) [69 – 71]. In one version of this approach, the bait protein is simply fused to an activated form of mammalian Ras that lacks its C-terminal farnesylation sequence. If the bait-Ras fusion interacts with a prey protein with an N-terminal farnesylation or a C-terminal myristoylation sequence, Ras is effectively localized to the plasma membrane. In a *cdc25–2* background, this localization activates Ras-dependent signal transduction pathways necessary for cell proliferation. This assay is straightforward. It has the advantages that it should be conformation-independent, that it does not require nuclear localization, and that it can be modified to be compatible with integral membrane proteins. One likely difficulty will be the design of readouts other than survival in a *cdc25-2* background, because the Ras-dependent signaling pathways in yeast are not completely defined. It is not hard to imagine, however, that one could design similar assays based on other localization events that can be coupled to transcription.

Another interesting approach is artificial induction of ubiquitin-mediated protein degradation (Fig. 7.8B and 7.8C). First introduced by Johnsson and Varshavsky in 1994 [72], the system has now been modified by Johnsson and co-workers so that

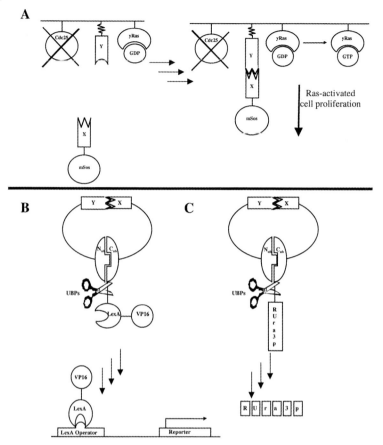

Fig. 7.8. Alternative methods for detecting protein-protein interactions, which do not rely on gene transcription. **A.** Protein-protein interactions can be detected based simply on membrane localization [69, 70]. This system is based on the isolation of a mutant form of the yeast not expressing factor Cdc25 that makes yeast temperature-sensitive for Ras-dependent cell proliferation. In this system, protein X is fused to the mammalian guanyl nucleotide exchange factor Sos (mSos) while protein Y is targeted to the plasma membrane by an N-terminal myristoylation tag. Thus, interaction between X and Y recruits mSos to the plasma membrane, converting Ras into the active GTP bound form, and activating Ras-dependent cell proliferation in a temperature sensitive *cdc25-2* background. **B.** A ubiquitin complementation assay has been developed in which protein-protein interactions are detected as cleavage of a reporter protein from the ubiquitin complex [72]. The ubiquitin complementation assay can be run as a

positive growth selection [73]. When the N- and C-terminal fragments (N_{ub} and C_{ub}) of ubiquitin are brought together by the interaction of proteins X and Y, ubiquitin-specific proteases (UBPs) cleave the LexA-VP16 protein which is fused to the C-terminus of the C_{ub} fragment. LexA-VP16 can then translocate to the nucleus and activate various reporter genes. **C.** The ubiquitin complementation assay can also be run as a negative growth selection [74]. RUra3 is fused to the C-terminal end of the C_{ub} fragment, which makes the cells expressing this fusion sensitive to growth in media containing 5-FOA; however, if ubiquitin is reconstituted by bringing together N_{ub} with C_{ub} through the interaction of proteins X and Y, then RUra3 is cleaved from C_{ub} by the UBPs. Cleavage of RUra3 exposes the arginine that has been engineered at the first position, rendering the protein sensitive to degradation by the N-end rule pathway and, thus, making the cells insensitive to 5-FOA.

it can be run as an *in vivo* selection [73, 74]. Briefly, the bait is fused to the N-terminus of a C-terminal fragment of ubiquitin, which has the Ura3 protein, an essential enzyme in the uracil biosynthesis pathway, fused to its C-terminus. The prey is fused to the N-terminus of ubiquitin, such that protein association leads to degradation of Ura3. The *URA3* gene is one of the most commonly used counter-selectable markers in yeast and can be selected for in the absence of uracil and against based on conversion of 5-FOA to 5-FU, which is toxic. In principle, there should be several advantages to this system. The assay should be compatible with membrane, as well as cytosolic, proteins. Significantly, it should be compatible with virtually any eukaryotic cell line, allowing mammalian proteins to be tested in their native environment, ensuring correct post-translational modification.

7.3
Applications

7.3.1
Protein-protein interactions

The first applications of the Y2H assay to molecular evolution were to identify peptides that bound target proteins with high affinity [75 – 79]. Conveniently, the components of the traditional Y2H assay could be used without modification for this type of experiment (Fig. 7.1A). Six years after the Y2H assay was introduced, peptide ligands for the retinoblastoma protein (Rb) were selected from a library of random peptides fused to the Gal4 AD [75]. Using a *HIS3* growth selection followed by a *lacZ* screen, seven peptides that bound Rb were isolated from a library of 3×10^6 transformants. The affinity of two of these peptides, and two peptides identified from a second round of selection, for Rb were measured by surface plasmon resonance and found to be in the low micromolar range.

A year later, Brent and co-workers demonstrated that the Y2H assay could be used to identify peptide aptamers that inhibit cyclin-dependent kinase 2 (Cdk2) from a library of random peptide sequences (Fig. 7.9) [76]. The 20 residue peptide library was displayed, not as a simple extension to the AD, but rather in the active site of *E. coli* thioredoxin (TrxA). The TrxA loop library was fused to the B42 AD, and Cdk2 was fused to the LexA$_{202}$ DBD. In a single round of selection, 6×10^6 B42-TrxA transformants, a very small percentage of the 10^{27} 20mers possible, were tested for binding to LexA$_{202}$-Cdk2. From this selection, they isolated 66 colonies that activated transcription of both a *LEU2* and a *lacZ* reporter gene. Impressively, 14 different peptide sequences that bound Cdk2 with high affinity were isolated in this one experiment. Using surface plasmon resonance, the peptide aptamers were shown to bind Cdk2 with K_Ds of 30 – 120 nM. In kinase inhibition assays, the peptide aptamers had IC$_{50}$s for the Cdk2/

A

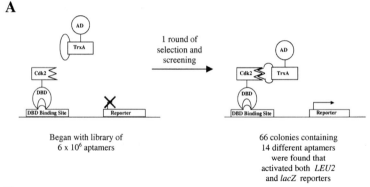

Began with library of
6 x 10⁶ aptamers

66 colonies containing
14 different aptamers
were found that
activated both *LEU2*
and *lacZ* reporters

B

Aptamer	K_D (nM)	Amino Acid Sequence
pep1	ND	ELRHRLGRAL SEDMVRGLAW GPTSHCATVP GRSDLWRVIR FL
pep2	64±16	LVCKSYRLDW EAGALFRSLF
pep3	112±17	YRWQQGVVPS NMASCSFRCQ
pep4	ND	SSFSLWLLMV KSIKRAAWEL GPSSAWNTSG WASLSDFY
pep5	52±3	SVRMRYGIDA FFDLGGLLHG
pep6	ND	RVKLGYSFWA QSLLRCISVG
pep7	ND	QLYAGCYLGV VIASSLSIRV
pep8	38±5	YSFVHHGFFN FRVSWREMLA
pep9	ND	QQRFVFSPSW FTCAGTSDFW GPEPLFDWTR D
pep10	105±10	QVWSLWALGW RWLRRYGWNM
pep11	87±7	WRRMELDAEI RWVKPISPLE
pep12	ND	RPLTGRWVVW GRRHEECGLT
pep13	ND	PVCCMMYGHR TAPHSVFNVD
pep14	ND	WSPELLRAMV AFRWLLERRP

Fig. 7.9. The Y2H system can be used to evolve peptides that bind target proteins with high affinities. **A.** Brent and co-workers adapted the Y2H system to select for short peptides (20 amino acids) that bind specifically to cyclin-dependent kinase 2 (Cdk2), in hopes of finding an inhibitor to this enzyme [76]. *E. coli* thioredoxin (TrxA) is fused to an AD and used as a rigid scaffold for displaying the peptide libraries. After one round of selection and screening from a library containing 6×10^6 peptide aptamers, 66 colonies were isolated that activated both a *LEU2* and a *lacZ* reporter gene. **B.** The 66 colonies identified contained peptides with 14 different amino acid sequences. Six of these peptides were selected for purification, and their K_D for Cdk2 was measured using surface plasmon resonance spectroscopy (ND = not determined).

cyclin E kinase complex of 1 – 100 nM. It is not surprising that a peptide library constrained in a protein loop gave higher affinity ligands than a simple linear peptide library. What is impressive, however, is that in both cases high affinity ligands are being isolated in a single round of selection. Similar results have been obtained using both B2H and protein complementation assays [80 – 82].

Building from the initial Cdk2 inhibitors, Kolonin and Finley derived peptide aptamers that inhibit two Cdks from *Drosophila melanogaster*, DmCdk1 and DmCdk2 [83]. They then showed that expression of these peptides in *Drosophila melanogaster* mimicked a defect in eye development associated with mutations that effect the cell cycle. In addition, Brent and co-workers showed that one of these peptide aptamers, when expressed in a mammalian cell line, inhibited the G_1 to S transition [84]. If high affinity binding proteins can be generated routinely in a single round of selection, monoclonal antibody technology could be outpaced by directed evolution of other generic scaffolds. Optimistically, six months are required from the start of immunization, through immortalization, and finally screening to generate a monoclonal antibody. On the other hand, if several peptide aptamer libraries were maintained for routine use, the libraries could be screened against a new target, false positives could be sorted out, and biochemical assays could validate a target in less than a month and at considerably less expense. In fact, it has been shown recently that antibody-antigen interactions can be detected using both the Y2H assay and the DHFR protein complementation assay [85, 86].

Two-hybrid experiments are not limited to peptides. For example, a two-hybrid selection was used to generate PDZ domains with new specificities [79]. In a recent twist, a B2H assay has been used to evolve peptides that inhibit protein dimerization [87]. The C-terminal domain of λ-repressor was replaced with a catalytically inactive variant of HIV protease. HIV protease exists predominantly as a homodimer at equilibrium. Thus, transcription of a *lacZ-tet* reporter gene was repressed by the dimerization of the λ-repressor-HIV protease fusion proteins. Random peptides, nine amino acids in length, were expressed from the C-terminus of thioredoxin. Peptides that inhibited dimerization of HIV protease were selected as *lacZ*$^+$*Tet*^{R}. From a library of 3×10^8 transformants, approximately 300 showed the desired phenotype. Several of the peptides were purified and shown to inhibit dimerization of HIV protease at low micromolar concentrations. Again, high affinity peptide ligands were isolated in a single round of selection. The authors note that, since the hit rate was only 1 in 10^7, the experiment might not have succeeded in yeast, where transformation efficiencies are only on the order of 10^6.

7.3.2
Protein-DNA interactions

Early on, it was realized that, just as the Y2H assay could be used to detect protein-protein interactions, transcriptional activators could be used directly, in a "one-hybrid" assay, to detect DNA-protein interactions (Fig. 7.1B). In truth, this type of experiment was done before the one-hybrid assay was conceptualized as such. For example, as early as 1983 a His^6→Pro Mnt variant was generated that preferentially binds a mutant Mnt operator using a transcription-based selection [88]. A plasmid encoding Mnt was mutagenized both by irradiation with UV light and by passage through a mutator strain. The mutant plasmids were then introduced into *E. coli* and selected against binding to the wt operator and for binding to the mutant operator. Because there are a variety of convenient reporter genes, the *E. coli* could be engineered to link DNA recognition to cell survival in both the negative (selection against binding to the wt operator) and the positive (selection for binding to the mutant operator) direction. Binding to the wt Mnt operator was selected against by placing a *tet* resistance (*tet*) gene under negative control of the wt Mnt operator. If a Mnt mutant bound the wt operator, it would block synthesis of the *tet* gene, and the *E. coli* cells would die in the presence of tetracycline. Then Mnt variants with altered DNA-binding specificity were selected for based on immunity to infection by P22 phage containing a mutant Mnt operator. The mutant Mnt operator-controlled synthesis of the proteins responsible for lysing the bacterial host. If a Mnt variant could bind to this mutant operator, it would turn off the lytic machinery, and the bacteria would survive phage infection. Four independent colonies were isolated from the two selections. Again, only a single round of selection was required for each step. All four colonies encoded the same $His^6 \rightarrow Pro$ mutation, two by a CAC→CCC and two by a CAC→CCT mutation. Not only did these mutants bind to the mutant operator, but they did not bind efficiently to the wt operator. This example illustrates the importance of screening both for the new activity and against the old activity when attempting to change the specificity of an existing protein.

More recently, Pabo and co-workers adapted the Hochschild RNA Pol assay to evolve zinc finger variants with defined DNA-binding specificities (Fig. 7.10) [57]. In this assay, protein-protein interactions are detected based on dimerization of a DNA-binding protein and the a-domain of RNA Pol and activation of RNA Pol-dependent transcription. To create a read-out for protein-DNA interactions, three tandem zinc-fingers were fused to Gal11 and Gal4, which binds with high affinity to Gal11, was fused to the a-domain of RNA Pol. In addition, a reporter system was engineered, such that if the zinc fingers bound with high affinity to the desired DNA sequence, they would activate transcription of a *HIS3* reporter gene. As with the peptide aptamers, zinc fingers with new DNA-binding specificities could be isolated from a library of ca. 10^9 zinc finger variants in a single round of selection. The authors, who have used phage display routinely in the past, note that the ease of selection is in contrast to phage display, where multiple rounds of selection and amplification were required in their lab for the identical evolution experiment (Fig. 7.10).

7.3.3
Protein-RNA interactions

Selecting for RNA-protein interactions is less straightforward because RNA-protein fusions cannot be generated directly *in vivo* and because routine biochemical assays that turn RNA-binding events into an amplified signal are convoluted to engineer [89]. This difficulty was circumvented by adding a third component to the two-hybrid sys-

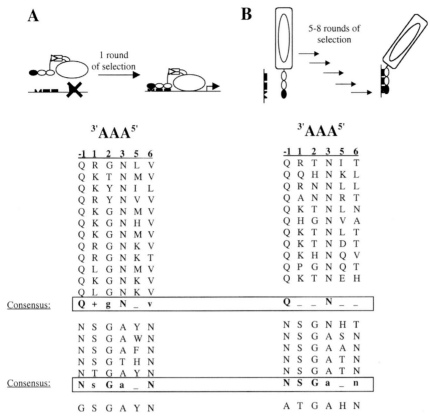

A

1 round of selection →

B

5-8 rounds of selection →

³'AAA⁵'

-1	1	2	3	5	6
Q	R	G	N	L	V
Q	K	T	N	M	V
Q	K	Y	N	I	L
Q	R	Y	N	V	V
Q	K	G	N	M	V
Q	K	G	N	H	V
Q	K	G	N	M	V
Q	R	G	N	K	V
Q	R	G	N	K	T
Q	L	G	N	M	V
Q	K	G	N	K	V
Q	L	G	N	K	V

Consensus: **Q + g N _ v**

N	S	G	A	Y	N
N	S	G	A	W	N
N	S	G	A	F	N
N	S	G	T	H	N
N	T	G	A	Y	N

Consensus: **N s G a _ N**

G S G A Y N

³'AAA⁵'

-1	1	2	3	5	6
Q	R	T	N	I	T
Q	Q	H	N	K	L
Q	R	N	N	L	L
Q	A	N	N	R	T
Q	K	T	N	L	N
Q	H	G	N	V	A
Q	K	T	N	L	T
Q	K	T	N	D	T
Q	K	H	N	Q	V
Q	P	G	N	Q	T
Q	K	T	N	E	H

Consensus: **Q _ _ N _ _**

N	S	G	N	H	T
N	S	G	A	S	N
N	S	G	A	A	N
N	S	G	A	T	N
N	S	G	A	T	N

Consensus: **N S G a _ n**

A T G A H N

Fig. 7.10. Pabo and co-workers compared the evolution of zinc fingers specific for the DNA sequence AAA using a B2H selection and a phage display selection [57, 111]. **A.** The bacterial RNA Pol system developed by Hochschild was used for two-hybrid selection [55]. Zinc figures 1, 2, and 3 from the Zif268 protein were fused to the Gal11 protein. The Gal4 protein, which binds Gal11 with high affinity, was fused to the N-terminal domain of the α-subunit of RNA Pol [57]. AAA sequence was engineered at the binding site for the first zinc finger, and six residues in zinc finger 1 were randomized. Thus, if the mutant zinc finger bound to the AAA site with high affinity, the RNA Pol complex was stabilized, activating transcription of a *HIS3* reporter gene. Significantly, in just one round of selection, several proteins were identified that bound specifically to the target DNA sequence. These sequences, which fall into two distinct sequences categories, are shown. The consensus sequences that can be drawn from this screen are also shown (capital letters indicate that all sequences showed this amino acid; lower case letters indicate a strong preference for this amino acid; "+" represents a preference for a positively charged amino acid; while "−" indicates no preference could be determined). **B.** A similar experiment was carried out by phage display, where zinc fingers 1, 2, and 3 from the Zif268 were displayed on the phage surface as a fusion to coat protein III [111]. Again, fingers 2 and 3 were held constant, while the same six amino acids in finger 1 were randomized. In this case, five to eight rounds of selection and amplification were required to obtain high affinity binders. The protein sequences obtained from these experiments are shown, as well as the consensus sequences. Both experiments produced very similar results, but the two-hybrid method only required one round of selection, compared to five to eight rounds for phage display.

tem and making a Y3H assay (Fig. 7.1C) [90, 91]. The third component is a hybrid RNA molecule where one half is a well-studied RNA molecule that binds to a known protein with high affinity and the other half is the RNA molecule of interest whose protein-binding partner is in question. In total, then, the Y3H system consists of two protein chimeras, one RNA chimera, and a reporter gene. The hybrid RNA molecule bridges the DNA-binding and activation domain fusion proteins and activates transcription of a reporter gene.

Wickens and co-workers designed an RNA Y3H system that uses the MS2 coat protein-stem-loop RNA interaction as an anchor (Fig. 7.11). In their first proof of principle experiment, they showed that this system could detect the interaction between two well-studied protein-RNA pairs: the iron response protein-iron response element interaction and the Tat-TAR interaction [90]. A particularly impressive application of this system was the cloning of a regulatory protein from *C. elegans* that binds to the 3' untranslated region of the *FEM-3* gene and mediates the sperm/oocyte switch in hermaphrodites [92]. Fields and co-workers have even tested the feasibility of using the RNA Y3H system to screen random libraries of RNA molecules to identify unknown

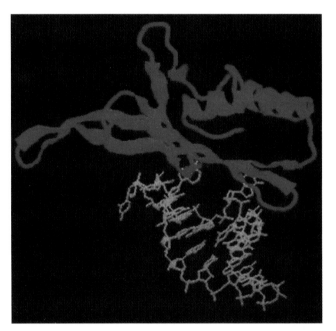

Fig. 7.11. Representation of the high-resolution structure of the bacteriophage MS2 coat protein (magenta) bound to an eighteen nucleotide RNA aptamer (cyan) at 2.8 Å resolution [112]. The MS2 coat protein is fused to the DBD in the RNA Y3H system and is used to anchor the RNA sequence at the promoter region [90].

RNA-protein interactions [93]. Fields built an RNA library by fusing fragments of yeast genomic DNA, each around 100 base pairs long, to the DNA encoding the RNA ligand for the MS2 coat protein. One and a half million yeast transformants were screened for their ability to activate both *HIS3* and *lacZ* reporter genes due to the interaction of the MS2-RNA fusion and a Gal4 AD-Snp1 fusion. Four separate RNA ligands to the Snp1 protein were found which activated both reporter genes. One of these RNA ligands was a fragment of the U1 RNA loop I, which is known to bind Snp1 in the U1 small nuclear RNA-protein complex. While this method holds promise for binding protein-RNA interaction, SELEX, an *in vitro* selection method, may be so powerful that other methods are left behind [94].

7.3.4
Protein-small molecule interactions

Just as an RNA molecule can be introduced to mediate the interaction between the DNA-binding and activation domains, so can a small molecule (Fig. 7.1D). These small molecules often have been termed chemical inducers of dimerization (CIDs). To date, both monomeric and dimeric small molecules have been employed. Dimeric CIDs have been used much in the same way as hybrid RNA molecules, with one half of the molecule serving as an anchor and the other half the compound in question. The first dimeric CIDs were dimers of the immunosuppressant FK506 [95]. Most papers describing the development of a new CID, including this first paper, included the application of the CID to transcription activation using a mammalian three-hybrid assay. Licitra and Liu built the first Y3H assay that employs two fusion proteins: the glucocorticoid receptor (GR) fused to the $LexA_{202}$ DBP, and FK506 binding protein (FKBP12) fused to the B42 AD [96]. The two fusion proteins are bridged by a heterodimeric dexamethasone-FK506 molecule. Dexamethasone binds to GR with a low nanomolar dissociation constant, as does FK506 to FKBP12. The key to the success of these systems is likely the ligand-receptor pairs, hence one major area of development is new CID pairs. Researchers at ARIAD Gene Therapeutics have developed several CIDs based on purely synthetic analogs of FK506 [97, 98]. Coumermycin, which is a naturally occurring asymmetric homodimer, has also been used as a CID [99]. We recently reported a dexamethasone-methotrexate CID that efficiently dimerizes GR and dihydrofolate reductase (DHFR) in the Y3H assay [42]. The main advantage to this system is that both dexamethasone (Dex) and methotrexate (Mtx) are readily available and that the heterodimeric derivative can be prepared readily. In addition the low picomolar affinity of Mtx for *E. coli* DHFR likely contributes to the efficacy of this CID. A methotrexate homodimer is being developed for use in the bacterial λ-repressor system [100]. Some of the small-molecule ligands used in CIDs are shown in Fig. 7.12.

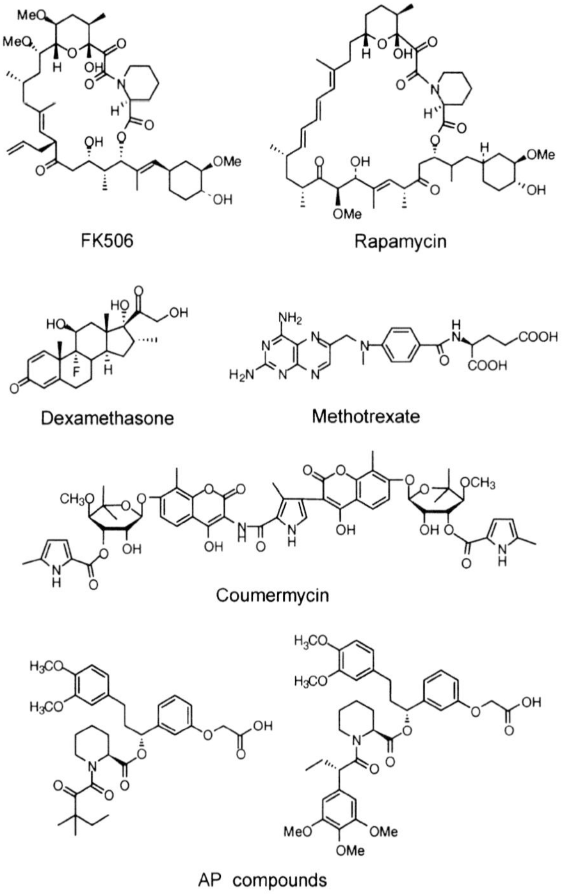

Fig. 7.12. Structures of several of the small molecules used as Chemical Inducers of Dimerizations (CIDs) in the three-hybrid systems to date.

The Y3H assay should be well suited as a tool for engineering proteins with new small molecule specificities. For example, a variant of the FKBP12-rapamycin-binding (FRB) domain of FKBP12-rapamycin-associated protein (FRAP) that is selective for a rapamycin analog was isolated from a library of FRB mutants by Liberles *et al.* using a three-hybrid assay carried out in a mammalian cell line [101]. The FRB mutants were designed to create a new pocket in their binding sites to accommodate an additional substituent installed on rapamycin to block binding to wt FRB. The most selective FRB mutant was identified by testing several FRB mutants in a three-hybrid assay. Just as the development of B2H assays is behind that of the yeast assay, only a few bacterial three-hybrid systems have been reported, and these systems still require optimization

[100, 102]. For molecular evolution applications, an important step will be the design of a robust bacterial small molecule three-hybrid system.

7.4 Conclusion

The two-hybrid assay should provide a convenient tool for performing molecular evolution experiments *in vivo*. Though it is used routinely to clone proteins based on protein-protein interactions, it has only recently begun to be applied to molecular evolution. The same features that have made two-hybrid systems powerful for testing cDNA libraries should hold for protein engineering. There have been significant technical improvements in the assay over the past decade. The assay can now be run as either a screen or a positive or negative growth selection. The availability of robust reporters with a range of sensitivities should facilitate the detection of low levels of activity in early rounds of selection and discrimination of the most active variants in the final rounds. Since the assay is carried out *in vivo*, the plasmids encoding the protein variants that survive at each round of a selection can be recovered readily either for sequencing or further rounds of mutation and selection. Bacterial "two-hybrid" systems are being introduced so that the experiments can be carried out in *E. coli* where libraries on the order of 10^8 can be tested. Early applications of the two-hybrid assay to the evolution of proteins with new protein- and DNA-recognition properties are very encouraging, suggesting that high affinity binders can be identified in a single round of selection. Three-hybrid systems expand these assays to include protein-RNA and protein-small molecule interactions. Efforts are even underway to convert these assays to generic complementation assays for enzyme catalysis [102 – 108].

Acknowledgement
V.W.C. is a recipient of a Beckman Young Investigator Award, a Burroughs-Wellcome Fund New Investigator Award in the Toxological Sciences, a Camille and Henry Dreyfus New Faculty Award, and a National Science Foundation Career Award. B.T.C. is a National Defense Science & Engineering Graduate Fellow.

References

[1] S. Fields, O. Song, *Nature* **1989**, *340*, 245–246.

[2] R. Brent, M. Ptashne, *Cell* **1985**, *43*, 729–736.

[3] I. Hope, K. Struhl, *Cell* **1986**, *46*, 885–894.

[4] S. J. Triezenberg, R. Kingsbury, S. McKnight, *Genes Dev* **1988**, *2*, 718–729.

[5] C.-T. Chien, P. Bartel, R. Sternglanz, S. Fields, *Proc Natl Acad Sci U S A* **1991**, *88*, 9578–9582.

[6] S. Fields, P. Uetz, ‹http://depts.washington.edu/sfields/yplm/data/index.html›.

[7] P. Chevray, D. Nathans, *Proc Natl Acad Sci U S A* **1992**, *89*, 5789–5793.

[8] T. Durfee, K. Becherer, P. Chen, S. Yeh, Y. Yang, A. Kilburn, W. Lee, S. Elledge, *Genes & Dev* **1993**, *7*, 555–569.

[9] P. James In *Methods in Molecular Biology: Two-Hybrid Systems*; P. MacDonald, Ed.; Humana Press: Totowa, 2001; Vol. 177.

[10] J. Boeke, F. LaCroute, G. Fink, *Mol Gen Genet* **1984**, *197*, 345–346.

[11] D. Mumberg, R. Muller, M. Funk, *Gene* **1995**, *156*, 119–112.

[12] A. Vojtek, S. Hollenberg, J. Cooper, *Cell* **1993**, *74*, 205–214.

[13] J. Gyuris, E. Golemis, H. Chertkov, R. Brent, *Cell* **1993**, *75*, 791–803.

[14] M. Johnston, *Microbiol Rev* **1987**, *51*, 458–476.

[15] M. Schnarr, P. Oertel-Buchheit, M. Kazmaier, M. Granger-Schnarr, *Biochimie* **1991**, *73*, 423–431.

[16] A. Efthymiadis, H. Shao, S. Hubner, D. Jans, *J Biol Chem* **1997**, *272*, 22134–22139.

[17] F. Beranger, S. Aresta, J. de Grunzberg, J. Camonis, *Nuc Acids Res* **1997**, *25*, 2035–2036.

[18] U. Yavuzer, C. Goding, *Gene* **1995**, *165*, 93–96.

[19] J. Ma, M. Ptashne, *Cell* **1987**, *51*, 113–119.

[20] A. Zervos, J. Gyuris, R. Brent, *Cell* **1993**, *72*, 223–232.

[21] M. Brown, R. MacGillivray, *Anal Biochem* **1997**, *247*, 451–452.

[22] E. Golemis, V. Khazak, *Methods Mol. Biol.* **1997**, *63*, 197–218.

[23] D. Shaywitz, P. Espenshade, R. Gimeno, C. Kaiser, *J Biol Chem* **1997**, *272*, 25413–25416.

[24] D. Ruden, J. Ma, Y. Li, K. Wood, M. Ptashne, *Nature* **1991**, *350*, 250–252.

[25] I. Serebriiskii, V. Khazak, E. Golemis, *J Biol Chem* **1999**, *274*, 17,080–17,087.

[26] B. Le Douarin, B. Pierrat, E. vom Baur, P. Chambon, R. Losson, *Nuc Acids Res* **1995**, *23*, 876–878.

[27] M. Ptashne *A Genetic Switch;* 2nd ed.; Blackwell Science & Cell Press: Cambridge, 1992.

[28] M. Vidal, R. Brachmann, A. Fattaey, E. Harlow, J. Boeke, *Proc Natl Acad Sci U S A* **1996**, *93*, 10315–10320.

[29] R. Cormack, K. Hahlbrock, I. Somssich, *Plant J* **1998**, *14*, 685–692.

[30] C. Leanna, M. Hannink, *Nuc Acids Res* **1996**, *24*, 3341–3347.

[31] J. Huang, S. Schreiber, *Proc Natl Acad Sci U S A* **1997**, *94*, 13396–13401.

[32] H.-M. Shih, P. Goldman, A. DeMaggio, S. Hollenberg, R. Goodman, M. Hoekstra, *Proc Natl Acad Sci U S A* **1996**, *93*, 13896–13901.

[33] M. Hirst, C. Ho, L. Sabourin, M. Rudnicki, L. Penn, I. Sadowski, *Proc Natl Acad Sci U S A* **2001**, *98*, 8726–8731.

[34] D. Tzamarias, K. Struhl, *Nature* **1994**, *369*, 758–761.

[35] I. Serebriiskii, E. Golemis In *Methods in Molecular Biology: Two-Hybrid System*; P. MacDonald, Ed.; Humana Press: Totowa, 2001; Vol. 177.

[36] I. Serebriiskii, E. Golemis, ‹http://www.fccc.edu/research/labs/golemis/InteractionTrapInWork.html›.

[37] E. Golemis, I. Serebriiskii, R. Finley, M. Kolonin, J. Gyuris, R. Brent In *Current Protocols in Molecular Biology*; F. Ausubel, R. Brent, R. Kingston, D. Moore, J. Seidman, J. Smith and K. Struhl, Eds.; John Wiley & Sons: New York, 1999; Vol. 3.

[38] P. James, J. Halladay, E. Craig, *Genetics* **1996**, *144*, 1425–1436.

[39] *Current Protocols in Molecular Biology*; John Wiley & Sons: New York, 1999; Vol. 3.

[40] J. Estojak, R. Brent, E. Golemis, *Mol Cell Biol* **1995**, *15*, 5820–5829.

[41] W. Abida, B. Carter, E. Althoff, H. Lin, V. Cornish, *submitted for publication*

[42] H. Lin, W. Abida, R. Sauer, V. Cornish, *J Am Chem Soc* **2000**, *122*, 4247–4248.

[43] J. Thillet, J. Absil, S. Stone, R. Pictet, *J Biol Chem* **1988**, *263*, 12500–12508.

[44] B. Schweitzer, A. Dicker, J. Bertino, *FASEB J* **1990**, *4*, 2441–2452.

[45] R. Kornberg, *Biol Chem* **2001**, *382*, 1103–1107.

[46] L. Myers, R. Kornberg, *Annu Rev Biochem* **2000**, *69*, 729–749.

[47] P. Cramer, D. Bushnell, R. Kornberg, *Science* **2001**,

[48] A. Gnatt, P. Cramer, J. Fu, D. Bushnell, R. Kornberg, *Science* **2001**,

[49] J. Hu, M. Kornacker, A. Hochschild, *Methods* **2000**, *20*, 80–94.

[50] J. Hu, E. O'Shea, P. Kim, R. Sauer, *Science* **1990**, *250*, 1400–1403.

[51] T. Schmidt-Dorr, P. Oertel-Buchheit, C. Pernelle, L. Bracco, M. Schnarr, M. Granger-Schnarr, *Biochemistry* **1991**, *30*, 9657–64.

[52] M. Dmitrova, G. Younes-Cauet, P. Oertel-Buchheit, D. Porte, M. Schnarr, M. Granger-Schnarr, *Mol Gen Genet* **1998**, *257*, 205–212.

[53] S. Bustos, R. Schleif, *Proc Natl Acad Sci U S A* **1993**, *90*, 5638–42.

[54] M. Kornacker, B. Remsburg, R. Menzel, *Mol Microbiol* **1998**, *30*, 615–24.

[55] S. Dove, J. Joung, A. Hochschild, *Nature* **1997**, *386*, 627–630.

[56] F. Whipple, *Nuc Acids Res* **1998**, *26*, 3700–3706.

[57] J. Joung, E. Ramm, C. Pabo, *Proc Natl Acad Sci U S A* **2000**, *97*, 7382–7387.

[58] M. C. Marsolier, M. N. Prioleau, A. Sentenac, *J Mol Biol* **1997**, *268*, 243–9.

[59] G. Karimova, J. Pidoux, A. Ullmann, D. Ladant, *Proc Natl Acad Sci U S A* **1998**, *95*, 5752–5756.

[60] F. Rossi, C. Charlton, H. Blau, *Proc Natl Acad Sci U S A* **1997**, *94*, 8405–8410.

[61] J. Pelletier, F. Campbell-Valois, S. Michnick, *Proc Natl Acad Sci U S A* **1998**, *95*, 12141–12146.

[62] I. Remy, S. Michnick, *Proc Natl Acad Sci U S A* **1999**, *96*, 5394–5399.

[63] I. Remy, S. Michnick, *Proc Natl Acad Sci U S A* **2001**, *98*, 7678–7683.

[64] S. Michnick, *Curr Opin Struct Biol* **2001**, *11*, 472–477.

[65] R. Subramaniam, D. Desveaux, C. Spickler, S. Michnick, N. Brisson, *Nat Biotechnol* **2001**, *19*, 769–772.

[66] A. Miyawaki, J. Llopis, R. Heim, J. McCaffery, J. Adams, M. Ikura, R. Tsien, *Nature* **1997**, *388*, 882–887.

[67] A. Miyawaki, R. Tsien, In *Methods in Enzymology*; J. Thorner, S. Emr and J. Abelson, Eds.; Academic Press: New York, 2000; Vol. 327.

[68] T. Ozawa, S. Nogami, M. Sato, Y. Ohya, Y. Umezawa, *Anal Chem* **2000**, *72*, 5151–5157.

[69] A. Aronheim, E. Zandi, H. Hennemann, S. Elledge, M. Karin, *Mol Cell Biol* **1997**, *17*, 3094–3102.

[70] Y. Broder, S. Katz, A. Aronheim, *Curr Biol* **1998**, *8*, 1121–1124.

[71] A. Aronheim, M. Karin In *Methods in Enzymology*; J. Thorner, S. Emr and J. Abelson, Eds.; Academic Press: New York, 2000; Vol. 328.

[72] N. Johnsson, A. Varshavsky, *Proc Natl Acad Sci U S A* **1994**, *91*, 10340–10344.

[73] I. Stagljar, C. Korostensky, N. Johnsson, S. te Heesen, *Proc Natl Acad Sci U S A* **1998**, *95*, 5187–5192.

[74] H. Laser, C. Bongards, J. Schuller, S. Heck, N. Johnsson, N. Lehming, *Proc Natl Acad Sci U S A* **2000**, *97*, 13732–13737.

[75] M. Yang, Z. Wu, S. Fields, *Nuc Acids Res* **1995**, *23*, 1152–1156.

[76] P. Colas, B. Cohen, T. Jessen, I. Grishina, J. McCoy, R. Brent, *Nature* **1996**, *380*, 548–550.

[77] C. Xu, A. Mendelsohn, R. Brent, *Proc Natl Acad Sci U S A* **1997**, *94*, 12473–12478.

[78] E. Fabbrizio, L. Le Cam, J. Polanowska, M. Kaczorek, N. Lamb, R. Brent, C. Sardet, *Oncogene* **1999**, *18*, 4357–4363.

[79] S. Schneider, M. Buchert, O. Georgiev, B. Catimel, M. Halford, S. Stacker, T. Baechi, K. Moelling, C. Hovens, *Nat Biotechnol* **1999**, *17*, 170–175.

[80] J. Pelletier, K. Arndt, A. Pluckthun, S. Michnick, *Nat Biotechnol* **1999**, *17*, 683–690.

[81] W. Zhu, R. Williams, T. Kodadek, *J Biol Chem* **2000**, *275*, 32098–32105.

[82] A. Dhiman, M. Rodgers, R. Schleif, *J Biol Chem* **2001**, *276*, 20017–20021.

[83] M. Kolonin, R. Finley, *Proc Natl Acad Sci U S A* **1998**, *95*, 14266–14271.

[84] B. Cohen, P. Colas, R. Brent, *Proc Natl Acad Sci U S A* **1998**, *95*, 14272–14277.

[85] M. Visintin, E. Tse, H. Axelson, T. Rabbitts, A. Cattaneo, *Proc Natl Acad Sci U S A* **1999**, *96*, 11723–11728.

[86] E. Mossner, H. Koch, A. Pluckthun, *J Mol Biol* **2001**, *308*, 115–122.

[87] S. Park, R. Raines, *Nat Biotechnol* **2000**, *18*, 847–851.

[88] P. Youderian, A. Vershon, S. Bouvier, R. Sauer, M. Susskind, *Cell* **1983**, *35*, 777–783.

[89] A. Cheng, V. Calabro, A. Frankel, *Curr Opin Struct Biol* **2001**, *11*, 478–484.

[90] D. SenGupta, B. Zhang, B. Kraemer, P. Pochart, S. Fields, M. Wickens, *Proc Natl Acad Sci U S A* **1996**, *93*, 8496–8501.

[91] B. Kraemer, B. Zhang, D. SenGupta, S. Fields, M. Wickens In *Methods in Enzymology*; J. Thorner, S. Emr and J. Abelson, Eds.; Academic Press: New York, 2000; Vol. 328.

[92] B. Zhang, M. Gallegos, A. Puoti, E. Durkin, S. Fields, J. Kimble, M. Wickens, *Nature* **1997**, *390*, 477–484.

[93] D. Sengupta, M. Wickens, S. Fields, *RNA* **1999**, *5*, 596–601.

[94] S. Klug, M. Famulok, *Mol Biol Rep* **1994**, *20*, 97–107.

[95] D. Spencer, T. Wandless, S. Schreiber, G. Crabtree, *Science* **1993**, *262*, 1019–24.

[96] E. Licitra, J. Liu, *Proc Natl Acad Sci U S A* **1996**, *93*, 12817–12821.

[97] J. Amara, e. al., *Proc Natl Acad Sci U S A* **1997**, *94*, 10618–23.

[98] T. Clackson, e. al., *Proc Natl Acad Sci U S A* **1998**, *95*, 10437–10442.

[99] M. Farrar, J. Alberola-Ila, R. Perlmutter, *Nature* **1996**, *383*, 178–181.

[100] S. Kopytek, R. Standaert, J. Dyer, J. Hu, *Chem Biol* **2000**, *7*, 313–321.

[101] S. Liberles, S. Diver, D. Austin, S. Schreiber, *Proc Natl Acad Sci U S A* **1997**, *94*, 7825-7830.

[102] S. Firestine, F. Salinas, A. Nixon, S. Baker, S. Benkovic, *Nat Biotechnol* **2000**, *18*, 544–547.

[103] E. Baum, G. Bebernitz, Y. Gluzman, *Proc Natl Acad Sci U S A* **1990**, *87*, 10023–27.

[104] B. Dasmahapatra, B. DiDomenico, S. Dwyer, J. Ma, I. Sadowski, J. Schwartz, *Proc Natl Acad Sci U S A* **1992**, *89*, 4159–4162.

[105] J. Stebbins, I. Deckman, S. Richardson, C. Debouck, *Anal Biochem* **1996**, *242*, 90–94.

[106] H. Sices, T. Kristie, *Proc Natl Acad Sci U S A* **1998**, *95*, 2828–2833.

[107] N. Dautin, G. Karimova, A. Ullmann, D. Ladant, *J. Bacteriol.* **2000**, *2000*, 7060–7066.

[108] K. Baker, C. Bleczinski, G. Salazar-Jimenez, D. Sengupta, H. Lin, S. Krane, V. Cornish, *submitted for publication*

[109] M. Wang, R. Reed, *Nature* **1993**, *364*, 121–126.

[110] P. Uetz, L. Giot, *e. al*, ((full list of names??))S. Fields, J. M. Rothberg, *Nature* **2000**, *403*, 623–627.

[111] H. Greisman, C. Pabo, *Science* **1997**, *275*, 657–661.

[112] S. Rowsell, N. Stonehouse, M. Convery, C. Adams, A. Ellington, I. Hirao, D. Peabody, P. Stockley, S. Phillips, *Nat Struct Biol* **1998**, *5*, 970–975.

[113] J. Harper, G. Adami, N. Wei, K. Keyomarsi, S. Elledge, *Cell* **1993**, *75*, 805–816.

[114] R. Stan, M. McLaughlin, R. Cafferkey, R. Johnson, M. Rosenberg, G. Livi, *J Biol Chem* **1994**, *269*, 32027–32030.

[115] M. Chiu, H. Katz, V. Berlin, *Proc Natl Acad Sci U S A* **1994**, *91*, 12574–12578.

8
Advanced Screening Strategies for Biocatalyst Discovery

Andreas Schwienhorst

8.1
Introduction

Since prehistoric times biocatalysis has played a significant role in chemical transformations such as baking, brewing, and cheese production [1 –3]. Today, with over 3000 enzymes identified so far, it is evident that many different types of chemical reactions can be performed with the help of enzymes (Tab. 8.1). Among various options in industrial catalysis, biocatalysts are generally chosen for their high activity, with reaction rate accelerations of typically 10^5–10^8. In addition, the chiral nature of enzymes results in a remarkable regio- and stereospecificity. As a consequence, biocatalysis usually works extremely well even in the absence of substrate functional-group protection. Furthermore, both natural and engineered enzymes can now be produced on a large scale in convenient host organisms using recombinant DNA technology. In recent years biocatalysis has therefore become increasingly attractive for industrial synthetic chemistry not least because of dramatic advances in recombinant DNA and high-throughput screening (HTS) technologies. Today, the combination of these approaches allows us to generate tailored biocatalysts for processes that generally work under mild conditions and generate few waste products, making biocatalysts an environmentally friendly alternative to conventional chemical catalysts.

Although the opportunities for applying biocatalysts are numerous, fast identification or engineering of appropriate enzyme catalysts remains one of the key limiting steps. Fortunately, a number of natural enzymes already work reasonably well as industrial biocatalysts. However, natural enzymes did not evolve over a billion years to work in industrial processes but to perform specific biological functions within the context of a living organism. Not surprisingly, hence, many natural enzymes exhibit properties that are rather unfavorable in the context of industrial processes. Problems comprise regulatory issues such as product inhibition and cofactor requirements. Other properties like stability under process conditions were neither required nor

Tab. 8.1. Examples of enzymes and concomitant screening methods

Enzyme	Assay method	Ref.
proteases	halo (A)	45 – 48
proteases	absorption (S)	70
proteases	FRET (S)	76 – 78
proteases	FP (S)	79
subtilisin	absorption (S)	59
methionine aminopeptidases	absorption (S)	71
D-Alanyl-D-Alanine dipeptidase	absorption (S)	60
penicillin G acylase	color (A, S)	106, 107
penicillin G acylase	fluorescence (S)	65
penicillin G acylase	fluorescence (S)	69
staphylococcal lipase	absorption (S)	108
lipases	fluorescence (S)	62
lipases (soil)	halo (A)	40, 49
pNB esterases	absorption (S)	22
P. fluorescens esterase	pH-indicator (A)	42
P. polymyxa glucosidase	color (A)	35
β-lactamases (soil)	color (A)	40
scytalone dehydratases	pH-indicator (A)	43
cellulase	halo, colony size (A)	30
sulfotransferases	absorption (S)	72
nitrate and nitrite reductases	absorption (S)	73
restriction endonucleases	FCS (S)	82
dehydrogenases (soil)	color (A)	39
P450 oxygenase	color (A)	44
B. flavum carotenogenic gene cluster	color (A)	41

A = agar plate-based assay; S = solution-based assay; FRET = fluorescence resonance energy transfer; FP = fluorescence polarization; FCS = fluorescence correlation spectroscopy

acquired during natural evolution and consequently impair technical applications of many isolated enzymes. In addition, the extent of substrate specificity sometimes limits the general synthetic utility of an enzyme. Although biocatalysts now cover a wide range of synthetic reactions, one still cannot use enzymes for the formation of every desired chemical bond or for every cleavage reaction. Hence, there is a constant need for novel biocatalysts with desired properties.

Traditionally, new biocatalysts were identified in labor-intensive screenings of microbial cultures. For this purpose, companies such as Novozymes amassed in-house collections with more than 25,000 classified fungal and bacterial cultures. However, taking together all collections worldwide, the number of identified species is presumably still a tiny fraction of all cultivable microbial organisms, which counts for less than 1 % of all microorganisms [4]. Since many of these microorganisms, in particular those that live in harsh environments, produce biocatalysts with potentially useful properties [5, 6], diverse approaches have been developed to also recruit these enzymatic activities and transfer them to more tractable organisms such as *E. coli* [7 – 10]. Finally, due to the revolutionary developments of molecular biology biocatalysts may also be evolved through engineering of known enzymes. Since our understanding of structure – function relationships is still very limited for proteins, rational attempts to design new enzymes were rather fruitless. The situation changed dramatically with the introduction of evolutionary optimization methods. The process of "laboratory evolution" is usually initiated by generating a large library ($>10^6$ clones) of enzyme variants, in particular by random mutagenesis [11 – 14] and homologous [15 – 25] as well as heterologous [26, 27] recombination. In a successive step of selection, enzyme features that are of interest for a certain application are linked to the survival or growth of a host microorganism and (hopefully) desired biocatalysts are genetically selected. Unfortunately, however, many enzyme properties cannot be subjected to genetic selection in this way. Therefore, enzyme libraries must be characterized in a clone-by-clone approach, i.e. by high-throughput screening. Typically, 10^3 – 10^6 variants are tested in this way.

Screening, however, is not only limited to the level of enzyme but may also be necessary on the level of the producer cell and the biocatalytic process itself, e.g. with respect to fermentation media. In all instances, the demand for screening large sample collections has stimulated technology developments in the areas of assay methodoogies with particular emphasis on automation, miniaturization, and sensitive detection technology.

8.2
Semi-quantitative Screening in Agar-plate Formats

Screening in agar plate formats is probably still the most widely used approach to identify biocatalysts with desired properties. Similar to genetic selection of whole cells, care must be taken to bring enzyme and substrate together. This can easily be achieved using *in vitro* translation systems [28]. With living systems, however, either substrates or enzymes have to penetrate the cell wall. At this point a very elegant approach is the display of enzymes on the surface of bacteriophage [29] or whole bacteria [30, 31]. The latter systems also proved to be suited for alternative FACS (Fluorescence Activated Cell Sorting)-based screening [31].

Usually, a library of microorganisms expressing enzymes of potential interest is spread on agar plates. Enzyme activity is assessed by monitoring a visible signal that is linked to enzyme function. Visual signals can, in principle, be generated by different mechanisms.

A peculiar case is certainly that of the green fluorescent protein (GFP). Functional GFP variants have been identified by visual inspection of colonies based on the intrinsic fluorescence of GFP under UV illumination [24, 32 – 34]. To increase accuracy of signal quantification as well as throughput, digital image spectroscopy has been applied in some cases [32].

More widely applicable are colorimetric assays. Here, the desired enzymatic activity is linked to a color change, e.g. on the level of bacterial colonies. A well-known example is the assay for β-galactosidase activity using its chromogenic substrate 5-bromo-4-chloro-3-indolyl β-D-galactoside (X-Gal). Depending on the concentration and activity of enzyme molecules colonies stain anything from dark blue to white. Meanwhile the method has been extended to a number of diverse applications including the screening for glucosidases [35] and even catalytic RNA molecules [36, 37]. Direct screening based on the level of colony color has been extended to a number of different enzymes including proteases amidohydrolases like penicillin G acylase (Fig. 8.1A), phosphatases [38], dehydrogenases [39] and β-lactamases [40]. Not only single enzyme activities but also complex metabolic functions are accessible by colorimetric colony-based methods. A recent example is the discovery of a carotenogenic gene cluster by functional complementation of a colorless *B. flavum* mutant, screening transformed cells for production of a yellow pigment [41].

Sometimes colorimetric screenings are inherently multistep with the enzymatic reaction of interest linked to a second chemical reaction which then produces a visible color. In its simplest setup the enzymatic reaction generates an alkaline or acidic product causing a shift in pH which is detected by a color change of a pH indicator dye. In this way esterases [42] as well as scytalone dehydratases [43] have recently been identified. The later method is of particular interest, since it couples the *in vivo* concentration of an enzyme's substrate to changes in the transcriptional level of a reporter operon and thus may represent a widely applicable screening technology. In a first example, the AraC DNA-binding domain was fused to a protein capable of binding both the substrate of scytalone dehydratase as well as a competing chemical inducer of dimerization. Dimerization caused by the later activity results in transcription of the chromosomal araBAD operon (reporter). Upon addition of substrate the chemical inducer of dimerization is displaced. Consequently, dimerization and thus reporter gene expression is prevented. In the presence of scytalone dehydratase, however, the substrate is cleaved and dimerization again leads to reporter gene activity.

A slightly more complex two-step colorimetric screening system was recently designed to discover oxygenases. Here, the oxygenase-catalyzed hydroxylation of certain aromatic substrates is combined with subsequent oxidative coupling. The later is achieved by coexpression of a variant of horseradish peroxidase (HRP) that generates colored products which are readily detected by digital imaging [44].

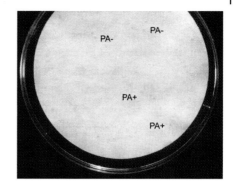

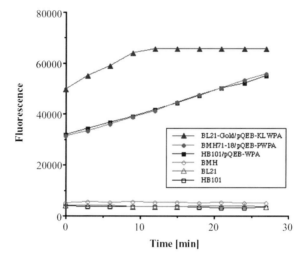

Fig. 8.1. Screening for penicillin G acylase activity. A) Screening in agar plate formats using 6-nitro-3-(phenylacetamido)-benzoic acid (NIPAB) [106]. Colonies secreting Penicillin G acylase activity stain a NIPAB-filter yellow. B) Screening in solution using phenylacetyl-MCA and periplasmic extracts without (open symbols) or with (▲,●, ■) penicillin G acylases from *Kluyvera citrophila*, *Proteus rettgeri* and *Escherichia coli* respectively [65].

A number of particular semi-quantitative screening methods in agar-plate formats depend on the secretion of (hydrolytic) enzymes. Clones that secrete active enzymes produce a so-called "halo", i.e. a zone of clearing in turbid solid growth medium with suspended substrate. In many cases the size of the halo is a good measure for the amount of enzymatic activity released by the growing colony.

Among the first applications of this type of experimental strategy were screenings for protease activity [45, 46]. Here, agar plates containing a turbid suspension of casein or skim milk proteins are used. Recent work on proteases include the discovery of subtilisin E variants with enhanced activity in aqueous organic solvent [47] and the screening of a library of *Streptomyces griseus* protease B variants to reveal sequence requirements for substrate specificity [48]. Functional screening based on halo formation is also used in the screening for lipases [40, 49], esterases, cellulases, keratinases, chitinases, amylases, β-lactamases and DNases [40]. Henne *et al.* used indicator plates containing triolein and rhodamine B to screen for lipolytic activities in an environmental DNA library [49]. Active clones could be identified by orange fluorescent halos around concomitant colonies. Kim *et al.* displayed a library of cellulase variants on the surface of *E. coli*. Indicator plates containing carboxymethyl cellulose (CMC) were used to identify active cellulase variants. Bacterial growth rates deduced from colony size were found to be correlated to the activities of displayed enzyme variants and served as a measure to select improved cellulase variants [30].

Screening based on halo formation is not only suited for the identification of single enzyme activities but may also be used to screen for whole metabolic pathways. Aarons *et al.* recently set up a system to screen for a functional enzyme cascade that is involved in the production of substances that act as antibiotics towards an indicator bacteria strain but not the producer strain. Using the *Bacillus* overlay-assay [50] they were able to isolate genes capable of complementing *P. fluorescens* F113 gacS and gacA mutants [51].

8.3
Solution-based Screening in Microplate Formats

Simple visual screens are widely used since they are often rapid and not extraordinary laborious. However, screenings on the basis of color or halo formation is at best semi-quantitative and rather insensitive to tiny changes in enzyme activity. In addition, screening procedures for biocatalyst properties that are of interest in industrial processes are often not compatible with agar-plate technology. In many of these cases screening has to be done in solution using whole microorganisms or isolated and sometimes purified enzyme preparations. Since large ensembles of biocatalysts have to be characterized, screening automation and computerized data acquisition are highly desirable. In general, standard assays are carried out in volumes of about 100 – 200 μl using 96-well microplates. For this purpose, clones have to be separated and transferred into microtiter formats using either cell sorting devices [52, 53] or colony pickers [54] (Fig. 8.2).

Enzymatic reactions can be assayed in different ways. The availability of automated high-pressure liquid chromatography (HPLC) and gas-liquid chromatography (GLC) has made the sequential analysis of several hundred reactions practical [55]. Virtually

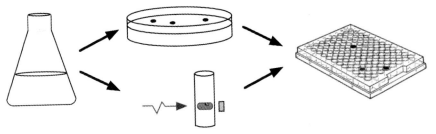

Fig. 8.2. Stages of enzyme screening. Microorganisms are grown in flasks or deep-well microplates. Separation of single clones is achieved either by FACS or by plating on agar plates and subsequent colony picking into microplates. Finally, assays are carried out in microplate formats using whole bacteria or cell extracts. Alternatively, bacterial clones can be directly screened on the level of colonies on indicator agar plates or during FACS.

any reaction can be analyzed in this way. Since the techniques require the serial analysis of individual reactions, however, analysis time increases linearly with library size. Consequently, there is a limitation on the size of libraries that can be screened in this fashion. An additional drawback is the need to purify a sample prior to analysis, e.g. to prevent contamination of the analytical columns. With an increased emphasis on high-throughput, automation and miniaturization, both the development of homogeneous assay formats and the use of high sensitivity detection techniques are desirable. Simple mix-and-measure assays are well suited for high-throughput screening as they avoid filtration, separation and wash steps that can be time-consuming and difficult to automate. There are, however, certain cases where e.g. filtration steps in parallel formats have been included into the assay [56]. The choice of detection technology employed is dependent on the particular class of assay target being investigated. Optical test systems are by far more important than assays based on radioisotopes which are employed only exceptionally [57]. As with semi-quantitative visual screens, assays based on absorbance or fluorescence intensity as read-outs are most widely used. Very often, specially designed chromogenic or fluorogenic substrates are employed that serve as representatives of the real substrate in the biocatalytic process. Chromogenic substrates are advantageous in that they may be employed both in the initial agar-based screening and in subsequent solution-based screening. Indeed, most colorimetric assay types that have been employed in agar plate formats have also been used in solution. Among the most common chromophores are *p*-nitrobenzyl-, *p*-nitrophenyl- and *p*-nitroanilide derivatives which have been used e.g. in the screening of esterases [22], epoxide hydrolases [58], proteases [59] and peptidases [60]. In contrast to absorbance, fluorescence is a more sensitive assay method and probably constitutes the majority of bioassays today. Fluorimetric assays are long known for a number of different enzyme activities [61]. A recent example is the dual wavelength assay of lipase activity that also allows direct assessment of stereoselectivity [62]. Protease assays were among the first fluorimetric enzyme assays to be developed [63]. The most important group of protease assays employs aromatic amines such as 7-amino-4-methylcoumarin as fluorophores [64]. Recently, the same technology has been extended to the analysis of amidohydrolases and in particular penicillin G acylase [65]. Phenylacetyl-4-methylcoumaryl-7-amide ($Km = 7.5 \mu$M) proved to mimic benzylpenicillin ($Km = 3.8 \mu$M), the industrial substrate of pencillin G acylase perfectly (Fig. 8.1B). As with most fluorimetric assays the consumption of substrate and enzyme is already low. An interesting option to further increase sensitivity and lower substrate concentrations has lately been described [66] for the detection of proteases, phosphatases, and glucosidases. The new assay monitors hydrolysis with the concurrent transfer of a solvatochromic dye across an oil – water barrier. In this way the fluorophore is concentrated several-fold in the organic phase and fluorescence is significantly enhanced.

With representative substrates that are designed to mimic the real substrate of interest, however, there is a considerable risk of not obtaining the desired result due to screening for processing of the "wrong" (chromogenic or fluorogenic) substrate which, in general, shows significant structural deviations with respect to the real sub-

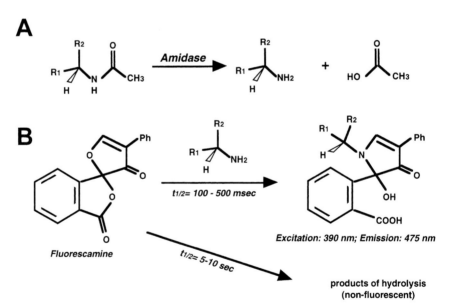

Fig. 8.3. Coupled assay for amidohydrolase activity. In the enzymatic reaction (A) an amide bond is cleaved releasing carboxylic acid and amine. The free amino group of the amine is then reacted with fluorescamine (B) in a rather fast reaction yielding fluorescent compounds. The excess of fluorescamine hydrolyzes spontaneously within seconds leaving only non-fluorescent compounds.

strate of interest. An effective strategy to escape this dilemma is to use the relevant substrate and specifically quantify one of the products of the enzymatic reaction using a suitable assay. In Fig. 8.3 a very well-known example of such a coupled assay is depicted: the titration of amino groups released by enzymatic cleavage of amide bonds using fluorescamine [67, 68]. The method has been extensively used to characterize amidohydrolases of different origin ([69]; R. Dietrich, A. Schwienhorst, unpublished results). Related methods include coupled assays for proteases [70], aminopeptidases [71], sulfotransferases [72] nitrate and nitrite reductases [73] and a new assay for esterases quantifying acetate as the leaving acyl moiety [74]. A peculiar example of the above strategy is based on an immunofluorometric assay. Here, products of the enzymatic reaction are specifically detected using a product-specific antibody, as applied in the assay for sialidase activity [75].

The process of measuring fluorescence intensity is not entirely selective and often causes unwanted fluorescence, e.g. autofluorescence caused by excitation and light-scattering signals from cellular and equipment sources. Therefore, a variety of alternative fluorescence-based detection methods have been developed and exploited to advantage in homogeneous assay formats. The fluorescence modes used include fluorescence resonance energy transfer (FRET), fluorescence polarization (FP), homogeneous time resolved fluorescence (HTRF), fluorescence correlation spectroscopy (FCS), fluorescence intensity distribution analysis (FIDA) and burst-integrated fluorescence lifetime measurements (BIFL).

FRET is the non-radiative transfer of energy between appropriate energy donor and acceptor molecules. The energy transfer efficiency varies inversely with the sixth power of the separation distance of donor and acceptor. Thus, small distance changes generate large changes in transfer efficiency. Important examples of recent applications of FRET concern protease assays in 96-well microplates [76, 77] as well as in miniaturized high-density formats [77, 78]. Typical enzyme substrates comprise short peptides corresponding to the protease cleavage site with appropriate donor and acceptor fluorophores at opposite ends of the molecule. In the intact substrate both donor and acceptor molecules are held in close proximity resulting in high energy transfer efficiencies as compared to the free fluorophores released upon enzymatic cleavage [31, 76]. FP measurements are based on changes in the rotational diffusion coefficient of small labeled probes upon enzymatic release from larger substrate molecules. Examples of the more interesting recent applications include assays for the detection of proteases [79]. As already mentioned sensitivity in fluorescence-based assays is often limited by a large background signal. This problem can be overcome by using time-resolved techniques such as HTRF. In particular, fluorescence lifetimes of rare earth chelates are often in the millisecond range whereas background "autofluorescence" is typically nanoseconds in duration. Use of pulsed laser excitation and gated detectors allows the background fluorescence to decay before sampling of the enzymatically produced assay signal takes place. Europium cryptates are frequently used due to their extremely long fluorescence lifetimes and the large Stokes shift. Recent applications of HTRF to screening of enzymatic activities include studies of tyrosine kinases [80] and reverse transcriptases [81]. Another emerging technology for high-throughput screening of biocatalysts is FCS. Here, measurements are carried out in femtoliter volumes using confocal optics. At nanomolar fluorophore concentrations only few if any molecules are present in the detection volume at a time. This diffusion of fluorescent molecules in and out the detection volume causes temporal fluctuations in the signal. Autocorrelation analysis of these time-dependent fluorescence signals provides information about the diffusion characteristics and hence about the size of the fluorescent molecules. Recent technological advances include the development of dual-color FCS which has been applied to assay restriction endonucleases [82]. Since working distances of the microscope objectives used are usually short, FCS detection technology is rather difficult to integrate with standard microplate handling systems. The same applies for FIDA measurements, where changes in fluorophore quantum yield or spectral shift and hence changes in the environment of the fluorophore are monitored [83]. In BIFL analysis, fluorescence bursts indicating traces of individual fluorescent molecules are registered and further subjected to selective burst analysis [84]. The two-dimensional BIFL data allow the identification and detection of different temporally resolved conformational states and has so far been primarily employed to study receptor – ligand interactions. Although highly sensitive, all non-standard fluorescence-based detection methods described above suffer from the fact that for many interesting enzyme classes convenient assays are still not available. In addition, most of these methods still have to be implemented into multichannel readers to function reliably in a high-throughput mode.

Among the more important recent advances in optical assay development are luminescence assays. Chemiluminescent assays utilize chemical reactions to produce light. The primary advantage of luminescent indicators is a low background signal which is usually limited to detector noise only. The best-known systems include luciferase [85] and the Ca^{2+}-sensitive photoprotein aequorin [86]. Luciferase was recently employed as a reporter gene in an interesting screening approach to identify thermoregulated genes [87]. Aequorin was used to develop an assay for proteolytic bond cleavage which in principle could be quite versatile in that a number of other cleavage reactions could also be monitored in this way. In the original assay, a cystein-containing peptidic substrate of HIV-1 protease was coupled with a cystein-free mutant of aequorin. The fusion protein was immobilized on a solid phase and employed as the substrate for HIV-1 protease. Proteolytic bond cleavage was detected by a decrease in the bioluminescence generated by the fusion protein on the solid phase [88]. Meanwhile, chemiluminescence-based assays are available for all major classes of enzymes except isomerases and ligases (see [89] for review). Generally, chemiluminescence-based enzyme assays are rather sensitive with many of them having subattomole detection limits.

For the assay of enzymes with products and reagents that have no absorption, fluorescence or luminescence in the ultraviolet or visible region, developments in analytical infrared spectroscopy can be used. In particular, mid-Fourier transform infrared (mFTIR) spectroscopy has been successfully applied to the determination of enzyme activities and kinetics, e.g. of β-fructosidase, phosphoglucose isomerase and polyphenol oxidase [90]. The method could very well be a tool that may also be applied to a variety of other enzyme classes. The potential of high-throughput applications, however, has yet to be demonstrated.

The demand for enzyme assays that not only monitor overall activity but also enantioselectivity stimulated the development of further assay systems that are still, however, in a rather experimental state with respect to high-throughput enzyme screening applications. These methods include assays based on electron spin resonance spectroscopy (ESR) [91], nuclear magnetic resonance (NMR) [92, 93], IR-thermography [94] or electrospray ionization spectrometry (ESI-MS) [95].

8.4
Robotics and Automation

The development of automation concepts for research use started in the early 80s with Zymark presenting the first robotic station that combined liquid handling technology with the movement of sample carriers. Today, a number of companies are producing fully automated robotic equipment designed for high-throughput screening. In addition, a large number of in-house system integrator groups in larger companies and scientific institutions exist which either make use of commercially available robot arms and peripherals (e.g. liquid-handling, readers) or even develop new tools. Principle architectures of automation systems can be divided into circular systems and linear systems (Fig. 8.4) [96].

Circular systems contain a stationary robot and provide efficient and compact work areas. Short distances between peripheral modules usually promote rather fast operations. In addition, robotic systems like the rotating arm system from CyBio ensure robust and largely unattended operation for months without teaching. Space, however, is generally too limited to integrate a large number of peripheral modules (Tab. 8.2). In linear systems, the robot has access to larger work areas by mounting the robot on a linear track. The movement of an item through a defined space, however, can be much slower than compared to circular systems.

Several devices of liquid handling are commercially available. Pipetting a volume of several microliter usually presents few problems. Below about 1 μl, due to obstacles

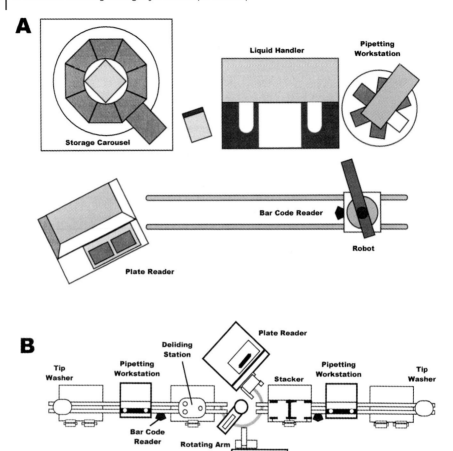

Fig. 8.4. Examples of robotic enzyme screening workstations.
A) Linear system (Zymark); B) Circular system (CyBio).

such as viscosity and surface tension, liquid handling in a pick-up mode is rather difficult or even impossible. New approaches, however, such as piezo-electric ink-jet technologies (Fig. 8.5A) or micromechanical devices, allow pipetting down to the pico-liter scale with speeds up to 10,000 drops per second [97]. Other methods, such as microsolenoid valve technology, microcapillary and solid-pin technologies are also available for transferring small droplets to nanoplates [98]. Parallelization is achieved by multi-channel pipetting. Pipettors with 2 to 1536 channels have been realized. For maximum flexibility, 96-channel pipettors that are able to move in x,y-direction are frequently used. In this way, microtiter plates with 96, 384, and 1536 wells can now be handled in the same system. Although many enzyme labora-

Tab 8.2. Important peripheral modules in robotic screening workstations

Peripheral	Function
Microplate stacker	storage of microplates
Active storage carousel	storage of microplates
UV/VIS-reader	signal detection (absorption)
Fluorescence reader	signal detection (fluorescence)
Luminescence reader	signal detection (luminescence)
Scintillation counter	signal detection (radioactivity)
1-channel pipettors	liquid handling
Multichannel pipettors	liquid handling
Washer	microplate washing (ELISA)
Dispensers	reagent addition
Microplate incubators	incubation
Bar-code readers	logistics
(De-)liding station	lid removal from microplates
Shaker	shaking (and incubation) of microplates
Centrifuge	centrifugation of microplates
Filtration module	filtration in microplate formats
Tip washer	tip washing after pipetting

tories still favor the standard 96-well microplate format [99], a considerable number of researchers have already switched to 384-well, 1536-well or even higher density nano-plate formats such as silicon wafers (Fig. 8.5B) with many thousands of compartments housing liquid volumes in the nanoliter range [28, 100 – 102]. The implementation of microplate-based screening in smaller volume, higher density formats, however, faces a number of technical hurdles. The need to control evaporation in open systems having high surface area-to-volume ratios becomes increasingly important as assay volumes are reduced. In addition, data acquisition easily becomes the rate-limiting step. This problem, however, may be avoided by replacing serially operating readers by imaging devices that simultaneously record signals from all compartments at once [77, 103]. Since throughput is essential for screening huge numbers of clones, efficient logistics, e.g. active microplate storage systems, bar-coding, computerized data acquisition and scheduling are also crucial factors. If, finally, assay conditions are designed properly to ensure robustness, reproducibility, and statistical significance of results, miniaturized biodevices may be used to screen large, combinatorial libraries of bio-catalysts to yield enzymes with tailored properties [104]. This may also be true for recent developments in the field of microfabricated fluidic devices which are currently being actively pursued in both academic and industrial laboratories [105].

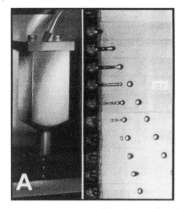

Fig. 8.5. Miniaturized biodevices for ultra-high-throughput screening. (A) Piezo-electric ink-jet pipettor that allows pipetting down to the pico-liter scale with speeds up to 10,000 drops per second and an accuracy of ~3 % [97]; (B) Sample carrier containing six segments of a silicon wafer. Each segment comprises 900 reaction compartments with a maximum volume of 120 nl each [28].

In summary, the basic tools for the discovery of new biocatalysts in high-throughput screening programs are abundantly available. Currently, several specialized companies exist that develop biocatalytic reagents, such as Novozymes, Genencor, Diversa and Maxygen. In addition, considerable growth in the number of biocatalytic processes can be observed in companies that have already developed valuable expertise in the area, such as BASF, Aventis, and Lonza Biotechnology. With the introduction of directed evolution and high-throughput screening technologies it has now become easier to improve biocatalysts. It thus can be anticipated that in near future many more biocatalytic processes will be exploited than are currently economically feasible.

References

[1] Rottländer, R. C. A.; Schlichterle, M. *Archaeo-Physica* **1980**, *7*, 61–70.

[2] Copeland, R. A. *Enzymes*, 1. ed.; Wiley-VCH: New York, 1996.

[3] Spicher, G. In *Handbuch Sauerteig: Biologie, Biochemie, Technologie*; Spicher, G., Stephan, H., Eds.; B. Behr's Verlag: Hamburg, 1999, pp 3–9.

[4] Demirjian, D. C.; Shah, P. C.; Moris-Varas, F. In *Biocatalysis – from discovery to application*; Fessner, W. D., Ed.; Springer-Verlag: Berlin, 1999; Vol. 200, pp 1–29.

[5] Robertson, D. E.; Mathur, E. J.; Swanson, R. V.; Marrs, B. L.; Short, J. M. *SIMS News* **1996**, *46*, 3–7.

[6] Adam, M. W.; Kelly, R. M. *Trends in Biotechnol.* **1998**, *16*, 329–332.

[7] Diversa ; Vol. US 6001574.

[8] Rondon, M. R.; Goodman, R. M.; Handelsman, J. *Trends Biotechnol.* **1999**, *17*, 403–409.

[9] Short, J. *Nature Biotechnol.* **1997**, *15*, 1322–1323.

[10] Bull, A. T.; Ward, A. C.; Goodfellow, M. *Microbiol. Mol. Biol.* **2000**, *64*, 573–606.

[11] Leung, D. W.; Chen, E.; Goeddel, D. V. *Technique* **1989**, *1*, 11–15.

[12] Cadwell, R. C.; Joyce, G. F. *PCR Methods Appl.* **1992**, 1–6.

[13] Spee, J. H.; de Vos, W. M.; Kuipers, P. *Nucleic Acids Res.* **1993**, *21*, 777–778.

[14] Fromant, M.; Blanquet, S.; Plateau, P. *Anal. Biochem.* **1995**, *224*, 347–353.

[15] Stemmer, W. P. C. *Nature* **1994**, *370*, 389–391.

[16] Crameri, A.; Raillard, S.-A.; Bermudez, E.; Stemmer, W. P. C. *Nature* **1998**, *391*, 288–291.

[17] Shao, Z.; Zhao, H.; Giver, L.; Arnold, F. H. *Nucleic Acids Res.* **1998**, *26*, 681–683.

[18] Zhao, H.; Giver, L.; Shao, Z.; Affholter, J. A.; Arnold, F. H. *Nat. Biotechnol.* **1998**, *16*, 258–261.

[19] Zhao, H.; Arnold, F. H. *Nucleic Acids Res.* **1997**, *25*, 1307–1308.

[20] Ninkovic, M.; Dietrich, R.; Aral, G.; Schwienhorst, A. *BioTechniques* **2001**, *30*, 530–536.

[21] Stemmer, W. P. *Proc. Natl. Acad. Sci. USA* **1994**, *91*, 10747–10751.

[22] Arnold, F. H.; Moore, J. C. *Adv. Biochem. Eng. & Biotechnol.* **1997**, *58*, 1–14.

[23] Proba, K.; Wörn, A.; Honegger, A.; Plückthun, A. *J. Mol. Biol.* **1998**, *275*, 245–253.

[24] Crameri, A.; Whitehorn, E. A.; Tate, E.; Stemmer, W. P. C. *Nature Biotechnology* **1996**, *14*, 315–319.

[25] Zhang, J. H.; Dawes, G.; Stemmer, W. P. C. *Proc. Natl. Acad. Sci. USA* **1997**, *94*, 4504–4509.

[26] Dietrich, R.; Wirsching, F.; Opitz, T.; Schwienhorst, A. *Biotechnol. Techniques* **1998**, *12*, 49–54.

[27] Ostermeier, M.; Shim, J. H.; Benkovic, S. J. *Nature Biotechnol.* **1999**, *17*, 1205–1209.

[28] Hempel, R.; Wirsching, F.; Schober, A.; Schwienhorst, A. *Anal. Biochem.* **2001**, *297*, 177–182.

[29] Legendre, D.; Laraki, N.; Gräslund, T.; Bjørnvad, M. E.; Bouchet, M.; Nygren, P.-A.; Borchert, T. V.; Fastrez, J. *J. Mol. Biol.* **2000**, *296*, 87–102.

[30] Kim, Y.-S.; Jung, H.-C.; Pan, J.-G. *Appl. Env. Microbiol.* **2000**, *66*, 788–793.

[31] Olsen, M. J.; Stephens, D.; Griffiths, D.; Daugherty, P.; Georgiou, G.; Iverson, B. L. *Nature Biotech.* **2000**, *18*, 1071–1074.

[32] Youvan, D. C. *Science* **1995**, *268*, 264.

[33] Heim, R.; Tsien, R. Y. *Current Biology* **1996**, *6*, 178–182.

[34] Siemering, K. R.; Golbik, R.; Sever, R.; Haseloff, J. *Curr. Biol.* **1996**, *6*, 1653–1663.

[35] González-Blasco, G.; Sanz-Aparicia, J.; González, B.; Hermoso, J. A.; Polaina, J. *J. Biol. Chem.* **2000**, *275*, 13708–13712.

[36] Schwienhorst, A. *Z. Phys. Chem.* **2001**, *in press.*

[37] Borrego, B.; Wienecke, A.; Schwienhorst, A. *Nucleic Acids Res.* **1995**, *23*, 1834–1835.

[38] Messer, W.; Vielmetter, W. *Biochem. Biophys. Res. Com.* **1965**, *21*, 182–186.

[39] Henne, A.; Daniel, R.; Schmitz, R. A.; Gottschalk, G. *Appl. Env. Microbiol.* **1999**, *65*, 3901–3907.

[40] Rondon, M. R.; August, P. R.; Bettermann, A. D.; Brady, S. F.; Grossman, T. H.; Liles, M. R.; Loiacono, K. A.; Lynch, B. A.; MacNeil, I. A.; Minor, C.; Tiong, C. L.; Gilman, M.; Osburne, M. S.; Clardy, J.; Handelsman, J.; Goodman, R. M. *Appl. Env. Microbiol.* **2000**, *66*, 2541–2547.

[41] Krubasik, P.; Sandmann, G. *Mol. Gen. Genet.* **2000**, *263*, 423–432.

[42] Bornscheuer, U. T.; Altenbuchner, J.; Meyer, H. H. *Biotech. Bioeng.* **1998**, *58*, 554–559.

[43] Firestine, S. M.; Salinas, F.; Nixon, A. E.; Baker, S. J.; Benkovic, S. J. *Nature Biotech.* **2000**, *18*, 544–547.

[44] Joo, H.; Arisawa, A.; Lin, Z.; Arnold, F. H. *Chem. Biol.* **1999**, *6*, 699–706.

[45] Cowan, D. A.; Daniel, R. M. *J. Biochem. Biophys. Meth.* **1982**, *6*, 31–37.

[46] Fossum, K. *Act. Pathol. Microbiol. Sc.–Section B* **1970**, *78*, 350–362.

[47] You, L.; Arnold, F. H. *Prot. Eng.* **1996**, *9*, 77–83.

[48] Sidhu, S. S.; Borgford, T. J. *J. Mol. Biol.* **1996**, *257*, 233–245.

[49] Henne, A.; Schmitz, R. A.; Bömeke, M.; Gottschalk, G.; Daniel, R. *Appl. Environ. Microbiol.* **2000**, *66*, 3113–3116.

[50] Fenton, A. M.; Stephens, P. M.; Crowley, J.; O'Callaghan, M.; O'Gara, F. *Appl. Environ. Microbiol.* **1992**, *58*, 3873–3878.

[51] Aarons, S.; Abbas, A.; Adams, C.; Fenton, A.; O'Gara, F. *J. Bacteriol.* **2000**, *182*, 3913–3919.

[52] Jung, H.-C.; Lebeault, J.-M.; Pan, J.-G. *Nature Biotechnol.* **1998**, *16*, 576–580.

[53] Georgiou, G.; Stathopoulos, C.; Daugherty, P. S.; Nayak, A. R.; Iverson, B. L.; Curtiss, R. I. *Nat. Biotechnol.* **1997**, *15*, 29 – 34.

[54] Uber, D. C.; Jaklevic, J. M.; Theil, E. H.; Lishanskaya, A.; McNeely, M. R. *BioTechniques* **1991**, *11*, 642 – 647.

[55] Müller, R.; Gerth, K.; Brandt, P.; Blöcker, H.; Beyer, S. *Arch. Microbiol.* **2000**, *173*, 303 – 306.

[56] Tortorella, M. D.; Arner, E. C. *Inflamm.* **1997**, *46, Suppl. 2*, S122 – S123.

[57] Donovan, R. S.; Dattin, A.; Baek, M.-G.; Wu, Q.; Sas, I. J.; Korczak, B.; Berger, E. G.; Roy, R.; Dennis, J. W. *Glycoconj. J.* **1999**, *16*, 607 – 615.

[58] Zocher, F.; Enzelberger, M. M.; Bornscheuer, U. T.; Hauer, B.; Wohlleben, W.; Schmid, R. D. *J. Biotechnol.* **2000**, *77*, 287 – 292.

[59] Zhao, H.; Arnold, F. H. *Proc. Natl. Acad. Sci. USA* **1997**, *94*, 7997 – 8000.

[60] Brandt, J. J.; Chatwood, L. L.; Yang, K.-W.; Crowder, M. W. *Anal. Biochem.* **1999**, *272*, 94 – 99.

[61] Roth, R. N. *Meth. Biochem. Anal.* **1969**, *17*, 189 – 285.

[62] Zandonella, G.; Haalck, L.; Spener, F.; Faber, K.; Paltauf, F.; Hermetter, A. *Chirality* **1996**, *8*, 481 – 489.

[63] Knight, C. G. *Meth. Enzymol.* **1995**, *248*, 18 – 34.

[64] Wirsching, F.; Opitz, T.; Dietrich, R.; Schwienhorst, A. *Gene* **1997**, *204*, 177 – 184.

[65] Ninkovic, M.; Riester, D.; Wirsching, F.; Dietrich, R.; Schwienhorst, A. *Anal. Biochem.* **2001**, *292*, 228 – 233.

[66] Cotenescu, M.-G.; LaClair, J. J. *J. Biotechol.* **2000**, *76*, 33 – 41.

[67] Takeshita, N.; Kakiuchi, N.; Kanazawa, T.; Komoda, Y.; Nishizawa, M.; Tani, T.; Shimotohno, K. *Anal. Biochem.* **1997**, *247*, 242 – 246.

[68] Tu, S.-I.; Grosso, L. *Biochem. Biophys. Res. Com.* **1976**, *72*, 9 – 14.

[69] Baker, W. L. *Antimicrobial Agents & Chemotherapy* **1983**, *23*, 26 – 30.

[70] Stebbins, J.; Debouck, C. *Anal. Biochem.* **1997**, *248*, 246 – 250.

[71] Zhou, Y.; Guo, X.-C.; Yi, T.; Yoshimoto, T.; Pei, D. *Anal. Biochem.* **2000**, *280*, 159 – 165.

[72] Burkart, M. D.; Wong, C.-H. *Anal. Biochem.* **1999**, *274*, 131 – 137.

[73] McNally, N.; Liu, X. Y.; Choudary, P. V. *Appl. Biochem. Biotechnol.* **1997**, *62*, 29 – 36.

[74] Baumann, M.; Stürmer, R.; Bornscheuer, U. T. *Angew. Chem. Int. Ed.* **2001**, *in press.*

[75] Räbina, J.; Pikkarainen, M.; Miyasaka, M.; Renkonen, R. *Anal. Biochem.* **1998**, *258*, 362 – 368.

[76] Grahn, S.; Kurth, T.; Ullmann, D.; Jakubke, H. *Biochim. Biophys. Acta* **1999**, *1431*, 329 – 337.

[77] Abriola, L.; Fuerst, P.; Schweitzer, R.; Sills, M. *J. Biomol. Screening* **1999**, *4*, 121.

[78] Mere, L.; Bennett, T.; Coassin, P.; England, P.; Hamman, B.; Rink, T.; Zimmerman, S.; Neglulescu, P. *Drug Discov. Today* **1999**, *4*, 363 – 369.

[79] Levine, L. M.; Michener, M. L.; Toth, M. V.; Holwerda, B. C. *Anal. Biochem.* **1997**, *247*, 83 – 88.

[80] Kolb, A.; Kaplita, P.; Hayes, D.; Park, Y.; Pernell, C.; Major, J.; Mathis, G. *Drug Discov. Today* **1998**, *3*, 333 – 342.

[81] Zhang, J.-H.; Chen, T.; Nguyen, S. H.; Oldenburg, K. R. *Anal. Biochem.* **2000**, *281*, 182 – 186.

[82] Koltermann, A.; Kettling, U.; Bieschke, J.; Winkler, T.; Eigen, M. *Proc. Natl. Acad. Sci. USA* **1998**, *95*, 1421 – 1426.

[83] Palo, K.; Mets, U.; Jager, S.; Kask, P.; Gall, K. *Biophys. J.* **2000**, *79*, 2858 – 2866.

[84] Eggeling, C.; Fries, J. R.; Brand, L.; Günther, R.; Seidel, C. A. M. *Proc. Natl. Acad. Sci. USA* **1998**, *95*, 1556 – 1561.

[85] Stewart, S. A. B.; Williams, P. *J. Gen. Microbiol.* **1992**, *138*, 1289 – 1300.

[86] Johnson, F. H.; Shimomura, O. *Meth. Enzymol.* **1978**, *57*, 271 – 291.

[87] Regeard, C.; Merieau, A.; Guespin-Michel, J. F. *J. Appl. Microbiol.* **2000**, *88*, 183 – 189.

[88] Deo, S. K.; Lewis, J. C.; Daunert, S. *Anal. Biochem.* **2000**, *281*, 87 – 94.

[89] Kricka, L. J.; Voyta, J. C.; Bronstein, I. *Meth. Enzymol.* **2000**, *305*, 370 – 390.

[90] Cadet, F.; Pin, F. W.; Rouch, C.; Robert, C.; Baret, P. *Biochim. Biophys. Acta* **1995**, *1246*, 142 – 150.

[91] Marcazzan, M.; Vianello, F.; Scarpa, M.; Rigo, A. *J. Biochem. Biophys. Methods* **1999**, *38*, 191 – 202.

[92] Louie, A. Y.; Hüber, M. M.; Ahrens, E. T.; Rothbächer, U.; Moats, R.; Jacobs, R. E.; Fraser, S. E.; Meade, T. J. *Nat. Biotech.* **2000**, *18*, 321 – 327.

[93] Hajduk, P. J.; Gerfin, T.; Boehlen, J.-M.; Häberli, M.; Marek, D.; Fesik, S. W. *J. Med. Chem.* **1999**, *42*, 2315–2317.

[94] Reetz, M. T.; Becker, M. H.; Kühling, K. M.; Holzwarth, A. *Angew. Chem. Int. Ed.* **1998**, *37*, 2647–2650.

[95] Reetz, M. T.; Jaeger, K.-E. *Chem. Eur. J.* **2000**, *6*, 407–412.

[96] Brandt, D. W. *Drug Discov. Today* **1998**, *3*, 61–68.

[97] Schober, A.; Günther, R.; Schwienhorst, A.; Döring, M.; Lindemann, B. F. *Bio-Techniques* **1993**, *15*, 324–329.

[98] Krul, K. G. *Spectrum* **1997**, *March 20*, 143/1-143-16.

[99] Zhao, H.; Arnold, F. H. *Curr. Op. Struct. Biol.* **1997**, *7*, 480–485.

[100] Köhler, J. M.; Schober, A.; Schwienhorst, A. *Experimental Technique of Physics* **1994**, *40*, 35–56.

[101] Schober, A.; Schwienhorst, A.; Köhler, J. M.; Fuchs, M.; Günther, R.; Thürk, M. *Microsystems Technologies* **1995**, *1*, 168–172.

[102] Schullek, J. R.; Butler, J. H.; Ni, Z.-J.; Chen, D.; Yuan, Z. *Anal. Biochem.* **1997**, *246*, 20–29.

[103] Schroeder, K. S.; Neagle, B. D. *J. Biomol. Screening* **1996**, *1*, 75–80.

[104] Michels, P. C.; Khmelnitsky, Y. L.; Dordick, J. S.; Clard, D. S. *Trends Biotechnol.* **1998**, *16*, 210–215.

[105] Sundberg, S. A. *Curr. Op. Biotechnol.* **2000**, *11*, 47–53.

[106] Kutzbach, C.; Ravenbush, E. *Hoppe-Seyler Z. Physiol.* **1974**, *354*, 45–53.

[107] Zhang, Q.; Zhang, L.; Han, H.; Zhang, Y. *Anal. Biochem.* **1986**, *156*, 413–416.

[108] van Kampen, M. D.; Egmond, M. R. *Eur. J. Lipid Sci. Technol.* **2000**, *102*, 717–726.

9
Engineering Protein Evolution

*Stefan Lutz and Stephen J. Benkovic**

9.1
Introduction

Proteins are involved in virtually every aspect of a living organism, yet they are composed of only twenty natural amino acids. The protein's tremendous diversity originates from its structural and functional versatility, achieved through polymerization of the chemically distinct amino acid building blocks into sequences of varying composition and length. Upon folding into complex three-dimensional structures, these macromolecules turn into molecular machines with the ability to accelerate even the most complex chemical reactions by many orders of magnitudes with unsurpassed selectivity and specificity.

The superior performance of proteins in respect to functional diversity, specificity, and efficiency has attracted scientists and engineers alike to create tailor-made catalysts by protein engineering. Customizing a protein towards practical applications such as industrial catalysts, biosensors, and therapeutics offers many economic incentives. At the same time, protein engineering can address fundamental questions of protein design in respect to their structure and function. This can then lead to the development of new technologies facilitating the more directed exploration of protein sequence space.

A readily available and abundant resource for examples on how to engineer proteins is Nature itself. The mechanisms of natural protein evolution have repeatedly succeeded in adapting protein function to ever changing environments. Understanding the principles and motives whereby new protein structures and functions emerge can provide useful guidelines for protein engineering. This chapter is directed towards the mechanisms of natural protein evolution, their implications on structure and function, and their experimental implementation *in vitro*.

9.2
Mechanisms of Protein Evolution in Nature

Two fundamental approaches can be employed to generate novel proteins; i) the *de novo* arrangement of small, structured peptides or ii) the remodeling of an existing protein framework. In computational simulations, the *de novo* assembly pathway demonstrated its effectiveness for the creation of novel functionality in proteins [1]. However, successful *in vitro* experiments are lacking and evidence for the occurrence of such a pathway in Nature has been limited to a family of antifreeze glycoproteins, found in arctic and antarctic fish ([2] and references therein). Having an Ala-Ala-Thr tripeptide (with minor sequence variations) as the fundamental building blocks, the polymeric structures presumably evolved by repeated duplication of a 9–basepair segment. Although antifreeze proteins are believed to have emerged during the geologically recent glaciation period, *de novo* assembly per se lost its significance after the early days of protein evolution. The sparse distribution of functionality in sequence space makes the repeated *de novo* creation of complex structures very inefficient and extremely difficult if not impossible within the given time constrains (the last 4 billion years).

Instead, remodeling of existing protein scaffolds represents a more efficient way to evolve novel functions and properties. Minor changes to explore the local sequence space by mutagenesis of individual positions within an established structure have

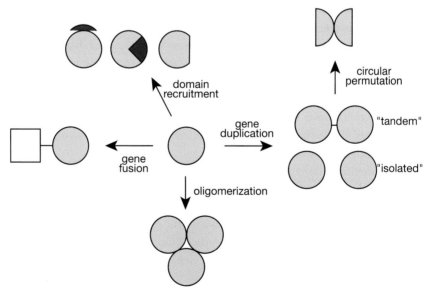

Fig. 9.1. Schematic overview of mechanisms for major structural reorganization during protein evolution. (Adapted from [49])

proven highly successful in accommodating small environmental changes and in op-timizing existing functions. The efficiency of such an approach is exemplified by the fine-tuning of antibody binding specificity by hyper-mutation of a set of precursors. Similarly, rationally designed single-amino acid changes in isocitrate dehydrogenase [3], lactate dehydrogenase from *Trichomonas vaginalis* [4], and T4 lysozyme [5] each resulted in a specific alteration in substrate specificity. Finally, a several-fold improve-ment of thermostability of subtilisin was observed upon introduction of two point mutations [6]. On the other hand, major rearrangements of proteins through duplica-tion, circular permutation, fusion, as well as deletion and insertion of large fragments create entirely novel structural combinations with potentially novel function [7] (Fig. 9.1). Such radical reorganization of genetic material results in the exploration of ex-tended sequence space and accelerated evolution.

9.2.1
Gene duplication

Reorganization, optimization, and development of novel function by duplication of genetic material is one of the fundamental mechanisms of protein evolution. Evi-dence for gene duplication can be found in small repetitive structures such as the previously discussed antifreeze proteins up to entire chromosomes and genomes. The unexpectedly high estimates of duplication frequency in eukaryotes, based on the analysis of several genomic databases, underline its significance in evolutionary terms [8]. Mechanistically, gene duplication can originate during chromosomal DNA replication or as a result of unequal crossing-over during meiosis [9]. Depending on the mechanism, either a tandem repeat or two isolated copies of the parental se-quences are generated (Fig. 9.1). In both cases, the copies are confronted with three possible outcomes: one of the two genes is i) inactivated by accumulation of mutation, ii) released from functional constrains which facilitates the accumulation of local mu-tations and leads to functional differentiation, or iii) both gene products maintain wild-type activity [8]. While gene inactivation has no obvious beneficial effect for the host organism, the presence of multiple functional copies of the same gene can amplify its expression level which may be advantageous by accelerating metabolic flux or increas-ing production of defensive compounds [10]. Finally, negotiating the fine line between complete inactivation and conservation of wild-type activity, duplication followed by functional differentiation can lead towards the evolution of novel function. How-ever, considering the sensitivity of protein structure and function to mutagenesis, divergence of the duplicate by random drift most likely results in enzyme inactiva-tion. The acquisition of a different function in contrast would be a relatively rare event. A recent modeling study, however, suggested that the accumulation of comple-mentary degenerative mutations in both gene duplicates could actually benefit the host by facilitating the long-term preservation of the duplicated genes. The extended conservation time of both copies increases the probability of evolving novel function [11].

An alternative view of the evolution of functionally novel proteins includes gene-recruitment [12] or gene sharing [13]. In this model, a period of bifunctionality of a single gene product precedes the duplication event upon which one copy maintains the ancestral function while the other adopts the new function. The various aspects of bifunctionality can hereby range from conservation of structural properties as demonstrated for crystallins [14] to what O'Brian and Herschlag [15] call promiscuous activity in ancestral structures.

Independent of the actual mechanism, functional divergence between two proteins with a common ancestor is exemplified in the βa-barrel structures of N-(phosphoribosyl-formimino)-aminoimidazole-carboxamide ribonucleotide isomerase (HisA) and imidazole glycerol phosphate synthase (HisF) of the histidine biosynthetic pathway [16]. The hypothesis of a common origin of the two enzymes was formulated based on sequence comparison [17] and has recently found support by the identification of extensive structural similarities [18].

9.2.2
Tandem duplication

In contrast to locally separated duplication products, tandem duplication of a gene results in the expression of two or more copies of the gene product within a single polypeptide chain (Fig. 9.1). Such multiplicity has, as reflected by the high frequency of its natural occurrence, clear advantages [7]:

- increased stability,
- new cooperative functions,
- formation of binding sites in a newly-formed cleft
- growth of long repetitive structures such as fibrous proteins, and
- production of multiple binding sites in series that produce either more efficient or more specific binding effects.

Furthermore, tandem repeats can undergo circular permutation and thereby reorganize the structural and functional integrity of the parental sequence.

Proteases
Often-cited examples for tandem duplications are pepsin, a member of the aspartate protease family [19], and chymotrypsin, representative of the trypsin-like protease family [20]. Structurally, both enzymes consist of two homologous, stacked domains, β-

sheets for pepsin and β-barrels for chymotrypsin, with the active site residues located at the domain interface. Despite extensive sequence divergence between the individual domain pairs, their overall backbone structures are mutually superimposable, strongly suggesting a single-domain homodimeric ancestor [21]. Direct evidence for the feasibility of such a primordial enzyme is provided by homodimeric aspartic proteases, found in retroviruses [22]. Similarly, the homodimeric β-barrel structure of the serine protease from cytomegalovirus, a member of the herpes virus family, may resemble an evolutionary precursor of modern serine proteases. Although crystallographic analysis clearly indicates no direct relation to the latter as revealed by significant differences in structural arrangement and the catalytic triad [23], the underlying design concept of using a two-barrel structure to form a functional enzyme is identical. Interestingly, more thorough analysis of the individual domains for both model structures have recently led to the suggestion that they themselves originated from multiple duplication events traceable back to five- and six-residue peptides [20].

βα-barrels

Another important structure that presumably arose by tandem duplication is the $(\beta\alpha)_8$-barrel. Being the most commonly used structural motif in nature, it is found in approximately 10 % of all known enzyme structures [24]. The prototypical $(\beta\alpha)_8$-barrel consists of eight serial strand-turn-helix-turn repeats, the β-strands forming an interior parallel β-sheet through a continuous hydrogen bonding network, and the helices forming the surface. Studying the symmetric nature of the barrel itself, multiple se-

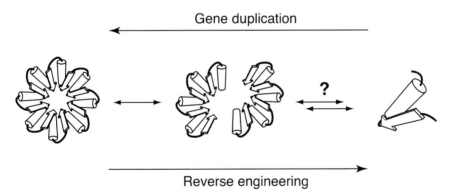

Fig. 9.2. Schematic illustration of the $(\beta\alpha)_8$-barrel structure. The highly symmetrical structure presumably arose by multiple gene duplication events from a βα-fragment. This hypothesis was tested using reverse engineering. Bisection of the barrel into two halves yielded correctly folded fragments which, upon heterodimerization, regained parental function. However, further fragmentation awaits experimental exploration.

quence alignments of the previously introduced gene products of *hisA* and *hisF* suggested the presence of an internal repetition within the fold [17]. Based on this data, the authors proposed that the $(\beta a)_8$-domain arose by tandem duplication of a $(\beta a)_4$-unit (Fig. 9.2). The validity of such a gene duplication hypothesis was recently examined from a structural perspective [18] and experimentally on HisF from *Thermotoga maritima* [25]. After bisection of the enzyme into two symmetrical $(\beta a)_4$-domains and expression of these in *E. coli*, the two subunits folded independently into their correct conformations. Compelling evidence in support of the hypothesis was found when coexpression of the two $(\beta a)_4$-units resulted in self-association into a functional heterodimer with activities similar to wild type. Although speculative, it has been suggest that the $(\beta a)_4$-barrel itself may have originated from even smaller elements such as individual (βa)-units ([26] and references therein). Experimentally, the rearrangement and substitution of (βa)-units in the context of the barrel structure are tolerated [27, 28].

9.2.3
Circular permutation

Circular permutation of a protein results in the relocation of its N- and C-termini within the existing structural framework. Initiated by a tandem duplication of a precursor gene, one mechanistic model proposes an in-frame fusion of the original termini, followed by the generation of a new start codon in the first repeat and a termination site in the second. In support of the model, tandem duplications are observed in prosaposins [29] and DNA methyltransferases [30], both genes for which circular permutated variants are also known.

Initial evidence for the structural diversification of the βa-barrel fold by circular permutation was found in sequence alignments of members of the a-amylase superfamily [31]. More recently, analysis of crystallographic data of transaldolase B from *E. coli* suggested that the enzyme was derived from circular permutation of a class I aldolase [32]. In either case, the shift of the two N-terminal βa-repeats (plus the β-strand of the third subunit for amylases) onto the C-terminus resulted in no apparent functional changes.

In general, the evolutionary advantages of circular permutation are not clear. However, *in vitro* experiments have revealed that although the final structure of a circular permuted protein is by and large the same, the rearrangement of the subunits changes the folding pathway and can affect the protein's conformational stability [33]. The potential benefit of such structural rearrangements from an engineering perspective was demonstrated by the increased anti-tumor activity of a circular permuted interleukin 4-*Pseudomonas* exotoxin fusion construct, resulting from decreased steric interference between the toxin and the interleukin-binding site [34].

9.2.4
Oligomerization

Noncovalent association of homo- and heterologous proteins plays an important role in enzyme function [35]. Oligomerization can modulate a protein's activity as demonstrated in allosteric enzyme regulation of hemoglobins [36] or by the dimer activation of transcription factors [37]. In addition, the agglomeration of structures at exposed hydrophobic surfaces increases the thermodynamic stability of the participating elements and can create cavities and grooves at the protein-protein interface that can accommodate substrate or regulator binding sites. Along the same line, oligomerization can be viewed as a primitive way of substrate shielding in an enzyme's active site.

A special case of oligomerization is the formation of homodimers by 3D-domain swapping, a term easily confused with "domain swapping". As shown in Fig. 9.3, the latter is a synonym for domain recruitment and describes the engineering of proteins by substitution of either gene fragments that encode for individual domains or, in the case of inteins, discrete protein fragments. This involves the breakage and reformation of covalent bonds in the peptide backbone. In contrast, 3D-domain swapping describes the structural reorganization between identical multi-domain proteins without breakage of covalent bonds. In the process, a portion of one protein is swapped with the same unit of an identical protein, forming an intertwined dimer or higher order oligomers. Several examples of naturally occurring 3D-domain swapping have been reviewed by Schlunegger *et al.* [38]. The exchanged structure elements range from secondary structures to entire domains and, although the advantages of 3D-domain swapping are not fully understood, one can envision regulatory purposes as well as increased structural stability [39].

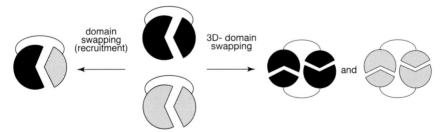

Fig. 9.3. Graphical distinction between domain swapping/recruitment and 3D-domain swapping.

9.2.5
Gene fusion

Fusion of entire genes, which translates into covalently linked multifunctional enzyme complexes, is a more conclusive commitment towards cooperativity between individual enzymes than non-covalent oligomerization. Often found for enzymes that catalyze subsequent steps in biosynthetic pathways, gene fusions can provide stability, advanced regulatory function, substrate channeling, the ability to direct concerted expression, and entropic advantages, giving the host organism a selective advantage.

Multifunctional gene fusion constructs in both the pyrimidine and purine biosynthetic pathways have been discussed in detail elsewhere [40]. Additional examples have been identified in the biosynthetic pathways for histidine [16] and the aromatic amino acids. In the latter, chorismic acid is the key intermediate from which tryptophan, as well as tyrosine and phenylalanine biosynthesis diverges. In *E. coli*, two bifunctional complexes catalyze the conversion of chorismic acid via the prephenate intermediate into the respective a-ketoacid precursor of tyrosine and phenylalanine. While the tyrosine precursor is synthesized by a chorismate mutase (CM) fused to prephenate dehydrogenase (CM-PDG), phenylpyruvate is produced by a CM-prephenate dehydratase complex (CM-PDT). Either complex is reportedly susceptible to feedback inhibition by its final product [41, 42]. After bisection of the fusion construct, the individual monomers remain functional, excluding the necessity of gene fusion for the purpose of substrate channeling. Instead, allosteric control seems the driving force behind dimer formation [41, 43]. These findings agree with experimental data from the bisection of *N*-acetylglucosamino-1-phosphate uridyltransferase, an enzyme complex that catalyzes the consecutive acetylation and uridylation of glucosamino-1-phosphate. While both enzymes are essential *in vivo*, their cleavage shows no apparent effect on kinetics of the individual enzymes, the pathway's overall flux, nor the organism's phenotype [44]. In contrast, enhanced substrate processing was observed for rationally designed β-galactosidase – galactose dehydrogenase and galactose dehydrogenase-bacterial luciferase fusion constructs [45, 46].

9.2.6
Domain recruitment

The generation of functional versatility in proteins by reorganization of individual protein subunits or domains is summarized under the term "domain recruitment". The idea behind domain recruitment has been vividly depicted by Ostermeier and Benkovic's adaptation of the cliché "a thousand monkeys typing at a thousand typewriters would eventually reproduce the works of Shakespeare" [40]. In reference to domain recruitment, these authors suggest that the writing would obviously work much more efficiently if the monkeys, once managing to write entire words or sentences, could duplicate these with a single keystroke.

Evidence for domain recruitment has been identified in a wide variety of proteins [47], mechanistically ranging from simple N- or C-terminal fusion to multiple internal insertions and possibly circular permutations [48]. A recent analysis of proteins in the protein structure database (PDB) has further indicated that structural rearrangements as a result of domain shuffling have significantly contributed to today's functional diversity [49]. A brief overview of the various modes of domain recruitment and their effects on function, is presented on examples of βa-barrel structures.

Substrate specificity

The family of glycosyl hydrolases utilizes a variety of oligo- and polysaccharides. Preserved across the family, the active site is located in a wide-open groove at the C-terminus of the βa-barrel core structure. Substrate specificity between the family members is modulated through additional domains, which effectively close down the active site, adjusting its size and shape towards the individual substrates. In lacZ from *E. coli*, four"added" domains were identified, orchestrated around the central βa-barrel [50]. These add-ons all represent individual folding units with distinct structures; jelly-roll barrel (domain 1), fibronectin III fold (domain 2 & 4), and β-sandwich (domain 5). Functionally, domain 1 was found to directly participate in shaping the active site pocket while domain 2 & 4 act as domain linker to correctly position domain 1 and 5. Although the role of domain 5 is less clear, it seems to contribute to substrate binding and may be relevant to oligomerization.

In contrast to the above examples, substrate specificity can also be acquired by deleterious modification of the protein framework. The absence of two consecutive a-helices in the barrel structure of *endo-β-N*-acetylglucosaminidase F_1 (Endo F_1) was first reported by Van Roey *et al.* [51]. While not affecting the structural or functional integrity of the protein, the unusual feature was interpreted as important to binding of the bulky glycoprotein substrates and their subsequent deglycosylation. Similar alterations of the βa-barrel core structure, although to a lesser extent, were observed in family relatives of Endo F_1 and could also be attributed to substrate binding [52].

Regulation

The catalytic activity of porcine pancreatic a-amylase, another member of the glucosyl hydrolase family, is metal cofactor-dependent [53]. A calcium ion-binding site is located at the interface of an antiparallel β-sheet, inserted in one of the loop regions of the βa-barrel, and the core structure. In the calcium-bound state, the insertion stabilizes the substrate-binding site, and indirectly constrains part of the active site in a catalytically competent conformation.

In phosphoenolpyruvate carboxylase, structural analysis indicates the arrangement of the total 40 helices around the active site in the βa-barrel core structure [54]. The additional domains create a binding site for aspartate, an allosteric inhibitor of the enzyme. Furthermore, the highly conserved hydrophobic C-terminal extension of the enzyme presumably is critical for oligomerization.

Chemistry

Flavocytochrome b_2 from *Saccharomyces cerevisiae*, a member of the FMN-dependent oxidoreductase superfamily, catalyzes the two-electron oxidation of lactate to pyruvate with subsequent electron-transfer to cytochrome c via the bound flavin [55]. What distinguishes the enzyme from other family members is the N-terminal fusion of a heme-binding domain to the βa-barrel structure, which hosts the primary active site. Rather than dumping the electrons from the reduced flavin hydroquinone onto molecular oxygen, they are transferred intramolecularly to the heme-binding domain and from there in a second intermolecular step to cytochrome c.

9.2.7
Exon shuffling

Structure analysis of several proteases involved in blood coagulation and fibrinolysis reveals a diverse, sometimes repetitive, assembly of discrete protein modules (Fig. 9.4) [56]. While these modules represent independent structural units with individual folding pathways, their concerted action contributes to function and specificity in the final protein product. On the genetic level, these individual modules are encoded in separate exons. Over the course of modular protein evolution, new genes are created by duplication, deletion, and rearrangement of these exons. Mechanistically, the exon shuffling actually takes place in the intervening intron sequences (intronic recombination – for further details see [10]).

The significance of exon shuffling to protein evolution, in particular in respect to the development of multicellularity, is signified by a short inventory of processes involving proteins created by modular assembly. Exon shuffling facilitates the construction of proteins involved in regulation of blood coagulation, fibrinolysis, and complement activation, plus most constituents of the extracellular matrix, cell adhesion proteins, and receptor proteins [10, 57].

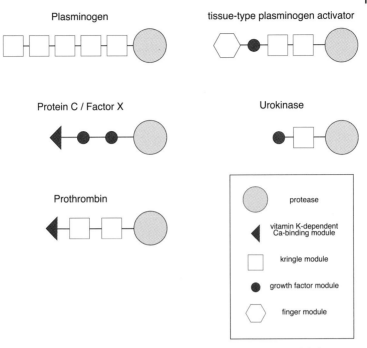

Fig. 9.4. Schematic illustration of modular protein evolution, exemplified on proteases, involved in blood coagulation and fibrinolysis. Varying type and number of modules, fused onto a common protease unit, generates a family of highly specific hydrolytic enzymes. (Adapted from [56])

9.3
Engineering Genes and Gene Fragments

In recent years, the protein engineering field has developed a series of *in vitro* methods that mimic natural processes in an attempt to improve existing enzymes and develop novel functions and properties. Particularly attractive in that respect have been modular approaches reminiscent of domain recruitment, gene fusion, and exon shuffling. By identifying and characterizing these individual modules or building blocks, protein dissection through fragmentation has substantially contributed to the fundamental understanding of protein design. Combined with rational and combinatorial methods to facilitate the recombination of these building blocks, the exploration of structural and functional properties of proteins has not only produced many improved biocatalysts of industrial importance but also has provided insight to protein function and evolution. The following sections introduce a variety of protein engineering techniques and discuss their applications.

9.3.1
Protein fragmentation

Methods for protein fragmentation were initially developed to study protein folding mechanisms and pathways [58, 59]. Since, their application has been extended towards the elucidation of structural organization in proteins and their evolutionary origins, as well as towards the design of novel two-hybrid systems.

The analysis of a protein's structural organization by protein fragment complementation facilitates the identification of locations in the peptide backbone, permissible for cleavage without loss in function. Knowledge of such sites is valuable since protein engineering frequently requires the substitution and insertion of modules such as secondary structure elements and domains into an existing protein scaffold. Intuitively, one would assume that the manipulation of a scaffold in regions of lower organization such as loops and linkers is better tolerated by the system. Indeed, the analysis of laboratory data and the Protein Data Bank indicated that the majority of insertions and deletions are actually located in loops [60, 61]. Nevertheless some experimental systems have demonstrated surprising flexibility in the accommodation of insertion in defined structural elements [62] and tolerance towards bisection of the peptide backbone in functionally critical regions [63 – 65].

Experimentally, the search for permissive sites by protein fragmentation can be attempted by various methods. In the absence of structural data, random fragmentation patterns of proteins can be generated by limited proteolysis. Although biased towards surface loop regions, the partial digestion by chemicals, as well as serine and cysteine proteases has been applied to a variety of proteins [66]. More recently, combinatorial approaches that enable a comprehensive mapping of the entire protein structure were described in the literature. Manoli and Bailey reported the transposon-mediated random insertion of a 31 amino acid fragment into a protein scaffold [67]. Alternatively, a protein scaffold can be sampled for permissive sites by circular permutation [68, 69]. Finally, N- and C-terminal truncation of overlapping DNA regions, using *Bal31* exonuclease [70, 71] or exonuclease III [72] can be used to search for bisection points.

Furthermore, protein fragmentation has successfully been employed to investigate the evolutionary origin of protein structures and folds. Alternatively to evolutionary studies that focus on the formation of complex protein structures, the disassembly of an existing structural framework may provide valuable insight into the nature of evolutionary significant building blocks (Fig. 9.2). Termed reverse engineering of proteins, the method calls for the fragmentation of an existing monomeric protein into pieces that ideally possess distinct structural properties, meaning that they fold independently, and exhibit affinity towards oligomerization.

Probably one of the most extensively studied folds in that respect is the $(\beta a)_8$-barrel. Earlier introduced in the context of tandem duplication, protein fragmentation experiments have provided data in support of a $(\beta a)_4$-domain precursor [25]. More extensive digestion experiments of the $(\beta a)_8$-barrel were reported by Vogel *et al.* [73] and Ray *et al.* [66]. In both cases, digestion of triosephosphate isomerases with subtilisin resulted in multiple fragments which maintained proper folding and could be religated by the protease under nonaqueous conditions, reconstituting the catalytic activity.

While providing valuable insight towards a better understanding of protein evolution, the bisection of proteins into functional heterodimers has also found practical applications to study protein-protein interactions. Heterodimeric variants of dihydrofolate reductase (DHFR), green fluorescent protein (GFP), GAR transformylase, and phosphotransferases were constructed to work in two-hybrid systems [64].

9.3.2
Rational swapping of secondary structure elements and domains

Rational protein engineering aims to exchange and fuse secondary structure elements and domains at precise locations, based on predictions through rational means, to form a functional hybrid structure. Despite the great complexity involved in rational protein engineering, two outstanding experimental successes were recently reported for β-barrel and β/a-barrel structures.

Engineering members of the serine protease family which structurally consist of two homologous, stacked β-barrels, Hopfner and coworkers designed a hybrid protease by fusing the N-terminal β-barrel from trypsin to the C-terminal β-barrel of factor Xa [74]. The location of the fusion point in the linker region between the two β-barrels was chosen following close inspection of the X-ray structures of trypsin and factor Xa. The resulting hybrid was shown to be fully functional as it hydrolyzed a broader but distinct spectrum of peptide substrates in comparison to the parental enzymes.

Using rational protein engineering in combination with DNA shuffling on the βa-barrel scaffold, Altamirano *et al.* [75] demonstrated the successful conversion of indole-3-glycerol phosphate synthase (IGPS) to *N*-(5'-phosphoribosyl)anthranilate isomerase (PRAI). The two enzymes, catalyzing consecutive reactions in the tryptophan biosynthetic pathway, are thought to have evolved by gene duplication of a common ancestor. Although they share binding affinity towards a common intermediate, their sequence identity (22 %) and chemistry is significantly different and no cross-reactivity is detectable. Based on initial modeling studies of the structural differences between the two enzymes, the authors deleted, substituted, and inserted gene fragments corresponding to secondary structure elements into the IGPS scaffold. While the resulting rational hybrid enzymes were virtually inactive, two subsequent rounds of "fine-tuning" the structural arrangements by DNA shuffling resulted in hybrid enzymes with 90 % sequence identity to IGPS, but PRAI-activity ten-fold higher than wild-type PRAI. The work is truly remarkable since neither of the two approaches alone could have generated the final hybrid protein. Instead, the authors successfully combined two methods into a new design strategy, rationally rough-drafting the protein scaffold and combinatorially fine-tuning it.

Generally, the rational construction and reorganization of hybrid enzymes relies on a comprehensive knowledge of structure and function of the proteins involved. The availability of such information for only a small percentage of proteins has so far limited the effective rational design on a wider range of targets.

9.3.3
Combinatorial gene fragment shuffling

Alternatively to rational design, combinatorial approaches resembling Darwinian protein evolution through random mutagenesis, recombination, and selection for the desired properties can be employed for protein engineering. These approaches do not require extensive structural and biochemical knowledge and their implementation has resulted in successful modifications and improvements of a wide diversity of targeted properties.

In the last decade, a rapidly increasing number of genetic methods for generating combinatorial libraries of proteins have been reported. The early *in vivo* systems, using chemical and UV-induced random mutagenesis and *E. coli* mutator strains, have largely been replaced by *in vitro* methods. With the introduction of error-prone PCR [76], the first technique for rapidly generating large families of related sequences of a specific target gene became available. The next milestone was the successful demonstration of *in vitro* evolution of proteins by sexual PCR or DNA shuffling [77, 78].

DNA shuffling

The directed evolution of proteins by random mutagenesis and DNA shuffling has proven extremely powerful to modify and optimize functional and physicochemical properties of existing proteins [6, 79 – 85] (for an up-dated list on targets see: http://www.che.caltech.edu/groups/fha/Enzyme/directed.html).

Experimentally, the method starts off with the random fragmentation of a target gene(s) by DNaseI treatment. The resulting mixture of double-stranded oligonucleotides, on average 20 – 100 basepairs in length, is then taken through several rounds of heat-denaturing and reannealing in the presence of a DNA polymerase. This iterative process generates a library of polynucleotides, consisting of fragments from various parental sequences. Today, multiple variations of Stemmer's original protocol have been reported in the literature [86 – 90].

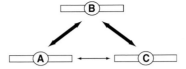

Fig. 9.5. Principle of family DNA shuffling, demonstrated on the identity network of three sequences. The thickness of the arrows corresponds to the pairwise sequence identities. The similarity of sequence A with C is low, disqualifying them from regular DNA shuffling. Addition of B acts as a bridging intermediate, having sufficient identity with either A or C to accommodate successful recombination of fragments from all three parents.

Despite the many successful applications of DNA shuffling, the methodology has one serious limitation. Sequence-dependent reassembly, the fundamental mechanism behind DNA shuffling, only works efficiently between sequences with >70 % sequence identity [91]. Below this threshold, DNA shuffling generates libraries biased towards regions of locally higher identity and reassembly of parental sequences becomes a dominant factor.

Family DNA shuffling or "molecular breeding" of multiple parental sequences lowers the approach's boundaries of sequence-dependence [82, 83, 92]. In contrast to traditional DNA shuffling, this strategy is based on homologous recombination of multiple parental sequences, some with pairwise DNA sequence identities as low as 50 %. The fundamental idea behind the approach is outlined in Fig. 9.5 on three sequences. Each member of the initial gene pool has a sufficiently high sequence identity to at least one other member. The homology between two specific sequences may be low, yet the remaining members of the pool, each of them with higher identities to either of the two, act as bridging elements and allocate the shuffling of elements from all parents.

The experimental implementation of family DNA shuffling on the four *amp C* lactamases from *Citrobacter freudii*, *Klebsiella pneumoniae*, *Enterobacter cloacea*, and *Yersinia enterocolitica* was highly successful, resulting in the identification of hybrid proteins with up to 540–fold increased activity [82]. Interestingly, a detailed sequence analysis of the functional hybrids revealed the absence of *amp C* fragments from the *Yersinia enterocolitica*. Contrary to the argument that its presence may not be required under the applied selection conditions, a recent computational analysis of the system suggested that a lack of homology in critical regions of *ampC* from *Yersinia* prevented the sequence from participating in the shuffling in the first place [93]. This alternative explanation highlights the importance of careful design of family DNA shuffling experiments with particular emphasis not just on global but also local DNA sequence homologies of all participants.

In recent years, a series of alternative methods to shuffle sequences of low-homology has been proposed. Kikuchi *et al.* [89] took advantage of multiple restriction sites distributed along the gene sequence, which in combination with DNA shuffling added crossovers independent of sequence identity to these region. Others suggested the spiking of the DNase-treated fragment mixture with artificial, designed intermediates; chemically synthesized oligonucleotides that introduce mutations or crossovers at specific location in a sequence [94].

Homology-independent fragment swapping

The limitations of DNA shuffling, arising from its dependency on sequence identity, have motivated the development of new technologies to facilitate an unbiased sampling of sequence space. At present, the reported approaches for creating sequence-independent hybrid libraries between two parental sequences include a family of methods collectively known as Incremental Truncation for the Creation of Hybrid enzYmes (ITCHY) [72, 95, 96] and Sequence-Homology Independent Protein RECombination (SHIPREC) [91].

Although the final products of all methods are the same, comprehensive single-crossover hybrid libraries that includes insertion and deletion sequences, the approaches employ different strategies of library generation (Fig. 9.6). The original IT-CHY method [72] involves separate exonuclease III-treatment of the two parental genes (Fig. 9.6a). The reaction conditions are chosen whereby the nuclease has a sufficiently low processivity ($\sim$10 nucleotides/min), allowing the generation of libraries of every single-base pair deletion by quenching of aliquots over time. Following removal of the single-stranded portion of the DNA fragments, the N-terminal and the C-terminal fragment libraries are mixed and ligated to generate the single-crossover library. Following the same fundamental idea but being experimentally significantly faster and less laborious is the use of nucleotide analogs such as a-phosphothioate nucleotides to create ITCHY libraries [96]. Furthermore, a variation of ITCHY to bias truncation libraries towards parental sizes has been developed [95].

SHIPREC, an alternative approach to create sequence-independent single-crossover libraries has been reported by Arnold and coworkers [91]. Utilizing an inverted fusion construct of the parental genes, partial digestion by DNase I cleaves N- and C-terminal fragments at random. Following a size selection of parental size fragments, the original termini are restored by sequence inversion (Fig. 6b).

The potential value of these techniques becomes apparent upon analysis of structure diversity and functional selection of ITCHY and SHIPREC libraries. For the model system of glycinamide ribonucleotide transformylase from *E. coli* (PurN) and human (hGART), comparison of the diversity of the naive libraries obtained by ITCHY with DNA shuffling show the expected even distribution of crossovers over the sampled sequence for the former but a clustering in the central, highly homologous region

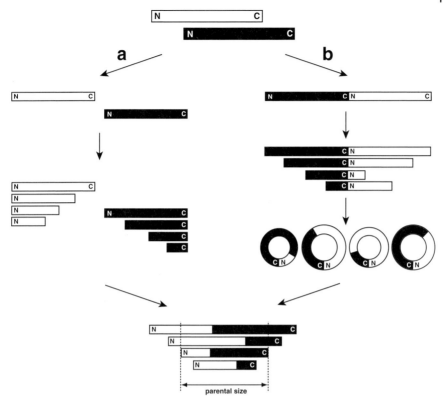

Fig. 9.6. Fundamental mechanisms of homology-independent fragment swapping by a) ITCHY and b) SHIPREC. In the ITCHY protocol, the two parental genes are truncated independently, creating two separate libraries that correspond to the N- and C-terminal fragments. Ligation of the mixed libraries then generates a comprehensive pool of hybrid sequences. SHIPREC on the other hand starts by limited DNaseI digestion of the fused genes. Intramolecular fusion, followed by restriction digest at the original fusion point, results in sequence inversion and generates the hybrid library.

for the latter [72]. Functional analysis of the ITCHY and SHIPREC libraries proves that interesting and catalytically active solutions for protein engineering can be found outside regions of high sequence identity.

The main drawback of these described methods is their limitation to a single crossover per hybrid sequence. In contrast, DNA shuffling and related methods have produced up to 14 intersects per nucleotide sequence [90]. Increasing numbers of crossovers are desirable because they have been shown to improve the evolution of proteins with desirable characteristics.

Homology-independent fragment shuffling

The creation of hybrid libraries with multiple sequence-independent crossovers through a combination of ITCHY and DNA shuffling (Fig. 9.7) was first proposed by Ostermeier et al. [97] and has recently been implemented experimentally [98]. Using the PurN/hGART model system, two complementary incremental truncation libraries (*purN-hGART* and *hGART-purN*) were generated (Fig. 9.7a). In preparation for the subsequent DNA shuffling, these libraries were then pre-selected for in-frame and parental-size hybrid sequences (Fig. 9.7b). Without in-frame selection, the probability of SCRATCHY library members encoding for nonsense sequences would increase exponentially with the number of introduced crossovers. Furthermore, hybrid enzymes with large insertions are prone to protein folding problems while deletions of extended portions of a protein are likely to terminate enzymatic activity.

The resulting libraries resemble large artificial "families" of intermediates, each carrying a particular crossover between the two parents. DNase I treatment of these families and subsequent reassembly by PCR results in the creation of naive hybrid libraries comprising crossovers from multiple "family" members (Fig. 9.7c). In experiments, analysis of the naive libraries revealed about 20 % of the members to have two and three crossovers. Their locations were randomly distributed over the entire expected sequence range and functional complementation of an auxotrophic *E. coli* strain provided evidence for the catalytic competence of multi-crossover hybrid constructs. In respect to catalytic performance of these hybrid enzymes, the method lags behind DNA shuffling approaches. This however can mostly be attributed to issues of protein folding rather than chemistry itself. In the following section, the aspects of protein folding and its increasing importance to protein engineering at low sequence identities will be discussed.

9.3.4
Modular recombination and protein folding

The biological activity of proteins generally depends on a unique three-dimensional conformation, which in turn is inherently linked to its primary sequence. Protein folding, the conversion of the translated polypeptide chain into the native state of a protein, is the critical link between gene sequence and three-dimensional structure. Mechanistically, folding is believed to proceed through a predetermined and ordered pathway, either via kinetic intermediates or by direct transition from the unfolded to the native state [99]. In both cases, local and non-local interactions alike stabilize transient structures along the pathway and funnel the intermediates towards the native state.

In the context of protein engineering, these folding and stability issues are of major importance. Modular protein engineering in general and homology-independent

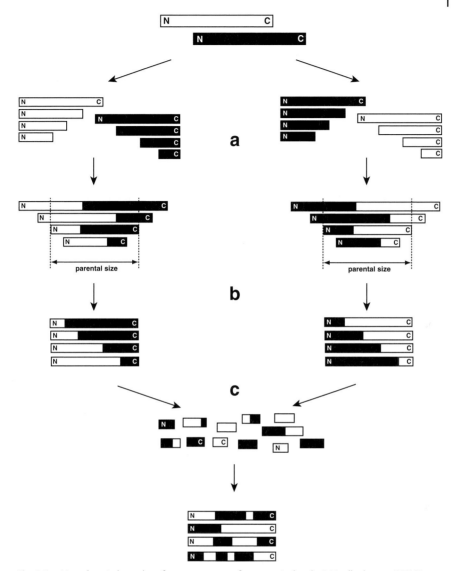

Fig. 9.7. Homology-independent fragment shuffling (SCRATCHY) is a combination of ITCHY and DNA shuffling. a) The approach starts by creating two complementary ITCHY libraries. b) To maximize the functional competence of the SCRATCHY library, hybrid sequences of approximately parental size that are in the correct reading frame are isolated. c) Finally the two ITCHY libraries are mixed, fragmented, and reassembled. The significant portion of the resulting hybrid libraries consist of members with multiple crossovers at locations independent of sequence identity.

methods in particular can introduce significant variation into a system's primary sequence, impacting the network of interactions that ultimately controls the folding trajectory. Proteins have generally demonstrated amazing flexibility and tolerance towards changes of individual amino acids and small peptide fragments, accommodating them by reorganization of the neighboring environment [62]. In the context of domain swapping and modular engineering, the situation is often more complex. Affecting mostly non-local interactions, this remodeling can disturb long-range interactions crucial to folding and structural integrity. Furthermore, possible steric and physicochemical interferences are expanded to entire surfaces of a protein module. Although the disruption of a network of stabilizing interactions between parental structures is possible on all levels of homology, the probability for structural inconsistencies that lead to misfolding rapidly increases with decreasing sequence similarity. When exploring distant regions in sequence space by homology-independent methods as discussed earlier, deleterious folding effects on hybrid protein libraries upon exchange of subunits of low homology can quickly nullify the potential benefits of structure-based protein engineering.

The problem of incorrect folding of hybrid proteins is illustrated by a system investigated in our laboratory. Attempting the substitution of domains from structures with very low sequence homology, incremental truncation libraries between the *E. coli* proteins GAR transformylase (PurN) and tRNA(fMet)-formyltransferase (FMT) were generated (Marc Ostermeier and Steve Tizio – unpublished data). The N-terminal portions of the two proteins share only 33 % sequence identity, yet their backbone tertiary structures are almost perfectly superimposable (Fig. 9.8). In addition, they both catalyze the same chemistries; the formyl group-transfer from N^{10}-formyl-tetrahydrofolate onto their respective substrates. In previous experiments, analogous substitution of fragments between the functionally related PurN and N^{10}-formyl-tetrahydrofolate hydrolase (PurU) from *E. coli* (~60 % DNA sequence identity), as well as PurN and hGART (~50 % DNA sequence identity) had yielded functional hybrid enzymes [63, 100]. A potential indicator for protein folding problems was the identification of several thermosensitive constructs of PurN/hGART, which failed to complement in an auxotrophic host under regular growth conditions at 37 °C, yet showed sufficient activity when incubated at ambient temperatures [95, 96]. Presumably, the perturbance of the network of non-local interactions at or near the fusion construct's intraface destabilizes the global structure and prevents proper folding at higher temperatures. In contrast, expression of the PurN/FMT hybrid library at either temperature did not yield any functional chimeras. While subsequent analysis of the libraries on the genetic level indicated the expected diversity, Western blots revealed that the expressed protein was found exclusively in the insoluble fraction of the cells (Steve Tizio – unpublished data). The same difficulties were reported for a series of rationally designed hybrid constructs between ornithine and aspartate transcarbamoylases, two structure homologs that share only 32 % sequence identity [101].

a)

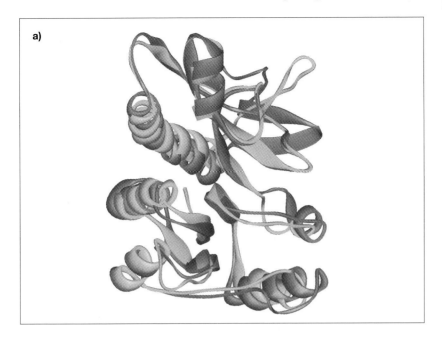

b)

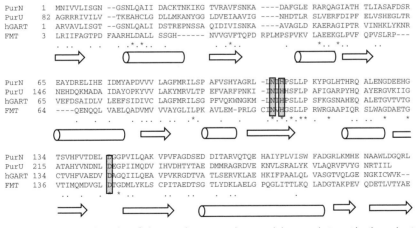

```
PurN     1  MNIVVLISGN  --GSNLQAII  DACKTNKIKG  TVRAVFSNKA  ----DAFGLE  RARQAGIATH  TLIASAFDSR
PurU    82  AGRRRIVILV  --TKEAHCLG  DLLMKANYGG  LDVEIAAVIG  ----NHDTLR  SLVERFDIPF  ELVSHEGLTR
hGART    1  ARVAVLISGT  --GSNLQALI  DSTREPNSSA  QIDIVISNKA  ----AVAGLD  KAERAGIPTR  VINHKLYKNR
FMT      3  LRIIFAGTPD  FAARHLDALL  SSGH------  NVVGVFTQPD  RPLMPSPVKV  LAEEKGLPVF  QPVSLRP---
            . . .    .   . .*. * . .         . . *.         .          *.. *. .       . .
```

PurN   65 EAYDRELIHE IDMYAPDVVV LAGFMRILSP AFVSHYAGRL -INIHPSLLP KYPGLHTHRQ ALENGDEEHG
PurU   146 NEHDQKMADA IDAYOPKYVV LAKYMRVLTP EFVARFPNKI -INIHSFLP AFIGARPYHQ AYERGVKIIG
hGART   65 VEFDSAIDLV LEEFSIDIVC LAGFMRILSG PFVQKWNGKM -INIHPSLLP SFKGSNAHEQ ALETGVTVTG
FMT   64 ----QENQQL VAELQADVMV VVAYGLILPK AVLEM-PRLG CINIHGSLLP RWRGAAPIQR SLWAGDAETG

PurN   134 TSVHFVTDEL DGGPVILQAK VPVFAGDSED DITARVQTQE HAIYPLVISW FADGRLKMHE NAAWLDGQRL
PurU   215 ATAHYVNDNL DEGPIIMQDV IHVDHTYTAE DMMRAGRDVE KNVLSRALYK VLAQRVFVYG NRTIIL
hGART 134 CTVHFVAEDV DAGQIILQEA VPVKRGDTVA TLSERVKLAE HKIFPAALQL VASGTVQLGE NGKICWVK--
FMT   136 VTIMQMDVGL DTGDMLYKLS CPITAEDTSG TLYDKLAELG PQGLITTLKQ LADGTAKPEV QDETLVTYAE
```

**Fig. 9.8.** a) Structural overlay of glycinamide ribonucleotide transformylase (PurN) [155] and the N-terminal domain of tRNA(fMet)-formyl-transferase (FMT) [156] from *E. coli*. Despite the low sequence identity (33 %), the two structures are almost perfectly superimposable. b) Structure-based multiple sequence alignment of PurN and FMT with human glycinamide ribonucleotide transformylase (hGART) and formyltetrahydrofolate hydrolase (PurU). Conserved residues are labeled with an asterisk and homologous positions are marked with a dot. The three active site residues are highlighted in shaded boxes.
```

We hypothesize that the observed instability of hybrid constructs is caused by two mechanisms: i) steric and functional clashes at the module surfaces and ii) disruption of folding pathways. For the former, interference at the module's intraface is illustrated by the "English muffin-experiment". Tearing two muffins in half and trying to fit the bottom of the first and the top of the second back together is not likely to match up precisely. Instead, the contact surface will have friction points and empty spaces. A similar picture can be drawn for the two fused protein fragments, causing the observed destabilization of the constructs. Furthermore, dissimilarities of module surfaces can result in the exposure of hydrophobic patches, increasing the tendency to agglomerate or misfold. While future studies will have to address these issues in detail, a practical solution to circumvent the problems may be site-directed or random mutagenesis. Mutagenesis of only few positions significantly improved protein solubility and as a result functionality in a variety of systems [102 – 106]. While the rationale for most changes is unclear, the majority of mutations were located on the protein surface and involved the substitution of hydrophobic amino acids with charged constituents, minimizing hydrophobic patches that would otherwise be prone to aggregation.

Furthermore, fundamental differences in the folding pathway between the two parental sequences could affect proper structural alignment in a chimera. This hypothesis is supported by recent results from two studies of evolutionary distant structural homologs. The first study focused on two members of the globin family, soybean leghemoglobin and myoglobin, two proteins with similar three-dimensional structure but only 13 % sequence identity [107]. Extensive analysis and comparison of the folding process revealed overall similar, compact helical folding intermediates. These intermediates however differ significantly in detail, consisting of helix A, G, H, and parts of helix B for myoglobin but helix E, G, and H for leghemoglobin. In the second study, the folding mechanisms of two proteins in the family of intracellular lipid binding proteins were investigated [108]. As in the previous report, the two proteins share very similar structure yet their sequences have diverged to only 23 % sequence identity. The spectroscopic analysis of the folding intermediates indicates very distinct folding pathways leading to the same final structures. Upon substitution of individual parts of two such structures, it seems reasonable to assume that neither of the two original folding pathways will be suitable to orchestrate the conformational organization of the structure as a whole. Instead, the system will likely follow a third, unknown trajectory with uncertain outcome. From an experimental standpoint, the issue can be addressed from two sides. As discussed earlier, mutagenesis can provide the means to make the necessary adjustments to restore solubility and functionality. In a recent report, protein-folding problems due to a C-terminal truncation in chloramphenicol acetyltransferase (CAT) could be corrected by the introduction of additional secondary site mutations [109]. Random mutagenesis of a deleterious, insoluble CAT variant followed by selection for chloramphenicol resistance in *E. coli* yielded a functional mutant carrying a single point mutation. The beneficial effect of the conservative leucine to phenylalanine substitution, located in the hydrophobic core of the protein, clearly could be attributed to an increase in folding efficiency.

An alternative concept can be the fusion of domains in the context of independent folding units, exemplified by the modular assembly of proteases involved in blood coagulation and fibrinolysis. Modules are commonly associated with domains and, on a genetic level, with exons [110 – 112] although the correlation is not dominant [113]. Recent work with zinc finger binding domains in conjunction with nucleases or regulatory elements demonstrates the successful design of hybrids enzymes based on modular engineering.

9.3.5
Rational domain assembly – engineering zinc fingers

A very popular target for exploring and exploiting the modular design of proteins is the zinc finger family. Zinc fingers consist of a $\beta\beta\alpha$ domain, stabilized by a zinc metal ion with tetrahedral planar coordination to two cysteine and two histidine side-chains. Interacting with the major groove in DNA, the zinc finger domain specifically recognizes three-nucleotide stretches. Serial assembly of individual finger domains can therefore create protein complexes with an extended DNA recognition pattern, a highly desirable property in the post genomic area. Repeatedly, studies have indicated that fusion of finger modules through their natural linker sequence is, however, limited to three domains. Each insertion of a zinc finger into the major groove of DNA causes a slight bending of the latter, putting any module past the third domain out of reach for its DNA recognition sequence. To specifically target a unique site in even a relatively small genome such as *E. coli*, the 9 nucleotide-recognition pattern of the triple finger is not sufficient. Furthermore, the repertoire of natural zinc finger recognition sequences, the diversity of the"codons", is limited to a few combinations, leaving only few sites that comprise the correct sequence and length. Recent advances in the field have addressed and provided insight to these questions and are discussed below.

Focusing on the limitations of codon diversity, random mutagenesis of existing zinc finger domains followed by screening for variations in the recognition sequence by phage display technology has led to the identification of domains recognizing all 16 possible 5'-GNN-3' combinations [114]. The ability to specifically target desired sites *in vivo*, using the expanded set of finger domains, was demonstrated. Fusing either three or six modules (two three-finger constructs, linked by a pentapeptide) to transcription factor domains, the expression of a luciferase reporter gene could be up or down-regulated [115]. An alternative approach to overcome the limitations of triplet diversity has been reported by Klug and coworkers [116, 117]. By introducing structured linkers between the individual binding domains, stretches of up to 10 nucleotides could be bridged without compromising binding affinities. Rather than trying to overcome the limitation of the number of naturally occurring codons in zinc fingers, their tactic is based on extending the length of the target sequence that can be covered by the DNA recognition domains. Having a set of linkers with flexible length allows the custom design of highly specific binding complexes using only the modules at hand. Combining the two approaches may even further increase the versatility of the zinc finger domains. Finally, the fusion of such engineered multi-module zinc finger subunits with a variety of catalytic subunits can provide very powerful tools for the engineering of entire genomes as demonstrated with transcriptional regulators [118] or with endonuclease domains [119].

9.3.6
Combinatorial domain recombination – exon shuffling

Advances in molecular biology and high-throughput screening, as well as the projected flexibility and diversity of modular assembly in natural protein evolution has inspired the development of various techniques to implement combinatorial domain recombination, better known as exon shuffling, *in vitro*.

Two early proposals focused on the exploitation of self-splicing introns to catalyze the fusion of RNA libraries. Fisch and coworkers [120] generated large 25-residue peptide libraries by constructing two random 30-base pair exon libraries, flanked either on their 5'- or 3'-end by a group I intron sequence with an inserted *lox*-Cre recombination site. Following electroporation of the first plasmid-based exon library into *E. coli* and transformation of the second library through phage infection, *in vivo* recombination of the two libraries by Cre-recombinase produced a repertoire of ~1.6 × 10^{11} members. In a final step, the *lox*-Cre recombination site between the two exon-libraries was excised by the adjacent intron sequences, retaining a 15-base pair stretch at the points of fusion. Screening of the resulting peptide library, comprising two hyper variable 10-amino acid regions with an intervening five-amino acid residue spacer, by phage display identified several sequences with up to nanomolar binding affinities against various protein targets.

Simultaneously, Mikheeva and Jarrell [121] proposed *in vitro* exon shuffling directly by engineered trans-splicing group II intron ribozymes. The method was implemented on a model system consisting of two fragments from human tissue plasminogen activator. RNA analogs were generated by separate reverse transcription of the two exons, fused to their corresponding intron sequences. A mixture of the two transcripts was then trans-spliced. Although the ligation efficiency was low, gel analysis clearly indicated product formation. In contrast to the previous approach, the engineered ribozymes facilitate the seamless fusion of the two exons.

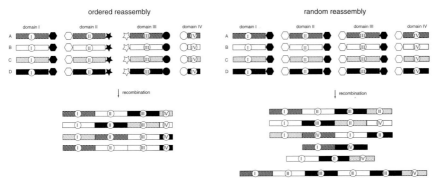

Fig. 9.9. Exon shuffling: combinatorial libraries of multidomain enzymes can be created through recombination of individual modules or domains (I – IV) from multiple parents (A – D) in an ordered fashion, maintaining the parental size (left diagram) or by random reassembly. This allows the insertion, deletion, substitution, or duplication of modules in a single step. The black and white hexagons, stars, and circles indicate unique complementary elements (complementary DNA strands or pairs of *trans*-splicing inteins).

An elegant alternative to the intron-catalyzed recombination has been derived from DNA shuffling [122]. For initial experiments on a human single-chain Fv framework, six complementarity-determining regions (CDRs) within the structure (which could be viewed as individual domains in a general sense) were identified. Fragments of the DNase-treated parental gene were then spiked with synthetic oligonucleotides comprising randomly mutated derivatives of these six CDRs and the mixture was reassembled by PCR. The analysis of the resulting naive library indicated a random mix of wild-type and mutant CDRs, evenly distributed over all six regions. As discussed in a recent review [57], a modified version of the approach now assembles libraries from mixed pools of homologous exons. The proposed method gives great flexibility to customized libraries based on the design of the chimeric oligonucleotide linkers. Presumably, the size and sequence of the linkers, as well as the concentration of individual exons in each pool and the mixing ratio of the different pools can be used to control the reassembly process (Fig. 9.9).

More hypothetically in the context of exon shuffling, post-translational protein-fragment recombination by *trans*-splicing inteins can be employed to recombine protein modules [123]. *Trans*-splicing inteins were first discovered in *Synechocystis sp.* PCC6803, catalyzing the specific and directed religation of the genetically separately encoded protein fragments that make up the replicative DNA polymerase III catalytic α-subunit (DnaE) [124]. Consisting of short protein sequences at the C-terminus of the first protein fragment and the N-terminus of the second protein fragment, the inteins, following translation, dimerize *in vivo* and catalyze the peptide-bond formation between the two protein fragments while simultaneously excising themselves (Fig. 9.10).

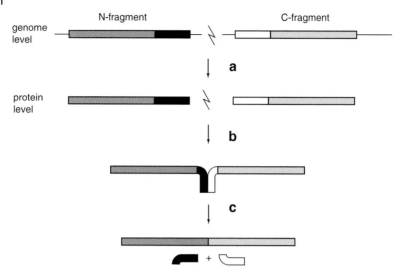

Fig. 9.10. Ligation of two protein fragments by *trans*-splicing inteins: a) on the DNA level, the two fragments can be encoded in separate ORFs and are flanked by either a 3'-extension encoding the N-intein (N-fragment) or 5'-extension encoding the C-intein (C-fragment). b) Following transcription and translation, the intein domains dimerize and c) excise themselves, fusing the N- and C-fragment by a peptide backbone bond.

Using the same basic approach for the successful random fusion of domains and modules of proteins on a post-translational rather than the genetic level would depend on two fundamental requirements. First, multiple pairs of *trans*-splicing inteins would have to be available and second, the cross-reactivity between the individual pairs of inteins would have to be either minimal or non-existent. So far, a single pair of naturally occurring *trans*-splicing inteins has been reported in the literature and an additional four pairs have been engineered from related *cis*-splicing inteins [125]. Assuming that the number of engineered intein pairs will further increase over time, the first requirement should not be a major limitation to the approach.

Cross-reactivity between N- and C-termini of *trans*-splicing inteins from different pairs is the other primary concern. Recently, Otomo *et al.* [126] reported the simultaneous *in vitro* recombination of three protein fragments, tagged with two separate pairs of *trans*-splicing inteins. The ligation, followed by *in vitro* refolding, generated a correctly assembled, functional full-length protein. Similar experiments in our laboratory, using heterodimeric-GFP, indicate that correct intein-mediated *in vivo* assembly of functional protein in the presence of multiple *trans*-inteins is also possible (C.P. Scott – personal communication).

9.4
Gene Fusion – From Bi- to Multifunctional Enzymes

While the rationale behind gene consolidation in natural systems is still disputed, protein engineering has made extensive use of the idea of linking genes. Some of the applications may be as simple as attaching a reporter module onto the N- or C-terminus of the target. Alternatively, joining parental genes by insertion of one sequence into the other has created bifunctional hybrids with, in some cases, allosteric properties. Finally, the assembly of multiple functional units into entire protein machineries has been directed towards the multi-step synthesis of complex organic molecules.

9.4.1
End-to-end gene fusions

Rationally designed end-to-end fusions of genes are employed in the purification, detection, selection, and localization of targeted gene products. Probably the widest-known applications of covalently connecting two proteins are for purification purposes. Fusion constructs between a target protein and a tagging gene such as the maltose-binding protein, chitin-binding domain, thioredoxin, β-galactosidase, or glutathione S-transferases are frequently employed to aid in purification and in some cases actually help stabilize and increase solubility of target proteins [127, 128]. In addition, fusion of a reporter gene such as GFP or bacterial luciferase can be used to identify and localize a protein of interest *in vivo* and *in vitro*. Issues of protein solubility and folding can also be addressed by gene fusions *in vivo*. C-terminal extensions of a target gene with GFP, chloramphenicol acetyltransferase, or β-galactosidase have successfully been applied to select for solubility and in-frame constructs [129 – 131]. Additional applications of fusion constructs include co-expression with transporter proteins that direct and anchor the target protein to the outer membrane, effectively displaying the protein on the host's surface. Implemented in phages using the pIII and pVIII coat proteins as fusion partners, as well as in bacterial hosts and yeast using surface-exposed loops or outer-membrane proteins [132 – 134], these systems are particularly useful for the screening of protein libraries.

9.4.2
Gene insertions

While most of the gene fusions discussed so far focused on N or C-terminal extensions, the insertion of an entire gene into a second gene is possible and may be advantageous. In a rather intriguing experiment, Betton *et al.* [135] inserted an entire β-

lactamase gene (*bla*, 708 nucleotides) into loop regions of the maltodextrin-binding protein (MalE). Inserting *bla* at position 130 or 303 in *malE* which were previously identified as permissive to smaller insertions and dissection [136 – 138], the authors were able to isolate fully active bifunctional fusion constructs. In addition, MalE was shown to increase the stability and modulate the catalytic activity of the fused lactamase. Similar experiments were reported, exploring the tolerance of the phosphoglycerate kinase scaffold towards insertion of dihydrofolate reductase or β-lactamase and its effect on folding [139], as well as creating bifunctional fusion proteins between 1,3-1,4-β-glucanase and 1,4-β-xylanase [140].

Further expanding the published protocol, Baird *et al.* [141] circularly permuted an internal fusion construct between the enhanced yellow fluorescent protein (EYFP) and a zinc finger domain or calmodulin. Conformational changes upon metal binding at the insertion domain are translated into structural rearrangements of EYFP. These structural changes in turn affect the fluorometric properties of EYFP; making these fusion constructs *in vivo* biosensors to monitor Zn^{2+} and Ca^{2+} concentration in cells. An allosterically regulated, GFP-based biosensor, sensitive to antibiotics, was engineered by inserting β-lactamase into one of the surface loops of GFP followed by random mutagenesis and selection for antibiotic-dependent changes in the fluorescence of the fusion libraries [142].

9.4.3
Modular design in multifunctional enzymes

One of the most exciting areas in the field of multifunctional enzyme systems is the synthesis of a wide array of organic molecules by polyketide and nonribosomal protein synthetases. These enzymes are generally characterized by multiple subunits which themselves consist of individual domains with distinct enzymatic activities (Fig. 9.11a). The range of natural products synthesized by these mega-synthetases includes a considerable number of important antibiotics, antifungals, antitumor and cholesterol-lowering compounds, immunosuppressants, and siderophores.

Three categories of synthetases are distinguished, based on their substrate specificity and mode of product synthesis. The two known types of polyketide synthetases (PKSs) (Type I and II) utilize acyl-coenzyme A (CoA) monomers while nonribosomal peptide synthetases (NRPSs) use amino acids and their analogs as substrates. Type I PKS and NRPS oligomerize these building blocks by a modular assembly-line arrangement while type II PKS iteratively assembles monomeric units.

The modular design of PKS and NRPS renders them convenient for protein engineering to generate hybrid enzymes that can synthesize a range of natural and unnatural metabolites. By independently manipulating what Cane *et al.* [143] call the four degrees of freedom in polyketide synthesis: variation of chain length, choice of ACP

a)

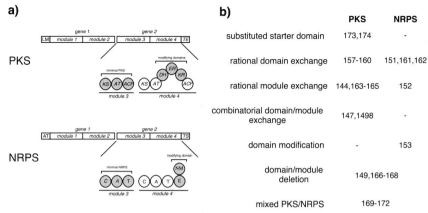

b)

	PKS	NRPS
substituted starter domain	173,174	-
rational domain exchange	157-160	151,161,162
rational module exchange	144,163-165	152
combinatorial domain/module exchange	147,1498	-
domain modification	-	153
domain/module deletion	149,166-168	
mixed PKS/NRPS		169-172

Fig. 9.11. Organizational overview and summary of protein engineering efforts on the polyketide synthetase (PKS) and nonribosomal peptide synthetase (NRPS) framework. a) In both systems, the complete synthetases consist of multiple subunits encoded on individual genes. The subunits themselves are divided in modules which each catalyze the addition of one acyl-building block (PKS) or amino acid (NRPS). The minimal module for PKSs is made up of three domains – a keto synthase (KS), an acyl transferase (AT), and an acyl carrier protein (ACP). In addition, modules can contain up to three modifying domains to derivatize β-carbons on the polyketide chain; a keto reductase (KR), a dehydratase (DH), and an enoylreductase (ER). Flanking the entire complex is a loading module (LM) and a termination module (TE). The minimal module for NRPSs consists of an adenylation (A), a thiolation (T), and a condensation (C) domain. Additional modifications on the peptide chain can be introduced by epimerization (E) or N-methylation (NM) domains. The loading module in NRPS consists of an adenylation/thiolation (AT) domain while the synthesis is complete by a thioesterase (TS). b) Protein engineering of PKSs and NRPSs, organized by structural objectives.

domains (substrate), degree of β-carbon modification, and stereochemistry, product libraries of extremely high diversity can hypothetically be synthesized. Likewise in NPRSs, the variation of chain length, type and stereochemistry of the substrates, and their level of post-synthetic derivatization can potentially yield an almost unlimited number of compounds with potentially novel or improved properties.

Initial efforts to rationally engineer PKS and NRPS have focused on the alteration of the biosynthetic pathway of metabolites by deletion, insertion, and substitution of individual domains (summarized in Fig. 9.11b). Although these experiments provided valuable information about the structure and function of the individual domains and modules, formation of functional hybrid enzymes proved difficult at times. In contrast, taking entire modules and rearranging them in the context of an enzyme complex was shown to be a more successful route towards the creation of functional hybrid enzymes. Of particular interest were data indicating the significance of the linker regions between covalently as well as non-covalently attached modules. By restoring the correct inter- or intra-polypeptide linker sequences, Khosla and coworkers swapped module positions in the polyketide synthetase 6–deoxyerythronolide B synthase (DEBS) and introduced modules from other PKSs, generating an array of functional hybrid enzymes [144 – 146].

Beyond the swapping of whole modules, efforts to prepare libraries of compounds have turned to randomization of the proteins within each module. With respect to these truly combinatorial engineering efforts, modular designed type I PKSs have so far been most amenable. Shortly after the report of the successful but elaborate generation of a combinatorial library by cassette replacement of domains in DEBS [147], a relatively simple and elegant multi-plasmid approach was described [148]. The three genes of DEBS were cloned in separate plasmid vectors, followed by introduction of mutations to diversify each individual gene. Upon triple transformation into a heterologous host, a small library of related polyketides was produced based on the synthesis by random members of the three vectors. Although these results are very exciting, the method is very much limited by the number of different plasmids that can be present in a single host cell. The efficiency of triple transformation is low and plasmid stability may be a major problem.

The engineering of iterative (type II) PKSs is considerably more difficult due to their repetitive use of a single module in the biosynthesis of natural products. So far, the only combinatorial libraries of polyketides generated with these types of PKS are created not by genetically encoded diversity but by construction of minimal PKS assemblies. Minimal PKSs are created by removal of all accessory proteins such as cyclases, ketoreductases, and aromatases, reducing the functional enzyme complex to three essential domains: a ketosynthase, a chain length factor, and an acyl carrier protein. Studying the minimal version of *whiE* (spore pigment) PKS, Moore and coworkers found more than two dozen unique polyketides, generated by a single enzyme multiplex [149]. The diversity was generated by premature product dissociation from the enzyme complex at various stages of the oligomerization, resulting in a variable product chain length and regiospecificity of the cyclization event. Similar variations in the natural product palette of PKS from *Streptomyces maritimus* suggest that the same mechanisms are employed by Nature to explore potentially beneficial effects of these secondary metabolites [150]. In contrast, Nature's mechanism of directing biosynthesis towards particular metabolites is accomplished by accessory proteins such as the cyclase in the natural *whiE* system. By non-covalently binding to the minimal PKS complex, the accessory protein stabilizes the product-enzyme intermediates, ensuring formation of full-length product and correct cyclization. While such a system is valuable for the generation of polyketide libraries, problems with reproducibility of library composition due to the unknown factors that influence the generation of the diversity may cause problems. Furthermore, upon identification of interesting or useful products in the library, no direct approach exists for resynthesis of these particular compounds. Instead, additional rational engineering is required to attempt the construction of a hybrid that exclusively produces the desired compound.

Finally, nonribosomal peptide synthetases represent the least explored members in this family of multienzyme complexes. They represent an important and versatile class of megasynthetases, involved in the biosynthesis of a wide spectrum of peptidylic secondary metabolites including penicillins, vancomycins, bleomycin, and cyclosporin A. The biosynthesis of these compounds is achieved by utilizing not only natural amino acid but also a wide range of D-configured and N-methylated amino acids and hydroxy acids. Variations in the peptide backbone structure result in linear, cyclic, and branched cyclic molecules and these structures can be modified by acylation, heterocyclization, reduction, and glycosylation. Very similar to the modular design of type I PKS, nonribosomal peptide synthetases consist of a linear arrangement of multiple modules linked either covalently or noncovalently (Fig. 9.11a). The individual modules are responsible for the incorporation of a specific amino acid in the growing peptide chain, therefore dictating the product sequence. The modules themselves are formed by three domains, an adenylation-domain (A-domain; activates incoming amino acid), a thiolation-domain (T-domain; peptidyl carrier protein), and a condensation-domain (C-domain; facilitates the reaction between peptidyl moiety and activated amino acid). As for PKSs, rational engineering experiments have focused on the exchange of individual domains within a module [151]. More recently, Marahiel and coworkers [152] reported the successful rearrangement and fusion of intact modules, generating novel peptide products based on the amino acid specificity of the individual building blocks.

Combinatorial approaches towards the creation of peptidylic libraries by NPRSs have not yet been reported. Considering the very similar overall design of NPRSs to modular PKSs, one can imagine applying similar methods as discussed above. In addition, the recently described site-directed mutagenesis of the substrate-recognizing A-domain of NRPSs, resulting in the alteration or relaxation of its substrate specificity, may be employed to generate a pool of heterogeneous products [153].

In summary, the structural and functional work on PKSs and NRPSs has provided a basic understanding of secondary metabolite biosynthesis. At the same time, these systems have proven extremely valuable to protein engineering efforts. However, the current methodology is not capable of coping with the demands of a large-scale combinatorial approach to generate polyketide- and peptide libraries. Although all the individual elements are readily available, the limitations of an effective shuffling and recombination protocol currently prevents the creation of libraries with >100 individual members.

9.5
Perspectives

Studies of protein evolution and protein engineering are progressing hand-in-hand. Advances in our understanding of protein evolution have unveiled strategies that have led to the development of novel methods in protein engineering. In return, mimicking natural processes in the laboratory has revealed some of the robust but at the same time delicate nature of complex protein structures, furthering our understanding of the fundamentals in protein evolution. The industrial aspects of protein engineering, as well as the exploration of natural protein evolution, can both profit from one another providing further insight into the fascinating world of proteins.

From a methodology standpoint, no single approach provides a universal solution to protein engineering. The nature of the individual target, the objective of the engineering project, and downstream factors such as the availability of a high-throughput screen or a selection procedure dictates the technique or techniques most suitable to address the problem at hand. While the development of fundamentally novel methods will progress, a particularly attractive approach has become the combination of individual techniques as demonstrated by Fersht and coworkers [75] using rational design and DNA shuffling, as well as by our laboratory [98] using ITCHY and DNA shuffling. Such pairwise approaches not only overcome the limitations of the individual methods but also provide the means for addressing increasingly complex issues of protein engineering.

Beyond the advances in experimental design, the development of *in silico* protein engineering is growing increasingly important. Progress in computational approaches now provide algorithms that simulate the nucleic acid fragment reassembly process during DNA shuffling [93] and assess the tolerance of individual amino acid substitution in the context of the global protein structure [154]. While the days of computer-based protein design are in the far future, the predictions from either modeling system can already provide guidance towards more effective experimental design.

Acknowledgements
The authors would like to thank their colleagues in the laboratory and in particular Dr. Walter Fast for many helpful discussion and critical review of the manuscript. We would also like to acknowledge Dr. Marc Ostermeier and Steve Tizio for providing the unpublished data on the PurN/FMT hybrid system.

References

[1] L. D. Bogarad, M. W. Deem, *Proc. Natl. Acad. Sci. USA* **1999**, 96, 2591–2595.

[2] C. H. Cheng, L. Chen, *Nature* **1999**, 401, 443–444.

[3] S. A. Doyle, S. Y. Fung, D. E. Koshland, Jr., *Biochemistry* **2000**, 39, 14348–14355.

[4] G. Wu, A. Fiser, B. ter Kuile, A. Sali, M. Muller, *Proc. Natl. Acad. Sci. USA* **1999**, 96, 6285–90.

[5] R. Kuroki, L. H. Weaver, B. W. Matthews, *Proc. Natl. Acad. Sci. USA* **1999**, 96, 8949–54.

[6] K. Miyazaki, F. H. Arnold, *J. Mol. Evol.* **1999**, 49, 716–720.

[7] A. D. McLachlan, *Cold Spring Harb. Symp. Quant. Biol.* **1987**, 52, 411–420.

[8] B M. Lynch, J. S. Conery, *Science* **2000**, 290, 1151–55.

[9] S. Ohno, *Evolution by Gene Duplication*, Springer Verlag, New York **1970**.

[10] L. Patthy, *Protein Evolution*, Blackwell Science, Malden, MA **1999**.

[11] A. Force, M. Lynch, F. B. Pickett, A. Amores, Y. L. Yan, J. Postlethwait, *Genetics* **1999**, 151, 1531–45.

[12] R. A. Jensen, *Annu. Rev. Microbiol.* **1976**, 30, 409–425.

[13] A. L. Hughes, *Proc. R. Soc. Lond. B Biol. Sci.* **1994**, 256, 119–124.

[14] J. Piatigorsky, G. Wistow, *Science* **1991**, 252, 1078–1079.

[15] P. J. O'Brien, D. Herschlag, *Chem. Biol.* **1999**, 6, R91–R105.

[16] R. Fani, P. Lio, A. Lazcano, *J. Mol. Evol.* **1995**, 41, 760–774.

[17] R. Fani, P. Lio, I. Chiarelli, M. Bazzicalupo, *J. Mol. Evol.* **1994**, 38, 489–495.

[18] D. Lang, R. Thoma, M. Henn-Sax, R. Sterner, M. Wilmanns, *Science* **2000**, 289, 1546–50.

[19] J. Tang, M. N. James, I. N. Hsu, J. A. Jenkins, T. L. Blundell, *Nature* **1978**, 271, 618–621.

[20] A. M. Baptista, P. H. Jonson, E. Hough, S. B. Petersen, *J. Mol. Evol.* **1998**, 47, 353–362.

[21] A. D. McLachlan, *Eur. J. Biochem.* **1979**, 100, 181–187.

[22] J. Tang, in A. J. Barrett, N. D. Rawlings, J. F. Woessner (Eds.): *Handbook of Proteolytic Enzymes, Vol. 272*, Academic Press, San Diego CA, **1998**, 805–814.

[23] X. Y. Qiu, J. S. Culp, A. G. DiLella, B. Hellmig, S. S. Hoog, C. A. Janson, W. W. Smith, S. S. AbdelMeguid, *Nature* **1996**, 383, 275–279.

[24] D. Reardon, G. K. Farber, *FASEB J.* **1995**, 9, 497–503.

[25] B. Hocker, S. Beismann-Driemeyer, S. Hettwer, A. Lustig, R. Sterner, *Nat. Struct. Biol.* **2001**, 8, 32–36.

[26] S. Janecek, S. Balaz, *J. Protein Chem.* **1993**, 12, 509–514.

[27] K. Luger, U. Hommel, M. Herold, J. Hofsteenge, K. Kirschner, *Science* **1989**, 243, 206–210.

[28] V. Mainfroid, K. Goraj, F. Rentier-Delrue, A. Houbrechts, A. Loiseau, A. C. Gohimont, M. E. Noble, T. V. Borchert, R. K. Wierenga, J. A. Martial, *Protein Eng.* **1993**, 6, 893–900.

[29] C. P. Ponting, R. B. Russell, *Trends Biochem. Sci.* **1995**, 20, 179–180.

[30] A. Jeltsch, *J. Mol. Evol.* **1999**, 49, 161–164.

[31] E. A. MacGregor, H. M. Jespersen, B. Svensson, *FEBS Lett.* **1996**, 378, 263–266.

[32] J. Jia, W. Huang, U. Schorken, H. Sahm, G. A. Sprenger, Y. Lindqvist, G. Schneider, *Structure* **1996**, 4, 715–724.

[33] K. Wieligmann, B. Norledge, R. Jaenicke, E. M. Mayr, *J. Mol. Biol.* **1998**, 280, 721–729.

[34] R. J. Kreitman, R. K. Puri, I. Pastan, *Cancer Res.* **1995**, 55, 3357–3363.

[35] G. D'Alessio, *Prog. Biophys. Mol. Biol.* **1999**, 72, 271–298.

[36] M. F. Perutz, *Nature* **1970**, 228, 726–739.

[37] K. A. Lee, *J. Cell Sci.* **1992**, 103, 9–14.

[38] M. P. Schlunegger, M. J. Bennett, D. Eisenberg, *Adv. Protein Chem.* **1997**, 50, 61–122.

[39] G. MacBeath, P. Kast, D. Hilvert, *Protein Sci.* **1998**, 7, 1757–1767.

[40] M. Ostermeier, S. J. Benkovic, *Adv. Protein Chem.* **2000**, 55, 29–77.

[41] S. Zhang, P. Pohnert, P. Kongsaeree, D. B. Wilson, J. Clardy, B. Ganem, *J. Biol. Chem.* **1998**, 273, 6248–53.

[42] J. Turnbull, J. F. Morrison, W. W. Cleland, *Biochemistry* **1991**, 30, 7783–7788.

[43] R. M. Romero, M. F. Roberts, J. D. Phillipson, *Phytochemistry* **1995**, 39, 263–276.

[44] F. Pompeo, Y. Bourne, J. van Heijenoort, F. Fassy, D. Mengin-Lecreulx, *J. Biol. Chem.* **2001**, 276, 3833–3839.

[45] P. Ljungcrantz, H. Carlsson, M. O. Mansson, P. Buckel, K. Mosbach, L. Buelow, *Biochemistry* **1989**, 28, 8786–92.

[46] C. Lindbladh, M. Persson, L. Buelow, K. Mosbach, *Eur. J. Biochem.* **1992**, 204, 241–247.

[47] R. B. Russell, *Protein Eng.* **1994**, 7, 1407–1410.

[48] M. Nardini, B. W. Dijkstra, *Curr. Opin. Struct. Biol.* **1999**, 9, 732–737.

[49] A. E. Todd, C. A. Orengo, J. M. Thornton, *J. Mol. Biol.* **2001**, 307, 1113–1143.

[50] D. H. Juers, R. E. Huber, B. W. Matthews, *Protein Sci.* **1999**, 8, 122–136.

[51] P. Van Roey, V. Rao, T. H. Plummer, Jr., A. L. Tarentino, *Biochemistry* **1994**, 33, 13989–96.

[52] A. C. Terwissscha van Scheltinga, M. Hennig, B. W. Dijkstra, *J. Mol. Biol.* **1996**, 262, 243–57.

[53] G. Buisson, E. Duee, R. Haser, F. Payan, *EMBO J.* **1987**, 6, 3909–16.

[54] Y. Kai, H. Matsumura, T. Inoue, K. Terada, Y. Nagara, T. Yoshinaga, A. Kihara, K. Tsumura, K. Izui, *Proc. Natl. Acad. Sci. USA* **1999**, 96, 823–828.

[55] Z. X. Xia, N. Shamala, P. H. Bethge, L. W. Lim, H. D. Bellamy, N. H. Xuong, F. Lederer, F. S. Mathews, *Proc. Natl. Acad. Sci. USA* **1987**, 84, 2629–33.

[56] L. Patthy, *Cell* **1985**, 41, 657–663.

[57] J. A. Kolkman, W. P. Stemmer, *Nat. Biotechnol.* **2001**, 19, 423–428.

[58] M. L. Tasayco, J. Carey, *Science* **1992**, 255, 594–597.

[59] A. G. Ladurner, L. S. Itzhaki, G. de Prat Gay, A. R. Fersht, *J. Mol. Biol.* **1997**, 273, 317–329.

[60] S. Pascarella, P. Argos, *J. Mol. Biol.* **1992**, 224, 461–471.

[61] N. Doi, H. Yanagawa, *Cell. Mol. Life Sci.* **1998**, 54, 394–404.

[62] M. Sagermann, W. A. Baase, B. W. Matthews, *Proc. Natl. Acad. Sci. USA* **1999**, 96, 6078–83.

[63] M. Ostermeier, A. E. Nixon, J. H. Shim, S. J. Benkovic, *Proc. Natl. Acad. Sci. USA* **1999**, 96, 3562–67.

[64] S. W. Michnick, I. Remy, F. X. Campbell-Valois, A. Vallee-Belisle, J. N. Pelletier, *Methods Enzymol.* **2000**, 328, 208–230.

[65] P. T. Beernink, Y. R. Yang, R. Graf, D. S. King, S. S. Shah, H. K. Schachman, *Protein Sci.* **2001**, 10, 528–537.

[66] S. S. Ray, H. Balaram, P. Balaram, *Chem. Biol.* **1999**, 6, 625–637.

[67] C. Manoil, J. Bailey, *J. Mol. Biol.* **1997**, 267, 250–263.

[68] R. Graf, H. K. Schachman, *Proc. Natl. Acad. Sci. USA* **1996**, 93, 11591–96.

[69] J. Hennecke, P. Sebbel, R. Glockshuber, *J. Mol. Biol.* **1999**, 286, 1197–1215.

[70] R. J. Eustance, S. A. Bustos, R. F. Schleif, *J. Mol. Biol.* **1994**, 242, 330–338.

[71] D. M. Nguyen, R. F. Schleif, *J. Mol. Biol.* **1998**, 282, 751–759.

[72] M. Ostermeier, J. H. Shim, S. J. Benkovic, *Nat. Biotechnol.* **1999**, 17, 1205–1209.

[73] K. Vogel, J. Chmielewski, *J. Am. Chem. Soc.* **1994**, 116, 11163–11164.

[74] K. P. Hopfner, E. Kopetzki, G. B. Kresse, W. Bode, R. Huber, R. A. Engh, *Proc. Natl. Acad. Sci. USA* **1998**, 95, 9813–18.

[75] M. M. Altamirano, J. M. Blackburn, C. Aguayo, A. R. Fersht, *Nature* **2000**, 403, 617–622.

[76] R. C. Cadwell, G. F. Joyce, *PCR Methods Appl.* **1994**, 3, S136–S140.

[77] W. P. Stemmer, *Nature* **1994**, 370, 389–391.

[78] W. P. Stemmer, *Proc. Natl. Acad. Sci. USA* **1994**, 91, 10747–10751.

[79] K. Chen, F. H. Arnold, *Proc. Natl. Acad. Sci. USA* **1993**, 90, 5618–22.

[80] J. C. Moore, F. H. Arnold, *Nat. Biotechnol.* **1996**, 14, 458–467.

[81] A. Crameri, G. Dawes, E. Rodriguez, S. Silver, W. P. Stemmer, *Nat. Biotechnol.* **1997**, 15, 436–438.

[82] A. Crameri, S. A. Raillard, E. Bermudez, W. P. Stemmer, *Nature* **1998**, 391, 288–291.

[83] J. E. Ness, M. Welch, L. Giver, M. Bueno, J. R. Cherry, T. V. Borchert, W. P. Stemmer, J. Minshull, *Nat. Biotechnol.* **1999**, 17, 893–896.

[84] C. Schmidt–Dannert, D. Umeno, F. H. Arnold, *Nat. Biotechnol.* **2000**, 18, 750–753.

[85] S. J. Allen, J. J. Holbrook, *Protein Eng.* **2000**, 13, 5–7.

[86] Z. Shao, H. Zhao, L. Giver, F. H. Arnold, *Nucleic Acids Res.* **1998**, 26, 681–683.

[87] H. Zhao, L. Giver, Z. Shao, J. A. Affholter, F. H. Arnold, *Nat. Biotechnol.* **1998**, 16, 258–261.

[88] Y. Zhang, F. Buchholz, J. P. Muyrers, A. F. Stewart, *Nat. Genet.* **1998**, 20, 123–128.

[89] M. Kikuchi, K. Ohnishi, S. Harayama, *Gene* **1999**, 236, 159–167.

[90] W. M. Coco, W. E. Levinson, M. J. Crist, H. J. Hektor, A. Darzins, P. T. Pienkos, C. H. Squires, D. J. Monticello, *Nat. Biotechnol.* **2001**, 19, 354–359.

[91] V. Sieber, C. A. Martinez, F. H. Arnold, *Nat. Biotechnol.* **2001**, 19, 456–460.

[92] N. W. Soong, L. Nomura, K. Pekrun, M. Reed, L. Sheppard, G. Dawes, W. P. Stemmer, *Nat. Genet.* **2000**, 25, 436–439.

[93] G. L. Moore, C. D. Maranas, S. Lutz, S. J. Benkovic, *Proc. Natl. Acad. Sci. USA* **2001**, 98, 3226–31.

[94] J. Minshull, W. P. Stemmer, *Curr. Opin. Chem.. Biol.* **1999**, 3, 284–290.

[95] M. Ostermeier, S. J. Benkovic, *Biotech. Lett.* **2001**, 23, 303–310.

[96] S. Lutz, M. Ostermeier, S. J. Benkovic, *Nucleic Acids Res.* **2001**, 29, E16.

[97] M. Ostermeier, A. E. Nixon, S. J. Benkovic, *Bioorg. Med. Chem.* **1999**, 7, 2139–2144.

[98] S. Lutz, M. Ostermeier, G. Moore, C. Maranas, S. J. Benkovic, *Proc. Natl. Acad. Sci. USA* **2001**, submitted,.

[99] S. E. Radford, *Trends Biochem. Sci.* **2000**, 25, 611–618.

[100] A. E. Nixon, M. Ostermeier, S. J. Benkovic, *Trends Biotechnol.* **1998**, 16, 258–264.

[101] L. B. Murata, H. K. Schachman, *Protein Sci.* **1996**, 5, 719–728.

[102] R. Wetzel, L. J. Perry, C. Veilleux, *Biotechnology (NY)* **1991**, 9, 731–737.

[103] Z. Lin, T. Thorsen, F. H. Arnold, *Biotechnol. Prog.* **1999**, 15, 467–471.

[104] L. Nieba, A. Honegger, C. Krebber, A. Pluckthun, *Protein Eng.* **1997**, 10, 435–444.

[105] G. E. Dale, C. Broger, H. Langen, A. D'Arcy, D. Stuber, *Protein Eng.* **1994**, 7, 933–939.

[106] M. Murby, E. Samuelsson, T. N. Nguyen, L. Mignard, U. Power, H. Binz, M. Uhlen, S. Stahl, *Eur. J. Biochem.* **1995**, 230, 38–44.

[107] C. Nishimura, S. Prytulla, H. Jane Dyson, P. E. Wright, *Nat. Struct. Biol.* **2000**, 7, 679–686.

[108] P. M. Dalessio, I. J. Ropson, *Biochemistry* **2000**, 39, 860–871.

[109] J. Van der Schueren, J. Robben, G. Volckaert, *Protein Eng.* **1998**, 11, 1211–1217.

[110] M. G. Rossmann, P. Argos, *Annu. Rev. Biochem.* **1981**, 50, 497–532.

[111] C. C. Blake, *Nature* **1979**, 277, 598.

[112] M. Go, *Nature* **1981**, 291, 90–92.

[113] T. Tsuji, K. Yoshida, A. Satoh, T. Kohno, K. Kobayashi, H. Yanagawa, *J. Mol. Biol.* **1999**, 286, 1581–96.

[114] D. J. Segal, B. Dreier, R. R. Beerli, C. F. Barbas, 3rd, *Proc. Natl. Acad. Sci. USA* **1999**, 96, 2758–63.

[115] R. R. Beerli, D. J. Segal, B. Dreier, C. F. Barbas, 3rd, *Proc. Natl. Acad. Sci. USA* **1998**, 95, 14628–33.

[116] M. Moore, Y. Choo, A. Klug, *Proc. Natl. Acad. Sci. USA* **2001**, 98, 1432–1436.

[117] M. Moore, A. Klug, Y. Choo, *Proc. Natl. Acad. Sci. USA* **2001**, 98, 1437–1441.

[118] R. R. Beerli, B. Dreier, C. F. Barbas, 3rd, *Proc. Natl. Acad. Sci. USA* **2000**, 97, 1495–1500.

[119] J. Smith, J. M. Berg, S. Chandrasegaran, *Nucleic Acids Res.* **1999**, 27, 674–681.

[120] I. Fisch, R. E. Kontermann, R. Finnern, O. Hartley, A. S. Soler-Gonzalez, A. D. Griffiths, G. Winter, *Proc. Natl. Acad. Sci. USA* **1996**, 93, 7761–7766.

[121] S. Mikheeva, K. A. Jarrell, *Proc. Natl. Acad. Sci. USA* **1996**, 93, 7486–90.

[122] A. Crameri, S. Cwirla, W. P. Stemmer, *Nat. Med.* **1996**, 2, 100–102.

[123] F. B. Perler, *Trends Biochem. Sci.* **1999**, 24, 209–211.

[124] H. Wu, Z. Hu, X. Q. Liu, *Proc. Natl. Acad. Sci. USA* **1998**, 95, 9226–31.

[125] F. B. Perler, *Nucleic Acids Res.* **2000**, 28, 344–345.

[126] T. Otomo, N. Ito, Y. Kyogoku, T. Yamazaki, *Biochemistry* **1999**, 38, 16040–44.

[127] N. Sheibani, *Prep. Biochem. Biotechnol.* **1999**, 29, 77–90.

[128] R. B. Kapust, D. S. Waugh, *Protein Sci.* **1999**, 8, 1668–74.

[129] L. Maxwell, A. K. Mittermaier, J. D. Forman-Kay, A. R. Davidson, *Protein Sci.* **1999**, 8, 1908–11.

[130] S. Waldo, B. M. Standish, J. Berendzen, T. C. Terwilliger, *Nat. Biotechnol.* **1999**, 17, 691–95.

[131] C. Wigley, R. D. Stidham, N. M. Smith, J. F. Hunt, P. J. Thomas, *Nat. Biotechnol.* **2001**, 19, 131–136.

[132] Georgiou, C. Stathopoulos, P. S. Daugherty, A. R. Nayak, B. L. Iverson, R. Curtiss, *Nat. Biotechnol.* **1997**, 15, 29–34.

[133] T. Lattemann, J. Maurer, E. Gerland, T. F. Meyer, *J. Bacteriol.* **2000**, 182, 3726–3733.

[134] T. Boder, K. D. Wittrup, *Methods Enzymol.* **2000**, 328, 430–444.

[135] M. Betton, J. P. Jacob, M. Hofnung, J. K. Broome-Smith, *Nat. Biotechnol.* **1997**, 15, 1276–79.

[136] Martineau, J. G. Guillet, C. Leclerc, M. Hofnung, *Gene* **1992**, 113, 35–46.

[137] J. M. Betton, P. Martineau, W. Saurin, M. Hofnung, *FEBS Lett.* **1993**, 325, 34–38.

[138] J. M. Betton, M. Hofnung, *EMBO J.* **1994**, 13, 1226–1234.

[139] B. Collinet, M. Herve, F. Pecorari, P. Minard, O. Eder, M. Desmadril, *J. Biol. Chem.* **2000**, 275, 17428–33.

[140] J. Ay, F. Gotz, R. Borriss, U. Heinemann, *Proc. Natl. Acad. Sci. USA* **1998**, 95, 6613–18.

[141] G. S. Baird, D. A. Zacharias, R. Y. Tsien, *Proc. Natl. Acad. Sci. USA* **1999**, 96, 11241–46.

[142] N. Doi, H. Yanagawa, *FEBS Lett.* **1999**, 453, 305–307.

[143] D. E. Cane, C. T. Walsh, C. Khosla, *Science* **1998**, 282, 63–68.

[144] R. S. Gokhale, S. Y. Tsuji, D. E. Cane, C. Khosla, *Science* **1999**, 284, 482–485.

[145] R. S. Gokhale, C. Khosla, *Curr. Opin. Chem.. Biol.* **2000**, 4, 22–27.

[146] S. Y. Tsuji, N. Wu, C. Khosla, *Biochemistry* **2001**, 40, 2317–25.

[147] R. McDaniel, A. Thamchaipenet, C. Gustafsson, H. Fu, M. Betlach, G. Ashley, *Proc. Natl. Acad. Sci. USA* **1999**, 96, 1846–1851.

[148] Q. Xue, G. Ashley, C. R. Hutchinson, D. V. Santi, *Proc. Natl. Acad. Sci. USA* **1999**, 96, 11740–45.

[149] Y. Shen, P. Yoon, T. W. Yu, H. G. Floss, D. Hopwood, B. S. Moore, *Proc. Natl. Acad. Sci. USA* **1999**, 96, 3622–27.

[150] J. Piel, C. Hertweck, P. R. Shipley, D. M. Hunt, M. S. Newman, B. S. Moore, *Chem. Biol.* **2000**, 7, 943–955.

[151] P. J. Belshaw, C. T. Walsh, T. Stachelhaus, *Science* **1999**, 284, 486–489.

[152] H. D. Mootz, D. Schwarzer, M. A. Marahiel, *Proc. Natl. Acad. Sci. USA* **2000**, 97, 5848–53.

[153] T. Stachelhaus, H. D. Mootz, M. A. Marahiel, *Chem. Biol.* **1999**, 6, 493–505.

[154] C. A. Voigt, S. L. Mayo, F. H. Arnold, Z. G. Wang, *Proc. Natl. Acad. Sci. USA* **2001**, 98, 3778–83.

[155] R. J. Almassy, C. A. Janson, C. C. Kan, Z. Hostomska, *Proc. Natl. Acad. Sci. USA* **1992**, 89, 6114–18.

[156] E. Schmitt, S. Blanquet, Y. Mechulam, *EMBO J.* **1996**, 15, 4749–58.

[157] R. McDaniel, C. M. Kao, S. J. Hwang, C. Khosla, *Chem. Biol.* **1997**, 4, 667–674.

[158] R. McDaniel, S. Ebert-Khosla, D. A. Hopwood, C. Khosla, *Nature* **1995**, 375, 549–554.

[159] D. Bedford, J. R. Jacobsen, G. Luo, D. E. Cane, C. Khosla, *Chem. Biol.* **1996**, 3, 827–831.

[160] Oliynyk, M. J. Brown, J. Cortes, J. Staunton, P. F. Leadlay, *Chem. Biol.* **1996**, 3, 833–839.

[161] H. Symmank, W. Saenger, F. Bernhard, *J. Biol. Chem.* **1999**, 274, 21581–21588.

[162] S. Doekel, M. A. Marahiel, Chem. Biol. **2000**, 7, 373–384.

[163] C. W. Carreras, D. V. Santi, *Curr. Opin. Biotech.* **1998**, 9, 403–411.

[164] L. Tang, H. Fu, R. McDaniel, *Chem. Biol.* **2000**, 7, 77–84.

[165] A. Ranganathan, M. Timoney, M. Bycroft, J. Cortes, I. P. Thomas, B. Wilkinson, L. Kellenberger, U. Hanefeld, I. S. Galloway, J. Staunton, P. F. Leadlay, *Chem. Biol.* **1999**, 6, 731–41.

[166] C. M. Kao, G. L. Luo, L. Katz, D. E. Cane, C. Khosla, *J. Am. Chem. Soc.* **1995**, 117, 9105–06.

[167] C. M. Kao, G. L. Luo, L. Katz, D. E. Cane, C. Khosla, *J. Am. Chem. Soc.* **1996**, 118, 9184–85.

[168] J. Cortes, K. E. Wiesmann, G. A. Roberts, M. J. Brown, J. Staunton, P. F. Leadlay, *Science* **1995**, 268, 1487–89.

[169] A. M. Gehring, E. DeMoll, J. D. Fetherston, I. Mori, G. F. Mayhew, F. R. Blattner, C. T. Walsh, R. D. Perry, *Chem. Biol.* **1998**, 5, 573–586.

[170] D. Tillett, E. Dittmann, M. Erhard, H. von Dohren, T. Borner, B. A. Neilan, *Chem. Biol.* **2000**, 7, 753–764.

[171] Y. Paitan, G. Alon, E. Orr, E. Z. Ron, E. Rosenberg, *J. Mol. Biol.* **1999**, 286, 465–474.

[172] J. F. Aparicio, I. Molnar, T. Schwecke, A. Konig, S. F. Haydock, L. E. Khaw, J. Staunton, P. F. Leadlay, *Gene* **1996**, 169, 9–16.

[173] A. F. Marsden, B. Wilkinson, J. Cortes, N. J. Dunster, J. Staunton, P. F. Leadlay, *Science* **1998**, 279, 199–202.

[174] S. Kuhstoss, M. Huber, J. R. Turner, J. W. Paschal, R. N. Rao, *Gene* **1996**, 183, 231–236.

10
Exploring the Diversity of Heme Enzymes through Directed Evolution

Patrick C. Cirino and Frances H. Arnold

10.1
Introduction

Enzymes are capable of clean, specific catalysis with high turnover rates. They have already proven useful in numerous synthetic applications, particularly for the high value and often chiral compounds demanded by the pharmaceutical, agricultural, and food industries. Redox enzymes such as peroxidases and cytochrome P450 mono-oxygenases catalyze valuable reactions on a vast spectrum of substrates. Despite their impressive synthetic potential, these enzymes have enjoyed only limited use due to their relative complexity, instability and, in some cases, low catalytic efficiency. Demands for clean, economical oxidation processes and for increasingly complex and specific oxidation products all point in the direction of biocatalytic routes. Directed evolution may be able to eliminate some of the shortcomings of enzymes, while improving and harnessing their natural catalytic power.

Metalloporphyrins are synthesized naturally and utilized biologically as redox catalysts, and as such are essential to life. These metal complexes have different chemical functions (see [1]); nature has discovered the ability to modulate the function by incorporating them into proteins which allow for a tremendous diversity of architecture and chemical environments surrounding the prosthetic group. Within the protein framework the prosthetic group becomes a versatile tool with varying, highly specialized capabilities.

Heme serves as the active center in different families of proteins classified by structural similarity (e.g. heme-binding peroxidases, cytochromes P450, globins, catalases). Within these families the metalloporphyrin has a primary function (e.g. hydroxylation or oxygen binding), but there is also considerable functional overlap among them. The protein regulates the function, but it is not known whether the particular folds that characterize each class are required for optimal function of that class. One could argue that nature has had a long time to optimize an enzyme's structure-function relationship. But evolution is restricted in its exploration of structure space and is contingent on previous history. P450-type hydroxylation reactions, for example, might be efficient in scaffolds very different from the one nature has adopted, but we only see the one that was discovered first. The original function of the P450 enzymes, in fact, may not even have been oxygen insertion. Some folds may be more likely to occur than others and therefore have a higher probability of being encountered, even though they may not be optimal. There are likely to be other, and perhaps even better, solutions that nature for a number of reasons may not have adopted, but that can be created in the laboratory. Directed evolution allows us to explore the interconversion of function within a structural framework and therefore address the question of how easily functions that are primarily associated with one scaffold can be grafted into another.

Evolution reflects the demands of survival and reproduction. Any one function is only as good as it has to be; function must also accommodate biological needs (e.g. regulation) that may hamper individual potential. It is clearly possible to engineer existing enzymes to improve specific functions, especially when they are removed from the context of biological compatibility. Evolution in the laboratory allows us to explore function free from biological constraints and to access and optimize functions or combinations of functions that are not biologically relevant. Such evolutionary design experiments will provide new insight into structure-function relationships as they also develop useful catalysts.

10.2
Heme Proteins

Heme consists of a tetrapyrrole ring system complexed with iron. The four pyrrole rings are linked by methene bridges, resulting in a highly conjugated and planar porphyrin. Shown in Fig. 10.1 complexed with iron, protoporphyrin IX is one of the most common porphyrins and is the prosthetic group found in all heme enzymes discussed in this chapter. The heme iron is octahedrally coordinated by six heteroatoms. The four equatorial ligands are the porphyrin nitrogens; the remaining two axial ligands lie above and below the plane of the heme. In heme proteins that do not directly bind oxygen or hydrogen peroxide both axial coordination sites are occupied by heteroatoms from nucleophilic amino acid residues. In cytochrome *c*, for example, these atoms are a histidine imidazole nitrogen and a methionine sulfur. In heme enzymes that bind oxygen or hydrogen peroxide, only one axial site is occupied by a basic amino acid heteroatom. This 'proximal' ligand is conserved throughout each enzyme family. All heme peroxidases except for chloroperoxidase (CPO) have a histidine nitrogen as the proximal ligand. In CPO and in all P450s the proximal ligand is a cysteinate sulfur. In catalase, it is a tyrosine oxygen. The sixth coordination site, distal to the heme iron,

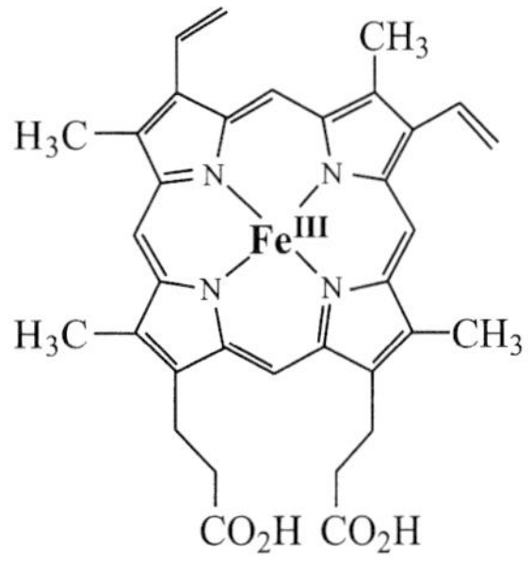

Fig. 10.1. Protoporphyrin IX complexed with iron.

is occupied by an oxygen atom from either O_2, H_2O or peroxide. This distal position is the catalytic center, and its coordination depends on the enzyme's status in the catalytic cycle.

Heme proteins collectively have three main functions: oxygen transport, electron transfer and catalysis of redox reactions using either peroxides or oxygen plus externally supplied electrons [2]. In this chapter we will limit our discussion to "b-type" enzymes, which contain a protoporphyrin IX [3]. We further focus on proteins comprising a single heme and a single polypeptide chain. This group of hemoproteins contains many well-studied enzymes whose (relative) simplicity and varying oxidative activities make them attractive catalysts. Available crystal structures of important b-type enzymes help us understand how the different heme-protein, heme-substrate, and protein-substrate interactions modulate active-site chemistry. Additionally, protein engineering studies of these enzymes have provided useful insights into function and the catalytic potential we might be able to achieve by directed evolution.

Most of the hemoproteins of one family also intrinsically possess functionality primarily belonging to that of a different family. For example, peroxidases have activities typically associated with the cytochromes P450, and vice versa. Enzymes from both families show catalase activity, and catalase has slight peroxidase activity. Myoglobin (Mb), whose primary function is to transport oxygen, is also capable of oxidizing substrates. Within one family there is still enormous diversity with regard to the primary reaction catalyzed, substrate specificity, catalytic rate, etc. It is the protein that controls factors such as the redox potential and stability of the oxidative iron species, the accessibility of substrates to the active site, and overall enzyme stability. With the same heme structure acting as the catalytic center for all these enzymes, it is interesting to try to understand how the protein serves to modulate heme catalysis.

10.3
Cytochromes P450

10.3.1
Introduction

A large volume of literature attests to the versatility of the cytochrome P450 monoox-ygenases with regard to substrate specificity, regio- and stereoselectivity and the breadth of reactions catalyzed. This family of monooxygenases exhibits a characteristic UV absorption maximum at 450 nm upon binding of carbon monoxide by the reduced enzymes. Entire books have been dedicated to P450 enzymes [4 – 6]. Many papers review what has been learned about these enzymes and their potential as catalysts [7 – 13]. Martinez and Stewart discuss P450 enzymes as catalysts for asymmetric olefin epoxidation [14], and Mansuy has reviewed the many reactions they catalyze [15]. Miles *et al.* [16] have reviewed protein engineering studies on P450s.

P450s are found in almost all organisms and primarily catalyze insertion of oxygen into carbon-hydrogen bonds. The general reaction equation can be written:

$$\text{RH} + \text{NAD(P)H} + \text{H}^+ + \text{O}_2 \xrightarrow{\text{P450}} \text{ROH} + \text{NAD(P)}^+ + \text{H}_2\text{O}.$$

P450s catalyze a number of other oxidative reactions, on substrates that range from alkanes to complex endogenous molecules such as steroids and fatty acids. Table 10.1 lists many of the oxidative reactions catalyzed by P450s. These enzymes are also known to catalyze non-oxidative dehydrase, reductase and isomerase reactions [15].

The P450s require a cofactor (NADH or NADPH) as a source of reducing equivalents to reduce oxygen and use a protein electron transport system. Depending on the P450, this system is composed of either two proteins (usually a reductase and a ferredoxin protein) or a single P450 reductase flavoprotein. The most widely studied of all P450 enzymes is the camphor hydroxylase (P450$_{cam}$) from the soil bacterium *Pseudomonas putida*, which uses two partner proteins (putidaredoxin reductase and putidaredoxin) for electron transfer from NADH to the heme enzyme [17]. Currently atomic structures for nine different P450s have been determined [18 – 26]. While their sequence identities are quite low (typically ~20 % on the amino acid level), all have similar structures.

Tab. 10.1. Cytochrome P450-catalyzed reactions [15].

Reaction	Substrate	Product
C-H Bond hydroxylation	$\overset{\mid}{\underset{\mid}{C}}-H$	$\overset{\mid}{\underset{\mid}{C}}-OH$
Epoxidation	$C=C$	epoxide
	arene (R)	arene oxide (R) → phenol (OH, R)
Oxidative N-dealkylation	$R_1R_2N-CH_2R_3$	$R_1R_2N-H + R_3CHO$
Oxidative O-dealkylation	$R-O-CH_2R'$	$ROH + R'CHO$
N-Hydroxylation	$\underset{R}{\overset{Ar}{N}}-H$	$\underset{R}{\overset{Ar}{N}}-OH$
Sulfoxidation	$\underset{R_2}{\overset{R_1}{S}}$	$\underset{R_2}{\overset{R_1}{S}}=O$
Peroxidase-type oxidation	phenol (OH, R)	diphenyl ether (R, OH) **OR** biphenyl (R, HO, OH)
NO Synthase-type oxidation (C=N bond cleavage)	$\underset{R_2}{\overset{R_1}{C}}=N-OH$	$\underset{R_2}{\overset{R_1}{C}}=O + NO \ (or \ N_xO_y)$
Oxidative deformylation	$\underset{R_2}{\overset{R_1}{H}}\underset{R_4}{\overset{R_3}{CHO}}$	$\underset{R_2}{\overset{R_1}{}}=\underset{R_4}{\overset{R_3}{}} + HCOOH$
Dehydrogenation	$H\overset{\mid}{\underset{\mid}{C}}-\overset{\mid}{\underset{\mid}{C}}H$	$C=C$

10.3.1
Mechanism

10.3.2.1 **The catalytic cycle**

Catalytic activity involves the generation of one or more short-lived, highly oxidizing intermediates at the heme iron and near the bound substrate. Figure 10.2 shows the catalytic reaction cycle. In the substrate-free, oxidized (ferric iron) state of the enzyme (**1**) the heme iron is in the low-spin six-coordinate form [27], with water as the sixth ligand. Binding of substrate results in dehydration of the active site so that the heme iron becomes five-coordinate (intermediate **2**) [10]. Additionally, the heme iron changes to predominantly high-spin and its reduction potential increases, thereby priming the enzyme for substrate turnover by allowing electron transfer to occur [10, 17, 28, 29]. It is believed that water exclusion from the active site is important not only for the change in coordination and reduction potential, but also to improve the coupling efficiency of electron transfer (see Section 10.3.2.2).

Oxygen binds to ferrous P450 after the first electron transfer, resulting in an unstable ferrous-oxy species (intermediate **4**) which then accepts the second electron. The electron transfer steps are rate-limiting [10]. The mechanism following the formation of the peroxo-iron species **5** involves incorporation of two protons and cleavage of the O–O bond, resulting in water formation. The two protons are pumped into the active site to the distal peroxo oxygen, with the initial formation of a hydroperoxo-iron intermediate **6**. The two electrons required for this step come from the heme, resulting in heme oxidation to an oxy-ferryl, or iron-oxo species **7**. Recent studies indicate that multiple intermediates effect substrate turnover, as described below [30, 32]. While the electronic structure(s) of the species performing oxidation remains a subject of debate, the iron-oxo species **7** is the most widely accepted active intermediate in P450-catalyzed reactions, and is normally depicted as $(Por) \cdot + Fe^{IV} = O$ by analogy with heme peroxidase *Compound I* (see Section 10.4.2).

Studies of site-directed mutants and kinetic solvent isotope effects have led to a proposed pathway for proton delivery to the heme [10, 33 – 37]. A highly conserved threonine residue near the heme seems to play a critical role in relaying protons from the solvent to the heme. It has been proposed that intermediates **5**, **6**, and **7** are all active oxygenating species with varying electrophilic or nucleophilic properties, contributing to the versatility of P450 enzymes (see Section 10.7.1.2) [30 – 32]. As shown in Fig. 10.3, each intermediate, peroxo-iron, hydroperoxo-iron, or iron-oxo, is believed to catalyze a different reaction. Accessibility of protons to the heme, as controlled by the surrounding protein structure, is apparently important in governing the oxidative activity of the enzyme.

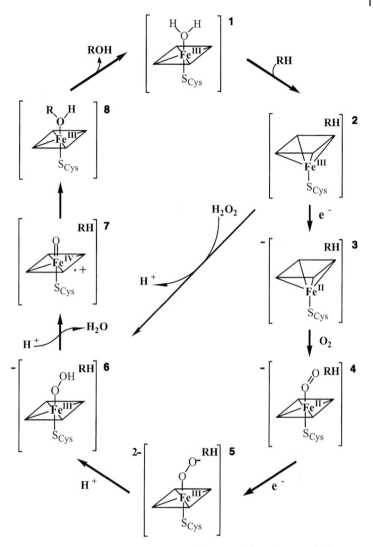

Fig. 10.2. Catalytic cycle of P450 including the peroxide shunt pathway. RH is substrate, and ROH is product. The porphyrin molecule is represented as a parallelogram. The overall charge on the structures is shown to the left of each bracket. Intermediates **1**, **2**, **7**, and **8** are neutral. Refer to text for a full description.

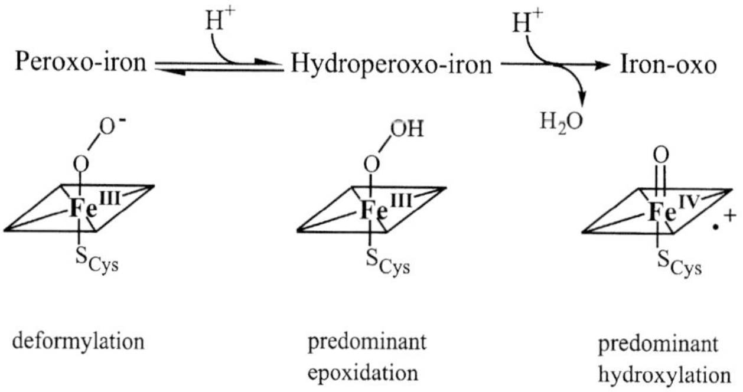

Fig. 10.3. Proposed reactive intermediates in the P450 cycle and the primary reactions they catalyze (adapted from [31]). Intermediates from left to right correspond to intermediates **5, 6** and **7**, respectively, in Fig. 10.2. Accessibility of protons to the heme plays an important role in P450 activity.

10.3.2.2 Uncoupling

Coupling efficiency refers to the percentage of reducing equivalents from NAD(P)H that are utilized for the oxidation of substrate. The oxidation of camphor by P450$_{cam}$ occurs with 100 % coupling efficiency: all electrons transferred from NADH are used in the stereospecific formation of 5-*exo*-hydroxycamphor. In contrast, the coupling efficiency of P450$_{cam}$ in styrene epoxidation is low, between 2 % [38] and 7 % [39]. In addition to the catalytic cycle shown in Fig. 10.2, there are several pathways in which reducing equivalents transferred to the heme can be consumed and transferred away from the substrate. Autooxidation of the ferrous-oxy intermediate **4**, H_2O_2 formation through the decomposition of intermediate **6** and H_2O formation by two-electron reduction plus diprotonation of intermediate **7** are all uncoupling mechanisms. Uncoupling by peroxide formation competes with substrate oxidation if the O–O bond cleavage step is inhibited or if peroxide dissociation is promoted [10]. Access of water molecules to the heme during catalysis increases polarity and could promote charge separation at the iron, thereby promoting release of hydrogen peroxide anion [17, 40]. This occurs when substrates are unable to exclude water from the active site, and would result in increased uncoupling by peroxide dissociation. Reduction of intermediate **7** could compete with substrate oxidation if the substrate is positioned too far from the ferryl oxygen.

10.3.2.3 Peroxide shunt pathway

P450s are capable of utilizing an oxygen atom from peroxide to catalyze oxygen insertion without electron transport proteins or the NAD(P)H cofactor, through the "peroxide shunt" pathway:

$$\text{RH} + \text{R'OOH} \xrightarrow{\text{P450}} \text{ROH} + \text{R'OH}.$$

As illustrated in Fig. 10.2, the shunt pathway bypasses a large portion of the enzyme's natural catalytic cycle, including the rate-limiting first electron transfer step (a rate constant of $\sim 15\,\text{s}^{-1}$ is reported for electron transfer in $P450_{cam}$ [41]). There have been many studies of the P450 peroxide shunt pathway [42 – 56]. Various peroxides and other oxidants (e.g. iodosobenzene, peracids and sodium periodate) will support the reaction, depending on the enzyme. This peroxygenase activity potentially adds to the versatility of cytochrome P450 catalysis.

10.4
Peroxidases

10.4.1
Introduction

Peroxidases are ubiquitous, and many are b-type heme proteins. Several good reviews summarize years of peroxidase research and describe peroxidase applications [57 – 61]. Some of the reactions catalyzed by peroxidases are listed in Tab. 10.2 and include oxidation of aromatic and heteroatom compounds, epoxidation, enantioselective reduction of racemic hydroperoxides, free radical oligomerizations and polymerizations of electron-rich aromatics, and the oxidative degradation of lignin [58, 60].

10.4.2
Mechanism

10.4.2.1 *Compound I* formation
Most of what is understood about heme peroxidases comes from studies of the plant enzyme, horseradish peroxidase (HRP). The characteristic peroxidase activity is one-electron oxidation coupled with reduction of H_2O_2 to H_2O. Figure 10.4 shows the peroxidase catalytic cycle. The native state of the heme is the same as that for the P450s, except the proximal ligand is a histidine nitrogen rather than a cysteine sulfur. In the first step of the reaction, H_2O_2 replaces H_2O at the axial position of heme Fe^{III}. The bound H_2O_2 is split heterolytically to form an iron-oxo derivative known as *Compound I*, which is formally two oxidation equivalents higher than the Fe^{III} resting state. *Compound I* is well-characterized and contains $Fe^{IV} = O$ and a π cation radical [62].

The mechanism of O–O bond cleavage is influenced by the protein environment around the heme, and is different for peroxidases and P450s. Cleavage of the O–O

Tab. 10.2. Peroxidase-catalyzed reactions (adapted from [58] and [60]).

Reaction (adapted from [58])		Typical substrates
Electron transfer	$2\,AH + ROOH \longrightarrow A\text{-}A + ROH + H_2O$	H_2O_2, ROOH, aromatic amines
Sulfoxidation	$R_1{-}S{-}R_2 + ROOH \longrightarrow R_1{-}S(=O){-}R_2 + ROH$	Thioanisole, H_2O_2, ROOH
Epoxidation	$R_1{-}CH{=}CH{-}R_2 + H_2O_2 \longrightarrow R_1\text{(epoxide)}R_2 + H_2O$	Alkenes, H_2O_2
Demethylation	$ROOH + R_1R_2N{-}CH_3 \longrightarrow R_1R_2N{-}H + ROH + HCHO$	N,N-Dimethylaniline, ROOH
Dehydrogenation	$2 \; (HO)(COOH)C{=}C(HOOC)(OH) \xrightarrow{O_2} 2 \; (O)(COOH)C{-}C(HOOC)(O) + 2\,H_2O$	Dihydroxyfumaric acid
α-Oxidation	$R_2{-}C(R_1)(H){-}CHO \xrightarrow{O_2} R_1{-}CH(OH){-}C(R_2)(O{-}O)H \longrightarrow R_1{-}C(=O){-}R_2 + HCOOH$	Aldehydes

CPO-catalyzed oxidations [60]

bond in peroxidases to form *Compound I* is promoted by a "push-pull" mechanism [63] (Fig. 10.5). Peroxidases have a catalytically critical histidine distal to the heme, along with an important cationic arginine [64]. The histidine pulls the proton from the heme-bound hydroperoxide, making the hydroperoxide a better nucleophile, while the arginine pulls on the O–O bond; together they work to cleave the O–O bond. In addition, a hydrogen-bonded carboxylate group near the proximal histidine increases electron density to the imidazole, creating an electron "push" that facilitates O–O cleavage. In hemoglobin and myoglobin, where O–O bond cleavage is not part of the primary function, there is no hydrogen-bonded carboxylate group near the proximal histidine. References [65 – 69] describe mutagenesis studies that have elucidated the roles of the

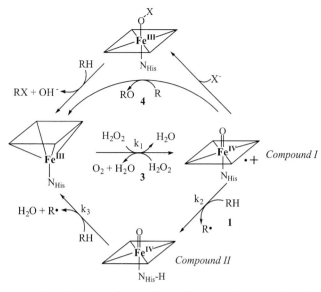

Fig. 10.4. Peroxidase catalytic cycle. Four pathways are shown for the return of *Compound I* to the resting state: **(1)** oxidative dehydrogenation, where RH is the substrate; **(2)** oxidative halogenation, where X^- is a halogen ion and RH is the substrate; **(3)** peroxide disproportionation, and **(4)** oxygen transfer, where R is the substrate.

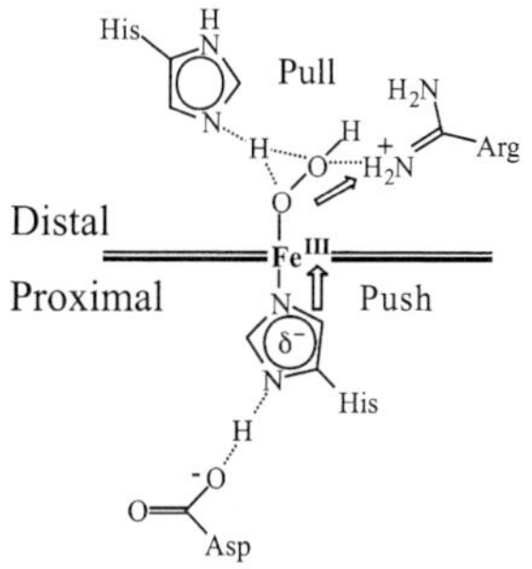

Fig. 10.5. "Push-pull" mechanism for *Compound I* formation in peroxidases (adapted from [12] and [63]). Distal residues (Arg and His) work together to cleave the O-O bond, while proximal residues (His and Asp) assist by supplying electron density.

critical distal residues in peroxidase and myoglobin. P450s generally have large substrate binding pockets, with no residues close to the heme on the distal side and therefore no "pull" effect. There, O–O cleavage is facilitated by the proton delivery system and by a very strong "push" by the cysteinate proximal ligand [2, 12, 70, 71].

Reduction of peroxidase *Compound I* back to the resting state can occur by one of four pathways, depending on the reaction catalyzed: oxidative dehydrogenation (pathway **(1)** in Fig. 10.4), oxidative halogenation **(2)**, peroxide disproportionation **(3)** or oxygen transfer **(4)** [60].

10.4.2.2 Oxidative dehydrogenation

Oxidative dehydrogenation is a primary biological function of peroxidases. *Compound I* is reduced in two one-electron transfers, as shown in Fig. 10.4. The overall reaction is

$$2 \text{ RH} + \text{H}_2\text{O}_2 \rightarrow 2 \text{ R} \cdot + 2 \text{ H}_2\text{O}.$$

The first reduced intermediate, *Compound II*, is formed when the π cation radical is reduced and a proton is transferred to the distal base (His). *Compound II* is then reduced to the Fe^{III} resting state with the simultaneous formation of water [70]. This second electron transfer step is one to two orders of magnitude slower than *Compound I* formation and is usually rate-limiting [72]. For HRP, the rate constant for *Compound I* formation (k_1) is $2.0 \times 10^7 \text{ M}^{-1}\text{s}^{-1}$ [73], while the rate-limiting step in phenol oxidation by HRP has a rate constant $k_3 \sim 3.0 \times 10^5 \text{ M}^{-1}\text{s}^{-1}$ [74, 75].

HRP catalyzes the oxidative dehydrogenation of a wide range of electron-rich aromatic compounds. The result of this radical formation pathway is dimerization and subsequent oligomerization of the substrates [76 – 78]. Peroxidases have been used to catalyze polymerizations of phenols (e.g. *p*-cresol and guaiacol) and aromatic amines (e.g. aniline, and *o*-phenyldiamine) [79, 80]. *N*- and *O*-dealkylations are also useful electron transfer reactions catalyzed by peroxidases. These reactions are used in industrial wastewater treatment and may have synthetic applications [81].

10.4.2.3 Oxidative halogenation

Heme haloperoxidases can also use peroxide and halide ions to halogenate an activated (benzylic/allylic) carbon. The halide is first oxidized to an active halogenating intermediate (Fig. 10.4, pathway (**2**)). The substrate is halogenated in the next step. The overall reaction is

$$\text{RH} + \text{H}_2\text{O}_2 + \text{H}^+ + \text{X}^- \rightarrow \text{RX} + 2 \text{ H}_2\text{O}.$$

10.4.2.4 Peroxide disproportionation

The H_2O_2 disproportionation reaction decomposes H_2O_2 into water and oxygen, as shown in Fig. 10.4 (pathway (**3**)). The overall reaction is

$$2 \text{ H}_2\text{O}_2 \rightarrow 2 \text{ H}_2\text{O} + \text{O}_2.$$

Catalase, a heme-containing enzyme with tyrosine as the proximal heme ligand, decomposes hydroperoxides and peracids by this reaction. Catalase is one of the most efficient enzymes known, with maximum turnover numbers on the order of 10^7 s^{-1}. In the absence of an electron donor, peroxidases and particularly CPO exhibit catalase activity [58, 82], as do some P450s [83].

10.4.2.5 Oxygen transfer

The oxygen transfer reactions catalyzed by peroxidases are very interesting for synthetic applications. The overall reaction for this peroxygenase activity is

$$R + H_2O_2 \rightarrow RO + H_2O.$$

Similar to the P450s, these reactions are often stereospecific and include hetero-atom oxidations, epoxidation, and oxidation of C-H bonds in allylic/benzylic compounds, alcohols, and indole [59].

10.5
Comparison of P450s and Peroxidases

P450s and peroxidases share key elements of their mechanisms. The proximal ligands and distal and proximal protein environments influence the mechanism of O–O bond cleavage, the stability of the intermediates and the accessibility of substrates to the heme. The fact that P450s are capable of oxidizing aliphatic hydrocarbons and olefins, in addition to the activated C-H bonds oxidized by peroxidases, reflects the stronger oxidative potential of the P450 iron-oxo species [84, 85]. In P450s the substrate binding pocket is relatively large and lies on the distal side of the heme, where the ferryl oxygen can be transferred to the substrate. Nature has tuned this binding pocket in the different P450s to make catalysts with a range of substrate specificities and reaction selectivities. Peroxidases, on the other hand, bind substrates near the heme edge, where electrons are transferred from the substrate to the heme center [59, 86]. This ability to oxidize substrates at the heme edge accounts for the broad substrate specificity of peroxidases [87]. While peroxidases have high one-electron oxidation activities, their oxygen transfer (peroxygenase) activities are low, reflecting limited substrate access to the heme iron (with the exception of CPO). Note that some P450s catalyze the coupling of phenols that is typical of peroxidases (shown in Tab. 10.1) [15]:

P450

$$2\ ArH + NAD(P)H + H^+ + O_2 \rightarrow Ar\text{-}Ar + NAD(P)^+ + 2\ H_2O.$$

The P450 shunt pathway and peroxygenase activity of peroxidases share identical overall reaction equations. P450s generally have high K_m values for H_2O_2; values of 15 mM

[42] and 250 mM [52] have been reported, while for HRP the K_m for H_2O_2 is $22\,\mu M$ [88]. Reaction of H_2O_2 with the P450 heme is several orders of magnitude slower than the equivalent reaction with peroxidase. Rapid reaction of H_2O_2 with the heme Fe^{III} requires a base to assist in binding of the peroxy anion [2]. The distal histidine serves this function in peroxidases. Replacing the distal histidine with a leucine in cytochrome c peroxidase (CCP) reduced the rate of reaction with peroxide by five orders of magnitude [89], to a rate similar to that of a P450.

An enzyme worthy of mention is the fatty acid α-hydroxylase (FAAH) from *Sphingomonas paucimobilis* [90]. This peroxygenase naturally utilizes H_2O_2 and sequence analysis indicates that it is a P450 (P450$_{SP\alpha}$). The environment around its active site is different from other P450s, presumably to accommodate H_2O_2 binding, resulting in a low K_m for H_2O_2 ($72\,\mu M$) [91, 92]. P450$_{SP\alpha}$ catalyzes H_2O_2-driven hydroxylation of fatty acids (carbon chain lengths between 11 and 18) with high stereoselectivity ($\sim$98 %) and turnover rates greater than $1000\,min^{-1}$ [93].

The reactions catalyzed by the cytochromes P450 are important in synthetic chemistry. The cofactor regeneration requirements of the P450s, however, severely limit their use outside of whole cells. The shunt pathway is a possible alternative to using cofactors, but this pathway is slow and the required peroxide levels are destructive to the enzyme. Directed evolution could improve the shunt pathway by increasing the stability of P450 in the presence of peroxides, increasing catalytic rates, and lowering the required peroxide concentration (reduce K_m) [94, 95]. Methods that use electrodes to drive P450 reactions are also being developed [96 – 99].

P450s are generally less stable than peroxidases. Recently a naturally thermostable P450 (stable up to 85 °C) was identified and characterized [25, 100 –103]. Both P450s and peroxidases are inactivated during catalysis, via heme alkylation by terminal olefins (see Section 10.8.3) and oxidative damage by peroxides. Eukaryotic P450s are associated with cell membranes and are therefore insoluble and difficult to use outside the cell. We believe that many of these limitations can be addressed by directed evolution.

10.6
Chloroperoxidase

CPO from the fungus *Caldariomyces fumago* catalyzes the oxidation of hydrochloric to hypochlorous acid with a turnover rate of $\sim$$10^3\,sec^{-1}$ [57]. While functionally categorized as a haloperoxidase, CPO possesses catalytic traits characteristic of peroxidases, P450s and catalase. The CPO proximal ligand is a cysteinate sulfur as in P450s. In the distal pocket, the catalytic base used for O–O cleavage is glutamic acid rather than the typical peroxidase histidine. Poulos and co-workers compared its crystal structure to those of peroxidases and P450s [104]. The CPO distal region is more hydrophobic than in other peroxidases, which allows it to bind substrates and promote P450-type reactions, although the presence of polar residues and restricted access to the distal face make the distal region more peroxidase-like than P450-like. The tertiary structure of CPO, however, resembles neither the P450s nor peroxidases [104], implying a separate evolutionary origin.

Several papers review the various reactions catalyzed by CPO [57–60, 105]. Some of these are shown in Tab. 10.2. CPO catalyzes the sulfoxidation of thioanisole at a turnover rate of $200 \, s^{-1}$ with >98 % enantiomeric excess [106]. This is orders of magnitude higher than typical peroxidases. CPO is also known to catalyze sulfoxidations on aliphatic sulfides at a similar rate [107]. CPO is the only peroxidase known to selectively hydroxylate hydrocarbons [60] and selectively catalyze the oxidation of primary alcohols to aldehydes [108, 109]. Chiral epoxides are useful in chemical synthesis because they can undergo stereospecific ring-opening to form bifunctional compounds. CPO catalyzes chiral epoxidations with high yields and high stereoselectivity [105]. Styrene epoxidation proceeds at $4.8 \, s^{-1}$ with CPO [110], versus $\sim 4 \, min^{-1}$ with P450$_{cam}$ [39]. Other CPO-catalyzed reactions include alkyne hydroxylation [111, 112] and heteroatom dealkylation [113, 114].

The rate of reaction of CPO with hydroperoxides is significantly lower than for other peroxidases, and closer to that for P450s [87]. Thus CPO is a rather slow one-electron oxidation catalyst, but it is the preferred heme enzyme for many sulfoxidation, epoxidation and hydroxylation reactions. CPO substrates, in contrast to P450 substrates, must have electron-rich groups. The major limitation to using CPO is its instability to peroxide. Whereas its half-life at pH 5 is 40 h without H_2O_2, the half-life drops to 0.5 h in 1 mM H_2O_2 [115]. To help overcome this, methods have been developed to maintain low levels of H_2O_2, either by *in situ* generation or controlled addition [115–119].

The P450 scaffold has evolved to serve as a ubiquitous monooxygenase. Nature manipulated this structure to create an immense library of P450s with finely-tuned activities and specificities. The CPO scaffold has not become a ubiquitous oxidation catalyst, but it supports high peroxygenase activity and may be superior to P450s for chemical transformations of certain substrates, particularly when it has been evolved in the laboratory for optimal performance in these applications.

10.7
Mutagenesis Studies

Mutagenesis studies demonstrate the "designability" of heme enzymes. For P450s, mutations that change substrate specificity, reaction specificity, activity and coupling efficiency have been reported. These studies primarily focused on the few P450s for which crystal structures are available, although they can be extended to other P450s by sequence alignment and homology modeling [16]. Many studies carried out on other heme enzymes have identified how amino acid substitutions can alter substrate specificity or explore the interconversion of function among various hemoproteins, based on structure comparisons. In this section we describe results that demonstrate the importance of protein structure in controlling heme activity and the ability to engineer function.

10.7.1
P450s

10.7.1.1 P450$_{cam}$

P450s have finely tuned substrate binding pockets and active sites. Loida and Sligar investigated how mutations of active site residues at varying distances from the heme affect turnover and the partitioning between uncoupling pathways in the oxidation of ethylbenzene by P450$_{cam}$ [40, 120]. Their results show that mutations can alter the coupling efficiency. As predicted, increasing the size of side chains close to the heme blocks substrate access and practically eliminates product formation. With substrates that do not fit tightly in the binding pocket, mutations further away that increased bulk help to force the substrate closer to the heme and increase product generation. The effects of single mutations were shown to be additive.

A tyrosine is positioned directly over the heme on the distal side in the P450$_{cam}$ active site. This Tyr96 residue interacts with the natural substrate camphor through a hydrogen bond to position it for regio- and stereospecific hydroxylation [18]. Various Tyr96 substitutions create active sites of varying volume and hydrophobicity that improve the oxidation of phenyl derivatives [121, 122]. For example, the oxidation of diphenylmethane, diphenylamine and 1,1-diphenylethylene by Tyr96Ala and Tyr96Gly mutants was regiospecific, with hydroxylation at the same *para* position for all three substrates, while wild-type P450$_{cam}$ showed no activity towards these substrates. Styrene oxide is produced from styrene by wild-type P450$_{cam}$ at $\sim$4 min^{-1}. Nickerson *et al.* [39] made Tyr96Ala and Tyr96Phe mutations to increase hydrophobicity and improve binding of styrene. Styrene oxidation was improved 25-fold by the Phe mutation and 9-fold by Tyr96Ala. In addition, the coupling efficiency was increased from 7 % to 32 % for the Tyr96Phe mutant, probably due to enhanced exclusion of water.

10.7.1.2 Eukaryotic P450s

Although the exact mechanism for P450-catalyzed epoxidation is not known, recent studies by Coon and co-workers indicate that at least one heme intermediate is active towards epoxidizing carbon-carbon double bonds [31]. To reduce proton delivery to the active site, they replaced the conserved threonine believed to facilitate proton delivery with alanine in two P450 enzymes (P450 2B4 and P450 2E1, both from rabbit liver). They then compared the rates of product formation and the ratios of epoxidized to hydroxylated products for the wild-type enzymes and the Thr→Ala mutants for several substrates. Figure 10.3 shows the proposed heme species and the oxidations catalyzed [31]. Coon proposes that, in addition to the iron-oxo species traditionally believed to be responsible for P450-catalyzed oxidations, the hydroperoxo-iron species could also epoxidize olefins. The experiments support this: for *cis*-butene, the ratio of epoxidized to hydroxylated product increased fivefold with the Thr→Ala mutant. These results once again demonstrate how the protein structure tunes catalytic activity. It is likely that P450-catalyzed epoxidation could be greatly enhanced by directed evolution.

10.7.2
HRP

The distal residues in peroxidases impair their ability to catalyze oxygen insertion reactions. In HRP, His42 is the conserved distal residue that aids in *Compound I* formation. Site-directed mutagenesis studies have shown that this histidine can be relocated to a nearby, less obstructive position and still promote *Compound I* formation [68]. The mutation His42Ala alone reduces the rate constant of *Compound I* formation (k_1 in Fig. 10.2) from $\sim 10^7\,M^{-1}s^{-1}$ to $\sim 10^1\,M^{-1}s^{-1}$, while a second mutation Phe41His brings this rate constant back up to $3 \times 10^4\,M^{-1}s^{-1}$. With this double mutant, the k_{cat} for styrene oxidation increased from $10^{-6}\,s^{-1}$ (for wild-type HRP) to $2.4 \times 10^{-2}\,s^{-1}$ (a 24,000-fold improvement), and thioanisole sulfoxidation increased from $0.05\,s^{-1}$ to $5.3\,s^{-1}$. The peroxidase activity of this double mutant, however, was drastically reduced with respect to wild-type. In later work Savenkova *et al.* [67] used the mutation Arg38His to recover the role of the distal histidine from His42Val. For the Arg38His, His42Val double mutant, the rate of thianisole sulfoxidation was increased 680-fold, with peroxidase activity still greatly sacrificed.

10.7.3
CPO

While CPO is highly attractive for synthetic applications [57], protein engineering has been hampered by the inability to express the fungal enzyme in bacteria or yeast. Hager and coworkers, however, have used gene replacement technology to allow functional expression and production of mutants in the enzyme's natural host, *Caldariomyces fumago* [110]. In an effort to understand the importance of the proximal thiolate ligand (Cys29), Yi *et al.* [110] generated the Cys29His mutant and compared the rates of halogenation, peroxidation, epoxidation (oxygen insertion) and catalase activity for the mutant and wild-type. Surprisingly, all four rates decreased only slightly (catalase activity diminished the most – but by only 40 %). Recently Consea *et al.* achieved functional expression of CPO in a different filamentous fungal host, *Aspergillus niger* [123].

10.7.4
Myoglobin (Mb)

The natural function of myoglobin is to reversibly bind dioxygen. Like peroxidases, myoglobin has proximal and distal histidine residues, although the distal histidine in myoglobin lies closer to the heme and helps to stabilize coordination of dioxygen or water. The active site of myoglobin is relatively hydrophobic, while in peroxidases it is relatively hydrophilic. Although the natural function does not require cleavage of the O–O bond, myoglobin is capable of peroxygenase catalysis, including olefin epoxidation and thioether sulfoxidation, although at much reduced rates compared to peroxidases [124 – 127]. The peroxygenase activity of myoglobin is well studied, including many site-directed mutagenesis studies [65, 128]. In one series of experiments with sperm whale Mb, turnover rates and stereospecificities for styrene epoxidation and thioanisole sulfoxidation were increased [129 – 132]. These studies were similar to those described for HRP in that the distal histidine (His64) was relocated to nearby positions. This destabilized coordination with O_2 or H_2O yet maintained critical hydrogen bonds with the ferryl oxygen, thereby allowing for increased peroxygenase activity. Attempting to engineer the distal region to look more like that of a peroxidase, Watanabe and co-workers created Mb mutants Leu29His/His64Leu and Phe34His/His64Leu [128]. The turnover rate for epoxidation of styrene was increased 300-fold (to 4.5 min^{-1}) for mutant Phe34His/His64Leu, while enantiomeric excess was improved from 3 % to 99 % for epoxidation of *cis-β*-methylstyrene by mutant Leu29His/His64Leu. Oxidation of thioanisole was increased 188-fold in rate (for Phe34His/His64Leu) and fourfold in enantiomeric excess (for Leu29His/His64Leu). More recently, the same group reported a single mutation, Phe43Trp, that increases guaiacol oxidation (one-electron oxidation activity) fourfold over wild-type [133]. Addition of the His64Leu mutation dramatically decreased activity.

10.8
Directed Evolution of Heme Enzymes

Heme enzymes are prime targets for biocatalyst engineering. Site-directed mutagenesis is helpful for testing mechanistic hypotheses on enzymes for which there are atomic structures, but are of limited utility when a better catalyst is the goal. Here, directed evolution methods will be particularly valuable. Few directed evolution studies have been performed on heme enzymes, and technical hurdles of heterologous enzyme expression and screening technology remain. However, recent developments in screening methods, expression and new methods of library generation are now making it possible to rapidly isolate specialized variants.

10.8.1
P450s

Cofactor requirements are a serious impediment to industrial use of biocatalysts such as cytochrome P450. In an effort to bypass the external cofactor and electron transfer proteins altogether, Joo et al. [95] evolved P450$_{cam}$ to better utilize hydrogen peroxide for hydroxylation via the shunt pathway. P450$_{cam}$ variants created by error-prone PCR followed by DNA-shuffling showed up to a 20-fold improvement in naphthalene hydroxylation utilizing H_2O_2. Additionally, mutants with different regiospecificities were identified from the different colored fluorescent products of HRP-catalyzed polymerization of the hydroxylated naphthalenes [95]. A few rounds of mutagenesis significantly improved the efficiency of the peroxide shunt pathway for this enzyme.

Key to the success of directed evolution are rapid, functional screens for identifying improved enzymes. Various methods have been developed for rapid screening of hydroxylation activity. The screen used by Joo et al. employed fluorescence digital imaging in a solid-phase assay, with which thousands of colonies can be assayed on a single plate [134]. Recently Joern et al. described an improved solid-phase assay suitable for monitoring P450- and other oxygenase-catalyzed hydroxylations of aromatic substrates using the Gibbs reagent [135]. Schwaneberg and coworkers have described a surrogate substrate assay that can be used to detect hydroxylation of aliphatic substrates in whole cells [136]. In this assay, ω-p-nitrophenoxycarboxylic acids (pNCAs) serve as surrogates for fatty acids [136], and hydroxylation produces the chromophore p-nitrophenolate. In order to screen for hydroxylation of alkanes, Farinas et al. [137] used the octane analogue ω-p-nitrophenoxyoctane (8-pnpane).

These screens have been used to direct the evolution of cytochrome P450 BM-3, a soluble enzyme from *Bacillus megaterium* that contains its reductase and hydroxylase domains on a single polypeptide chain. P450 BM-3 primarily catalyzes the hydroxylation of fatty acids (~12 to 18 carbons long) at the ω-1, ω-2, and ω-3 positions, but also hydroxylates long-chain amides and alcohols and epoxidizes long chain unsaturated fatty acids [138 – 140]. Schwaneberg et al. [141] showed that P450 BM-3 hydroxylates pNCAs, with a minimum required substrate chain length of 10 – 11 carbons.

The P450 BM-3 heme domain crystal structure shows a funnel-shaped shaft lined with hydrophobic residues for substrate binding [19, 142, 143]. Li et al. [144] used saturation mutagenesis at sites believed to be important for substrate binding in order to improve activity on shorter chain-length substrates. Mutants were screened for activity towards pNCA substrates, and a P450 BM-3 variant with five mutations that efficiently hydroxylates 8-pNCA was obtained. Li et al. [145] also discovered that some mutants formed a blue pigment, which they identified as indigo from the hydroxylation of indole. Indole hydroxylation could not be detected with wild-type P450 BM-3, but a triple mutant with a k_{cat} of 2.7 s^{-1} and a k_{cat}/K_m of 1365 M^{-1}s^{-1} was discovered. This work suggests that P450 BM-3 can be engineered to hydroxylate a wide range of substrates. This is particularly significant in light of the superior stability and activity of P450 BM-3 compared to most other P450s [146] and the fact that it is easily expressed and purified from E. coli [141, 147].

In another example in which directed evolution was used to alter the substrate range of P450 BM-3, Farinas et al. generated mutants with improved activity for alkane hydroxylation [137] by screening mutant libraries on the surrogate substrate p-nitrophenoxyoctane. The improved mutants also showed higher activity towards octane. Two generations of laboratory evolution yielded variants with up to fivefold higher octane oxidation activity compared to wild-type P450 BM-3.

Directed evolution has also increased the peroxide shunt pathway activity of the P450 BM-3 heme domain [94]. Mutant libraries were screened using 12-pNCA in the presence of 50 mM H$_2$O$_2$ and 1 mM H$_2$O$_2$ (K_m for H$_2$O$_2$ is ~18 mM). Activity in 1 mM H$_2$O$_2$ was increased more than eightfold, resulting in improved total substrate turnover. Additionally, it was found that expressing the heme domain without the P450 BM-3 reductase leads to better enzyme expression.

10.8.2
Peroxidases

Cherry and co-workers evolved a fungal peroxidase (CiP) from the ink cap mushroom *Coprinus cinereus* to resist high temperature, high pH and high concentrations of peroxide [148]. The goal was to develop a peroxidase to be used as a dye-transfer inhibitor in laundry detergents, an application requiring stability under harsh conditions. Site-directed mutagenesis improved the enzyme's stability to alkali and hydrogen peroxide. These mutations were combined with other mutations discovered by random mutagenesis (error-prone PCR) and saturation mutagenesis and screening. Combination of all the favorable mutations generated a mutant with 100-fold improved thermal stability and 2.8 times the oxidative stability of wild-type, although these improvements came at the cost of reduced overall activity. *In vivo* shuffling using a yeast homologous recombination system was then employed to improve activity. Mutants with high activity were shuffled with those showing improved thermostability but reduced activity. This generated a mutant with 174 times the thermal stability and 100 times the oxidative stability of wild-type CiP, with specific activity comparable to wild-type.

Cytochrome *c* peroxidase (CCP) catalyzes the oxidation of ferrocytochrome *c* (cyt *c*). Many site-directed mutagenesis studies have been performed on this enzyme in efforts to better understand and rationally engineer heme peroxidase function [149 – 156]. Various studies indicate that the large protein substrate cyt *c* binds to CCP in a different region than smaller substrates [155, 157]. Whereas small substrates such as phenol and aniline are believed to approach the heme from its distal side and bind at the heme edge, cyt *c* lies much further from the heme, and electron transfer seems to occur from the proximal side [158]. The low activity and selectivity exhibited by CCP compared to HRP and CPO is attributed to the limited access by small substrates to the heme [159].

Iffland *et al.* [160] used directed evolution to generate CCP mutants with novel substrate specificities. They used error-prone PCR and DNA shuffling to generate mutant libraries and screened for increased activity on guaiacol by detection of the brown-colored product tetraguaiacol. After three rounds of evolution mutants were isolated with 300-fold increased guaiacol activity and up to 1000-fold increased specificity for guaiacol relative to cyt *c*. It is interesting that all selected mutants contained the mutation Arg48His. Arg48 is a conserved distal residue that aids in *Compound I* formation and stabilization in the "push-pull" mechanism. The histidine in its place apparently reduces steric constraints for substrate access to the heme, but provides enough charge to maintain activity.

HRP is widely used as a reporter enzyme in bioanalytical chemistry and diagnostics; it also has potential applications in chemical synthesis. Protein engineering of HRP has been very limited, however, due to the lack of a convenient microbial host for functional expression and mutagenesis. Expression in *E. coli*, for example, leads primarily to formation of inactive inclusion bodies [161]. HRP contains four disulfide bridges and is ~21 % glycosylated by weight [161 – 164]. *S. cerevisiae* is known to facilitate glycosylation and disulfide formation [165], although the patterns for yeast protein modifications differ from those in fungi and plant cells [166, 167]. Morawski *et al.* used directed evolution to discover mutations that would promote secretion of functional HRP in yeast [168]. Having achieved functional expression in *S. cerevisiae*, they were able to further evolve the activity and stability of the enzyme [169]. A traditional colorimetric peroxidase assay was used for screening, in which the substrate 2,2'-azino-di-(3-ethyl)benzthiazoline-6-sulfonic acid (ABTS) is oxidized in one-electron steps to generate intensely colored radical products. With three rounds of random mutagenesis and screening, the total activity of yeast culture supernatant was increased 40-fold over wild-type. The best mutants were then expressed in *Pichia pastoris*, resulting in further improvements in total activity. Additional evolution generated HRP variants with improved thermostability and greater resistance to a series of chemical denaturants, as well as higher total activity.

10.8.3
CPO

One consequence of heme enzyme-catalyzed epoxidation of terminal olefins is the suicide inactivation of the enzyme due to *N*-alkylation of the prosthetic heme group [170 – 174]. Inactivation only occurs if the olefin is accepted as a substrate [175]. Neither the epoxide nor the substrate, but rather an active intermediate in the mechanism of substrate turnover, is responsible for heme alkylation [170], indicating that epoxidation and heme alkylation diverge at some point prior to epoxide formation. CPO epoxidizes olefins and is therefore subject to inactivation by primary olefins through this mechanism-based suicide reaction. Using *C. fumago* as their expression system, Rai *et al.* [176] created and screened random mutants of CPO to find ones that are more resistant to inactivation by allylbenzene. Halogenation activity was determined in a monochlorodimedone (MCD) absorbance assay, in which MCD is chlorinated to dichlorodimedone. Peroxidase activity was measured on the colorimetric assay substrate ABTS. Additionally, epoxidation activity was measured towards styrene. Three mutants that are resistant to suicide inactivation by allylbenzene (as well as 1-hexene and 1-heptene) were isolated after three and four rounds of mutagenesis. The fourth round mutant also exhibited improved styrene epoxidation activity. The results from this work demonstrate the power of directed evolution, with a handful of mutations, to make significant changes in a property that may at first seem a limitation inherent to the system.

10.8.4
Catalase I

Catalase I of *Bacillus stearothermophilus* shows 95 % specificity to the catalase reaction and 5 % specificity to peroxidase activity [177]. In early examples of evolution of a heme enzyme, Trakulnaleamsai *et al.* [178] and later Matsuura *et al.* [179] generated catalase libraries by random mutagenesis using sodium nitrite [180] and isolated variants with improved peroxidase activity. The second round of mutagenesis generated a triple mutant with 58 % specificity to peroxidase activity, but decreased thermostability. The authors further evolved their catalase mutant to improve thermostability back to that of wild-type by adding random peptide tails to the C-terminus [181].

10.8.5
Myoglobin

Directed evolution was used to improve the peroxidase activity of horse heart myoglobin [182]. The colorimetric ABTS assay was used to screen for activity. For HRP, the rate constant of *Compound I* formation (k_1 in Fig. 10.4) is $\sim 2.0 \times 10^7 \, \text{M}^{-1}\text{s}^{-1}$, while for horse heart myoglobin k_1 is $\sim 500 \, \text{M}^{-1}\text{s}^{-1}$ and is rate-limiting in the peroxidase reaction. Four rounds of mutagenesis and screening improved the value of k_1 ~ 25-fold and the value of k_3 ~ 1.6-fold, such that the two rate constants were about equal. All four substitutions in the most active mutant were in the heme pocket, within 5 Å of the heme group, demonstrating the importance of protein structure and charge in this region.

One set of site-directed mutagenesis studies improved the peroxygenase activity of Mb [128]; in a different random mutagenesis study the peroxidase activity was improved. Clearly these reactions are very different from one another, and neither is myoglobin's natural function. Here is another example where an alternative scaffold supports functions that are characteristic of other conserved structures. Heme enzyme function is flexible and highly evolvable.

10.8.6
Methods for recombination of P450s

Generating diversity is an important part of directed evolution. Nature supplies enormous genetic diversity within families of related genes, and we can use diversity for evolution by recombining, or shuffling, related genes [183]. However, family shuffling methods require a high degree of sequence identity for successful recombination ($\sim 70\,\%$). P450s from different species have low sequence identities (often $<20\,\%$), even though their tertiary structures are conserved. Sieber *et al.* [184] recently reported a method for producing libraries of hybrid sequences that does not rely on DNA hybridization and therefore can be used to recombine more distantly related sequences. The method, called SHIPREC (sequence homology-independent protein recombination), was used to create single-crossover hybrids of human P450 1A2 and bacterial P450 BM-3, which share only 16 % amino acid sequence identity. The hybrid genes were fused to the gene for chloramphenicol acetyl transferase, so that only variants that were translated in-frame and soluble in the cytoplasm would confer resistance to chloramphenicol. From these, two hybrids that retained the activity of the membrane-associated human enzyme but were more soluble in the *E. coli* cytoplasm were isolated and characterized. Iterative SHIPREC should be able to create more complex libraries with multiple crossovers.

Other technical problems for directed evolution by shuffling can be a high level of parental sequences in the recombined libraries and non-uniform representation of parental genes in the hybrids [191]. Abecassis *et al.* [185] used a family shuffling strategy (CLERY: combinatorial libraries enhanced by recombination in yeast) to generate a highly complex recombined library of human cytochromes P450 1A1 and 1A2 (74 % nucleotide sequence identity) composed of 86 % chimeric genes representing nearly equal contributions from each gene. This work combines *in vitro* shuffling methods [186 – 188] with previously developed *in vivo* yeast recombination methods [189, 190].

10.9
Conclusions

A rich source of potential industrial biocatalysts, the heme enzymes are also a superb testing ground for laboratory evolution. Directed evolution approaches are already generating customized heme enzymes and probing the limits of heme enzyme catalysis. Over the next few years, these same approaches will allow us to explore the interconversion of function among different protein scaffolds and thereby observe how the protein modulates heme chemistry and how new functions are acquired.

References

[1] Sheldon, R.A., Editor. *Metalloporphyrins in Catalytic Oxidations.* **1994**, Marcel Dekker, Inc.: New York.

[2] Erman, J.E., Hager, L.P., and Sligar, S.G., *Adv. Inorg. Biochem.*, **1994**, *10*, 71–118.

[3] Gray, H.B. and Ellis, W.R., *Electron Transfer*, in *Bioinorganic Chemistry*, I. Bertini, H.B. Gray, S.J. Lippard, and J.S. Valentine, Editors. **1994**, University Science Books: Sausalito, CA. p. 315 – 363.

[4] Phillips, I.R. and Shephard, E.A., Editors. *Cytochrome P450 Protocols*, Volume 107. **1998**, Humana Press: Totowa, NJ.

[5] Lewis, D.F.V., *Cytochromes P450: Structure, Function, and Mechanism.* **1996**, Taylor & Francis: Bristol, PA.

[6] Ortiz de Montellano, P.R., Editor. *Cytochrome P450: Structure, Mechanism, and Biochemistry*, 2nd Edition. **1995**, Plenum Press: New York and London.

[7] Roberts, G.C., *Chem. Biol.*, **1999**, *6*, R269–R272.

[8] Wong, L.L., *Curr. Opin. Chem. Biol.*, **1998**, *2*, 263–268.

[9] van den Brink, H.M., van Gorcom, R.F., van den Hondel, C.A., and Punt, P.J., *Fungal. Genet. Biol.*, **1998**, *23*, 1-17.

[10] Wong, L.L., Westlake, C.G., and Nickerson, D.P., *Struct. Bond.*, **1997**, *88*, 175 – 207.

[11] Kellner, D.G., Maves, S.A., and Sligar, S.G., *Curr. Opin. Biotechnol.*, **1997**, *8*, 274–278.

[12] Sono, M., Roach, M.P., Coulter, E.D., and Dawson, J.H., *Chem. Rev.*, **1996**, *96*, 2841–2887.

[13] Nelson, D.R., Koymans, L., Kamataki, T., Stegeman, J.J., Feyereisen, R., Waxman, D.J., Waterman, M.R., Gotoh, O., Coon, M.J., Estabrook, R.W., Gunsalus, I.C., and Nebert, D.W., *Pharmacogenetics*, **1996**, *6*, 1–42.

[14] Martinez, C.A. and Stewart, J.D., *Curr. Org. Chem.*, **2000**, *4*, 263–282.

[15] Mansuy, D., *Comp. Biochem. Physiol. C Pharmacol. Toxicol. Endocrinol.*, **1998**, *121*, 5–14.

[16] Miles, C.S., Ost, T.W.B., Noble, M.A., Munro, A.W., and Chapman, S.K., *Biochim. Biophys. Acta-Protein Struct. Molec. Enzym.*, **2000**, *1543*, 383–407.

[17] Mueller, E.J., Loida, P.J., and Sligar, S.G., *Twenty-five years of P450cam research*, in *Cytochrome P450 Structure, Mechanism, and Biochemistry*, Second Edition. P.R. Ortiz de Montellano, Editor. **1995**, Plenum Press: New York and London. p. 83–124.

[18] Poulos, T.L., Finzel, B.C., and Howard, A.J., *J Mol Biol*, **1987**, *195*, 687–700.

[19] Ravichandran, K.G., Boddupalli, S.S., Hasermann, C.A., Peterson, J.A., and Deisenhofer, J., *Science*, **1993**, *261*, 731–736.

[20] Hasemann, C.A., Ravichandran, K.G., Peterson, J.A., and Deisenhofer, J., *J. Mol. Biol.*, **1994**, *236*, 1169–1185.

[21] Cupp-Vickery, J.R. and Poulos, T.L., *Nat. Struct. Biol.*, **1995**, *2*, 144–153.

[22] Park, S.Y., Shimizu, H., Adachi, S., Nakagawa, A., Tanaka, I., Nakahara, K., Shoun, H., Obayashi, E., Nakamura, H., Iizuka, T., and Shiro, Y., *Nat. Struct. Biol.*, **1997**, *4*, 827–832.

[23] Ito, S., Matsuoka, T., Watanabe, I., Kagasaki, T., Serizawa, N., and Hata, T., *Acta Crystallogr. Sect. D-Biol. Crystallogr.*, **1999**, *55*, 1209–1211.

[24] Williams, P.A., Cosme, J., Sridhar, V., Johnson, E.F., and McRee, D.E., *Mol. Cell*, **2000**, *5*, 121–131.

[25] Park, S.Y., Yamane, K., Adachi, S., Shiro, Y., Weiss, K.E., and Sligar, S.G., *Acta Crystallogr. Sect. D-Biol. Crystallogr.*, **2000**, *56*, 1173–1175.

[26] Podust, L.M., Poulos, T.L., and Waterman, M.R., *Proc. Natl. Acad. Sci. USA*, **2001**, *98*, 3068–3073.

[27] Sligar, S.G. and Gunsalus, I.C., *Proc. Natl. Acad. Sci. USA*, **1976**, *73*, 1078–1082.

[28] Fisher, M.T. and Sligar, S.G., *J. Am. Chem. Soc.*, **1985**, *107*, 5018–5019.

[29] Sligar, S.G., *Biochemistry*, **1976**, *15*, 5399–5406.

[30] Toy, P.H., Newcomb, M., Coon, M.J., and Vaz, A.D.N., *J. Am. Chem. Soc.*, **1998**, *120*, 9718–9719.

[31] Vaz, A.D., McGinnity, D.F., and Coon, M.J., *Proc. Natl. Acad. Sci. USA*, **1998**, *95*, 3555–3560.

[32] Vaz, A.D., Pernecky, S.J., Raner, G.M., and Coon, M.J., *Proc. Natl. Acad. Sci. USA*, **1996**, *93*, 4644–4648.

[33] Kimata, Y., Shimada, H., Hirose, T., and Ishimura, Y., *Biochem. Biophys. Res. Commun.*, **1995**, *208*, 96–102.

[34] Aikens, J. and Sligar, S.G., *J. Am. Chem. Soc.*, **1994**, *116*, 1143–44.

[35] Raag, R., Martinis, S.A., Sligar, S.G., and Poulos, T.L., *Biochemistry*, **1991**, *30*, 11420–29.

[36] Martinis, S.A., Atkins, W.M., Stayton, P.S., and Sligar, S.G., *J. Am. Chem. Soc.*, **1989**, *111*, 9252–53.

[37] Imai, M., Shimada, H., Watanabe, Y., Matsushima-Hibiya, Y., Makino, R., Koga, H., Horiuchi, T., and Ishimura, Y., *Proc. Natl. Acad. Sci. USA*, **1989**, *86*, 7823–27.

[38] Fruetel, J.A., Collins, J.R., Camper, D.L., Loew, G.H., and Ortiz de Montellano, P.R., *J. Am. Chem. Soc.*, **1992**, *114*, 6987–93.

[39] Nickerson, D.P., Harford-Cross, C.F., Fulcher, S.R., and Wong, L.L., *FEBS Lett.*, **1997**, *405*, 153–156.

[40] Loida, P.J. and Sligar, S.G., *Biochemistry*, **1993**, *32*, 11530–38.

[41] Hintz, M.J., Mock, D.M., Peterson, L.L., Tuttle, K., and Peterson, J.A., *J. Biol. Chem.*, **1982**, *257*, 14324–32.

[42] Li, Q.S., Ogawa, J., and Shimizu, S., *Biochem. Biophys. Res. Commun.*, **2001**, *280*, 1258–61.

[43] Yu, X.C., Liang, C., and Strobel, H.W., *Biochemistry*, **1996**, *35*, 6289–6296.

[44] Anari, M.R., Khan, S., Liu, Z.C., and Obrien, P.J., *Chem. Res. Toxicol.*, **1995**, *8*, 997–1004.

[45] Coon, M.J., Blake, R.C., White, R.E., and Nordblom, G.D., *Method. Enzymol.*, **1990**, *186*, 273–278.

[46] Estabrook, R.W., Martin-Wixtrom, C., Saeki, Y., Renneberg, R., Hildebrandt, A., and Werringloer, J., *Xenobiotica*, **1984**, *14*, 87–104.

[47] McCarthy, M.B. and White, R.E., *J. Biol. Chem.*, **1983**, *258*, 9153–58.

[48] Blake, R.C. and Coon, M.J., *J. Biol. Chem.*, **1980**, *255*, 4100–4111.

[49] White, R.E., Sligar, S.G., and Coon, M.J., *J. Biol. Chem.*, **1980**, *255*, 11108–11.

[50] Capdevila, J., Estabrook, R.W., and Prough, R.A., *Arch. Biochem. Biophys.*, **1980**, *200*, 186–195.

[51] Rahimtula, A.D., O'Brien, P.J., Seifried, H.E., and Jerina, D.M., *Eur. J. Biochem.*, **1978**, *89*, 133–141.

[52] Nordblom, G.D., White, R.E., and Coon, M.J., *Arch. Biochem. Biophys.*, **1976**, *175*, 524–533.

[53] Hrycay, E.G., Gustafsson, J.A., Ingelman-Sundberg, M., and Ernster, L., *Biochem. Biophys. Res. Commun.*, **1975**, *66*, 209–216.

[54] Rahimtula, A.D. and O'Brien, P.J., *Biochem. Biophys. Res. Commun.*, **1974**, *60*, 440–447.

[55] Peterson, J.A., Ishimura, Y., and Griffin, B.W., *Arch. Biochem. Biophys.*, **1972**, *149*, 197–208.

[56] Hrycay, E.G. and O'Brien, P.J., *Arch. Biochem. Biophys.*, **1971**, *147*, 28–35.

[57] van Rantwijk, F. and Sheldon, R.A., *Curr. Opin. Biotechnol.*, **2000**, *11*, 554–564.

[58] Adam, W., Lazarus, M., Saha-Moller, C.R., Weichold, O., Hoch, U., Haring, D., and Schreier, P., *Adv. Biochem. Eng. Biotechnol.*, **1999**, *63*, 73–108.

[59] Colonna, S., Gaggero, N., Richelmi, C., and Pasta, P., *Trends Biotechnol.*, **1999**, *17*, 163–168.

[60] vanDeurzen, M.P.J., vanRantwijk, F., and Sheldon, R.A., *Tetrahedron*, **1997**, *53*, 13183–13220.

[61] Everse, J., Everse, K.E., and Grisham, M.B., Editors. *Peroxidases in Chemistry and Biology*, Volume 2. **1991**, CRC Press: Boca Raton, FL.

[62] Kyte, J., *Mechanism in Protein Chemistry.* **1995**, Garland Publishing, Inc.: New York and London.

[63] Poulos, T.L., *Adv. Inorg. Biochem.*, **1988**, *7*, 1–36.

[64] Smith, A.T. and Veitch, N.C., *Curr. Opin. Chem. Biol.*, **1998**, *2*, 269–278.

[65] Ozaki, S., Matsui, T., Roach, M.P., and Watanabe, Y., *Coord. Chem. Rev.*, **2000**, *198*, 39–59.

[66] Matsui, T., Ozaki, S., Liong, E., Phillips, G.N., and Watanabe, Y., *J. Biol. Chem.*, **1999**, *274*, 2838–2844.

[67] Savenkova, M.I., Kuo, J.M., and Ortiz de Montellano, P.R., *Biochemistry*, **1998**, *37*, 10828–10836.

[68] Savenkova, M.I., Newmyer, S.L., and Ortiz de Montellano, P.R., *J. Biol. Chem.*, **1996**, *271*, 24598–24603.

[69] Newmyer, S.L. and Ortiz de Montellano, P.R., *J. Biol. Chem.*, **1995**, *270*, 19430–38.

[70] Dawson, J.H., *Science*, **1988**, *240*, 433–439.

[71] Marnett, L.J., Weller, P., and Battista, J.R., *Comparison of the Peroxidase Activity of Hemeproteins and Cytochrome P-450*, in *Cytochrome P450: Structure, Mechanism, and Biochemistry*, First Edition. P.R. Ortiz de Montellano, Editor. **1986**, Plenum Press: New York and London. p. 29–77.

[72] Dunford, H.B. and Stillman, J.S., *Coord. Chem. Rev.*, **1976**, *19*, 187–251.

[73] Dunford, H.B. and Nadezhdin, A.D., *On the past eight years of peroxidase research*, in *Oxidases and Related Redox Systems*, T.E. King, H.S. Mason, and M. Morrison, Editors. **1982**, Pergamon: Elmsford, New York. p. 653–670.

[74] Sakurada, J., Sekiguchi, R., Sato, K., and Hosoya, T., *Biochemistry*, **1990**, *29*, 4093–4098.

[75] Dunford, H.B. and Adeniran, A.J., *Arch. Biochem. Biophys.*, **1986**, *251*, 536–42.

[76] Pietikainen, P. and Adlercreutz, P., *Appl. Microbiol. Biotechnol.*, **1990**, *33*, 455–58.

[77] Dordick, J.S., Klibanov, A.M., and Marletta, M.A., *Biochemistry*, **1986**, *25*, 2946–51.

[78] Hewson, W.D. and Dunford, H.B., *J. Biol. Chem.*, **1976**, *251*, 6043–52.

[79] Kobayashi, S., Shoda, S., and Uyama, H., *Adv. Polym. Sci.*, **1995**, *121*, 1–30.

[80] Dordick, J.S., *Trends Biotechnol.*, **1992**, *10*, 287–293.

[81] Klibanov, A.M., Tu, T.M., and Scott, K.P., *Science*, **1983**, *221*, 259–260.

[82] Sun, W., Kadima, T.A., Pickard, M.A., and Dunford, H.B., *Biochem. Cell. Biol.*, **1994**, *72*, 321–331.

[83] Truan, G. and Peterson, J.A., *Arch. Biochem. Biophys.*, **1998**, *349*, 53–64.

[84] Macdonald, T.L., Gutheim, W.G., Martin, R.B., and Guengerich, F.P., *Biochemistry*, **1989**, *28*, 2071–2077.

[85] Hayashi, Y. and Yamazaki, I., *J. Biol. Chem.*, **1979**, *254*, 9101–06.

[86] Casella, L., Gullotti, M., Ghezzi, R., Poli, S., Beringhelli, T., Colonna, S., and Carrea, G., *Biochemistry*, **1992**, *31*, 9451–9459.

[87] Marnett, L.J. and Kennedy, T.A., *Comparison of the Peroxidase Activity of Hemoproteins and Cytochrome P450*, in *Cytochrome P450 Structure, Mechanism and*

Biochemistry, 2nd Edition. P.R. Ortiz de Montellano, Editor. **1995**, Plenum Press: New York and London. p. 49–80.

[88] Smith, A.T., Sanders, S.A., Thorneley, R.N., Burke, J.F., and Bray, R.R., *Eur. J. Biochem.*, **1992**, *207*, 507–519.

[89] Erman, J.E., Vitello, L.B., Miller, M.A., and Kraut, J., *J. Am. Chem. Soc.*, **1992**, *114*, 6592–93.

[90] Matsunaga, I., Yokotani, N., Gotoh, O., Kusunose, E., Yamada, M., and Ichihara, K., *J. Biol. Chem.*, **1997**, *272*, 23592–23596.

[91] Imai, Y., Matsunaga, I., Kusunose, E., and Ichihara, K., *J. Biochem. (Tokyo)*, **2000**, *128*, 189–194.

[92] Matsunaga, I., Yamada, M., Kusunose, E., Miki, T., and Ichihara, K., *J. Biochem. (Tokyo)*, **1998**, *124*, 105–110.

[93] Matsunaga, I., Sumimoto, T., Ueda, A., Kusunose, E., and Ichihara, K., *Lipids*, **2000**, *35*, 365–371.

[94] Cirino, P.C., Schwaneberg, U., and Arnold, F.H., *unpublished results*. 2001.

[95] Joo, H., Lin, Z., and Arnold, F.H., *Nature*, **1999**, *399*, 670–673.

[96] Mayhew, M.P., Reipa, V., Holden, M.J., and Vilker, V.L., *Biotechnol. Prog.*, **2000**, *16*, 610–616.

[97] Reipa, V., Mayhew, M.P., and Vilker, V.L., *Proc. Natl. Acad. Sci. USA*, **1997**, *94*, 13554–13558.

[98] Estabrook, R.W., Faulkner, K.M., Shet, M.S., and Fisher, C.W., *Methods Enzymol.*, **1996**, *272*, 44–51.

[99] Faulkner, K.M., Shet, M.S., Fisher, C.W., and Estabrook, R.W., *Proc. Natl. Acad. Sci. USA*, **1995**, *92*, 7705–7709.

[100] Yano, J.K., Koo, L.S., Schuller, D.J., Li, H.Y., Ortiz de Montellano, P.R., and Poulos, T.L., *J. Biol. Chem.*, **2000**, *275*, 31086–31092.

[101] Koo, L.S., Tschirret-Guth, R.A., Straub, W.E., Moenne-Loccoz, P., Loehr, T.M., and Ortiz de Montellano, P.R., *J. Biol. Chem.*, **2000**, *275*, 14112–14123.

[102] Chang, Y.T. and Loew, G., *Biochemistry*, **2000**, *39*, 2484–2498.

[103] McLean, M.A., Maves, S.A., Weiss, K.E., Krepich, S., and Sligar, S.G., *Biochem. Biophys. Res. Commun.*, **1998**, *252*, 166–172.

[104] Sundaramoorthy, M., Terner, J., and Poulos, T.L., *Structure*, **1995**, *3*, 1367–1377.

[105] Hager, L.P., Lakner, F.J., and Basavapathruni, A., *J. Mol. Catal. B-Enzymatic*, **1998**, *5*, 95–101.

[106] vanDeurzen, M.P.J., Remkes, I.J., vanRantwijk, F., and Sheldon, R.A., *J. Mol. Catal. A-Chemical.*, **1997**, *117*, 329–337.

[107] Colonna, S., Gaggero, N., Carrea, G., and Pasta, P., *Chem. Commun.*, **1997**, 439–40.

[108] Geigert, J., Dalietos, D.J., Neidleman, S.L., Lee, T.D., and Wadsworth, J., *Biochem. Biophys. Res. Commun.*, **1983**, *114*, 1104–08.

[109] Thomas, J.A., Morris, D.R., and Hager, L.P., *J. Biol. Chem.*, **1970**, *245*, 3129–34.

[110] Yi, X., Mroczko, M., Manoj, K.M., Wang, X., and Hager, L.P., *Proc. Natl. Acad. Sci. USA*, **1999**, *96*, 12412–17.

[111] Hu, S.H. and Hager, L.P., *J. Am. Chem. Soc.*, **1999**, *121*, 872–873.

[112] Hu, S.H. and Hager, L.P., *Tetrahedron Lett.*, **1999**, *40*, 1641–1644.

[113] Kedderis, G.L. and Hollenberg, P.F., *J. Biol. Chem.*, **1984**, *259*, 3663–3668.

[114] Kedderis, G.L., Koop, D.R., and Hollenberg, P.F., *J. Biol. Chem.*, **1980**, *255*, 10174–82.

[115] VanDeurzen, M.P.J., Seelbach, K., VanRantwijk, F., Kragl, U., and Sheldon, R.A., *Biocatal. Biotransform.*, **1997**, *15*, 1–16.

[116] van de Velde, F., Lourenco, N.D., Bakker, M., van Rantwijk, F., and Sheldon, R.A., *Biotechnol. Bioeng.*, **2000**, *69*, 286–291.

[117] van de Velde, F., van Rantwijk, F., and Sheldon, R.A., *J. Mol. Catal. B-Enzymatic*, **1999**, *6*, 453–461.

[118] Seelbach, K., vanDeurzen, M.P.J., vanRantwijk, F., Sheldon, R.A., and Kragl, U., *Biotechnol. Bioeng.*, **1997**, *55*, 283–288.

[119] Vandeurzen, M.P.J., Groen, B.W., Vanrantwijk, F., and Sheldon, R.A., *Biocatalysis*, **1994**, *10*, 247–255.

[120] Loida, P.J. and Sligar, S.G., *Protein Eng.*, **1993**, *6*, 207–212.

[121] Bell, S.G., Rouch, D.A., and Wong, L.L., *J. Mol. Catal. B-Enzymatic*, **1997**, *3*, 293–302.

[122] Fowler, S.M., England, P.A., Westlake, A.C.G., Rouch, D.R., Nickerson, D.P., Blunt, C., Braybrook, D., West, S., Wong, L.L., and Flitsch, S.L., *J. Chem. Soc-Chem. Comm.*, **1994**, 2761–2762.

[123] Conesa, A., van De Velde, F., van Rantwijk, F., Sheldon, R.A., van Den Hondel, C.A., and Punt, P.J., *J. Biol. Chem.*, **2001**, *276*, 17635–17640.

[124] Matsui, T., Nagano, S., Ishimori, K., Watanabe, Y., and Morishima, I., *Biochemistry*, **1996**, *35*, 13118–13124.

[125] Tschirret-Guth, R.A. and Ortiz de Montellano, P.R., *Arch. Biochem. Biophys.*, **1996**, *335*, 93–101.

[126] Adachi, S., Nagano, S., Ishimori, K., Watanabe, Y., Morishima, I., Egawa, T., Kitagawa, T., and Makino, R., *Biochemistry*, **1993**, *32*, 241–252.

[127] Rao, S.I., Wilks, A., and Ortiz de Montellano, P.R., *J. Biol. Chem.*, **1993**, *268*, 803–809.

[128] Ozaki, S., Yang, H.J., Matsui, T., Goto, Y., and Watanabe, Y., *Tetrahedron-Asymmetry*, **1999**, *10*, 183–192.

[129] Matsui, T., Ozaki, S., and Watanabe, Y., *J. Am. Chem. Soc.*, **1999**, *121*, 9952–57.

[130] Ozaki, S., Matsui, T., and Watanabe, Y., *J. Am. Chem. Soc.*, **1997**, *119*, 6666–67.

[131] Matsui, T., Ozaki, S., and Watanabe, Y., *J. Biol. Chem.*, **1997**, *272*, 32735–38.

[132] Ozaki, S., Matsui, T., and Watanabe, Y., *J. Am. Chem. Soc.*, **1996**, *118*, 9784–85.

[133] Ozaki, S., Hara, I., Matsui, T., and Watanabe, Y., *Biochemistry*, **2001**, *40*, 1044–52.

[134] Joo, H., Arisawa, A., Lin, Z., and Arnold, F.H., *Chem. Biol.*, **1999**, *6*, 699–706.

[135] Joern, J.M., Sakamoto, T., Arisawa, A., and Arnold, F.H., *J. Biomol. Screen*, **2001**, *6*, 219–223.

[136] Schwaneberg, U., Otey, C., Cirino, P., Farinas, E., and Arnold, F.H., *J. Biomol. Screen*, **2001**, *6*, 111–118.

[137] Farinas, E., Schwaneberg, U., Glieder, A., and Arnold, F.H., *Adv. Synth. Catal.*, **2001**, *343*, 601–606.

[138] Graham-Lorence, S., Truan, G., Peterson, J.A., Falck, J.R., Wei, S., Helvig, C., and Capdevila, J.H., *J. Biol. Chem.*, **1997**, *272*, 1127–35.

[139] Capdevila, J.H., Wei, S., Helvig, C., Falck, J.R., Belosludtsev, Y., Truan, G.,

Graham-Lorence, S.E., and Peterson, J.A., *J. Biol. Chem.*, **1996**, *271*, 22663–71.

[140] Boddupalli, S.S., Estabrook, R.W., and Peterson, J.A., *J. Biol. Chem.*, **1990**, *265*, 4233–39.

[141] Schwaneberg, U., Schmidt-Dannert, C., Schmitt, J., and Schmid, R.D., *Anal. Biochem.*, **1999**, *269*, 359–366.

[142] Li, H. and Poulos, T.L., *Nat. Struct. Biol.*, **1997**, *4*, 140–146.

[143] Modi, S., Sutcliffe, M.J., Primrose, W.U., Lian, L.Y., and Roberts, G.C., *Nat. Struct. Biol.*, **1996**, *3*, 414–417.

[144] Li, Q.S., Schwaneberg, U., Fischer, M., Schmitt, J., Pleiss, J., Lutz-Wahl, S., and Schmid, R.D., *Biochim. Biophys. Acta – Protein Struct. Molec. Enzym*, **2001**, *1545*, 114–121.

[145] Li, Q.S., Schwaneberg, U., Fischer, P., and Schmid, R.D., *Chemistry*, **2000**, *6*, 1531–36.

[146] Guengerich, F.P., *J. Biol. Chem.*, **1991**, *266*, 10019–22.

[147] Schwaneberg, U., Sprauer, A., Schmidt-Dannert, C., and Schmid, R.D., *J. Chromatogr. A*, **1999**, *848*, 149–159.

[148] Cherry, J.R., Lamsa, M.H., Schneider, P., Vind, J., Svendsen, A., Jones, A., and Pedersen, A.H., *Nat. Biotechnol.*, **1999**, *17*, 379–384.

[149] Bhaskar, B., Bonagura, C.A., Jamal, J., and Poulos, T.L., *Tetrahedron*, **2000**, *56*, 9471–75.

[150] Hirst, J. and Goodin, D.B., *J. Biol. Chem.*, **2000**, *275*, 8582–91.

[151] Bonagura, C.A., Bhaskar, B., Sundaramoorthy, M., and Poulos, T.L., *J. Biol. Chem.*, **1999**, *274*, 37827–33.

[152] Wilcox, S.K., Putnam, C.D., Sastry, M., Blankenship, J., Chazin, W.J., McRee, D.E., and Goodin, D.B., *Biochemistry*, **1998**, *37*, 16853–62.

[153] Yeung, B.K., Wang, X., Sigman, J.A., Petillo, P.A., and Lu, Y., *Chem. Biol.*, **1997**, *4*, 215–221.

[154] Bonagura, C.A., Sundaramoorthy, M., Pappa, H.S., Patterson, W.R., and Poulos, T.L., *Biochemistry*, **1996**, *35*, 6107–15.

[155] Wilcox, S.K., Jensen, G.M., Fitzgerald, M.M., McRee, D.E., and Goodin, D.B., *Biochemistry*, **1996**, *35*, 4858–66.

[156] Roe, J.A. and Goodin, D.B., *J. Biol. Chem.*, **1993**, *268*, 20037–45.

[157] DePillis, G.D., Sishta, B.P., Mauk, A.G., and Ortiz de Montellano, P.R., *J. Biol. Chem.*, **1991**, *266*, 19334–41.

[158] Pelletier, H. and Kraut, J., *Science*, **1992**, *258*, 1748–1755.

[159] Casella, L., Monzani, E., Gullotti, M., Santelli, E., Poli, S., and Beringhelli, T., *Gazz. Chim. Ital.*, **1996**, *126*, 121–125.

[160] Iffland, A., Tafelmeyer, P., Saudan, C., and Johnsson, K., *Biochemistry*, **2000**, *39*, 10790–98.

[161] Smith, A.T., Santama, N., Dacey, S., Edwards, M., Bray, R.C., Thorneley, R.N., and Burke, J.F., *J. Biol. Chem.*, **1990**, *265*, 13335–43.

[162] Gray, J.S., Yang, B.Y., and Montgomery, R., *Carbohydr. Res.*, **1998**, *311*, 61–69.

[163] Dunford, H.B., *Horseradish Peroxidase: Structure and Kinetic Properties*, in *Peroxidases in Chemistry and Biology*, Volume 2. J. Everse, K.E. Everse, and M.B. Grisham, Editors. **1991**, CRC Press: Boca Raton, FL. p. 1–24.

[164] Welinder, K.G., *Eur. J. Biochem.*, **1979**, *96*, 483–502.

[165] Romanos, M.A., Scorer, C.A., and Clare, J.J., *Yeast*, **1992**, *8*, 423–488.

[166] Sudbery, P.E., *Curr. Opin. Biotechnol.*, **1996**, *7*, 517–524.

[167] Parekh, R., Forrester, K., and Wittrup, D., *Protein Expr. Purif.*, **1995**, *6*, 537–545.

[168] Morawski, B., Lin, Z., Cirino, P., Joo, H., Bandara, G., and Arnold, F.H., *Protein Eng.*, **2000**, *13*, 377–384.

[169] Morawski, B., Quan, S., and Arnold, F.H., *Biotechnol Bioeng*, **2001**, *66*, 99–107.

[170] Groves, J.T. and Han, Y.Z., *Models and Mechanisms of Cytochrome P450 Action*, in *Cytochrome P450: Structure, Mechanism, and Biochemistry*, Second Edition. P.R. Ortiz de Montellano, Editor. **1995**, Plenum Press: New York and London. p. 3–48.

[171] Tian, Z.Q., Richards, J.L., and Traylor, T.G., *J. Am. Chem. Soc.*, **1995**, *117*, 21–29.

[172] Groves, J.T., Fish, K.M., Avaria-Neisser, G.E., Imachi, M., and Kuczkowski, R.L., *Prog. Clin. Biol. Res.*, **1988**, *274*, 509–524.

[173] Ortiz de Montellano, P.R., Mangold, B.L., Wheeler, C., Kunze, K.L., and Reich, N.O., *J. Biol. Chem.*, **1983**, *258*, 4208–4213.

[174] Kunze, K.L., Mangold, B.L., Wheeler, C., Beilan, H.S., and Ortiz de Montellano, P.R., *J. Biol. Chem.*, **1983**, *258*, 4202–4207.

[175] Ortiz de Montellano, P.R., Kunze, K.L., and Mico, B.A., *Mol. Pharmacol.*, **1980**, *18*, 602–605.

[176] Rai, G.P., Zong, Q., and Hager, L.P., *Israel J. Chem.*, **2000**, *40*, 63–70.

[177] Yomo, T., Yamano, T., Yamamoto, K., and Urabe, I., *J. Theor. Biol.*, **1997**, *188*, 301–312.

[178] Trakulnaleamsai, S., Yomo, T., Yoshikawa, M., Aihara, S., and Urabe, I., *J. Ferment. Bioeng.*, **1995**, *79*, 107–118.

[179] Matsuura, T., Yomo, T., Trakulnaleamsai, S., Ohashi, Y., Yamamoto, K., and Urabe, I., *Protein Eng.*, **1998**, *11*, 789–795.

[180] Myers, R.M., Lerman, L.S., and Maniatis, T., *Science*, **1985**, *229*, 242–247.

[181] Matsuura, T., Miyai, K., Trakulnaleamsai, S., Yomo, T., Shima, Y., Miki, S., Yamamoto, K., and Urabe, I., *Nat. Biotechnol.*, **1999**, *17*, 58–61.

[182] Wan, L., Twitchett, M.B., Eltis, L.D., Mauk, A.G., and Smith, M., *Proc. Natl. Acad. Sci. USA*, **1998**, *95*, 12825–12831.

[183] Ness, J.E., Del Cardayre, S.B., Minshull, J., and Stemmer, W.P.C., *Adv. Protein Chem.*, **2001**, *55*, 261–292.

[184] Sieber, V., Martinez, C.A., and Arnold, F.H., *Nat. Biotechnol.*, **2001**, *19*, 456–60.

[185] Abecassis, V., Pompon, D., and Truan, G., *Nucleic Acids Res.*, **2000**, *28*, E88.

[186] Crameri, A., Raillard, S.A., Bermudez, E., and Stemmer, W.P., *Nature*, **1998**, *391*, 288–291.

[187] Stemmer, W.P., *Nature*, **1994**, *370*, 389–391.

[188] Stemmer, W.P., *Proc. Natl. Acad. Sci. USA*, **1994**, *91*, 10747–51.

[189] Pompon, D. and Nicolas, A., *Gene*, **1989**, *83*, 15–24.

[190] Mezard, C., Pompon, D., and Nicolas, A., *Cell*, **1992**, *70*, 659–670.

[191] Joern, J.M., Meinhold, P., and Arnold, F.H., *J. Mol. Biol.*, **2001**, *in print*.

11

Directed Evolution as a Means to Create Enantioselective Enzymes for Use in Organic Chemistry

Manfred T. Reetz and Karl-Erich Jaeger

11.1
Introduction

The stereoselective synthesis of chiral organic compounds is of considerable academic and industrial interest [1 – 3]. For example, the so-called "chiral market" of enantio-merically pure or enriched organic compounds is expanding rapidly, the total sales of chiral pharmaceuticals alone exceeding 100 billion US in 2000 [1e]. Many of these compounds are prepared in the laboratories of organic chemists. Presently, classical antipode separation is the method most often used in industry [1d]. However, this requires stoichiometric amounts of an appropriate optically active reagent as well as large amounts of organic solvents. For ecological and economic reasons, asym-metric catalysis can be expected to be more efficient, provided that active and highly enantioselective catalysts can be found. Catalytic asymmetric transformations can be carried out either in the form of kinetic resolution of racemates or in reactions invol-ving prochiral substrates. Two options are available, namely transition metal catalysis [2] and biocatalysis [3]. Success in the area of asymmetric transition metal catalysis requires efficient ligand tuning (Fig. 11.1a), which is a difficult goal [2]. Experi-ence, intuition, knowledge of the reaction mechanism and application of molecular modeling as well as time-consuming trial and error are required. Many successful examples have been reported in the academic literature, and some of these have been commercialized [1].

In the case of biocatalysis, enzymes [3] and catalytic antibodies [4] have attracted most attention. Since enzymes are inherently the more active catalysts, they have been used most often. Indeed, many industrial processes for the enantioselective production of certain chiral intermediates are based on the application of enzymes, as in the lipase-catalyzed kinetic resolution of an epoxy-ester used in the production of the anti-hypertensive therapeutic Diltiazem [5]. Recently, it has been noted that there seems to be a trend in industry to use enzymes more often than in the past

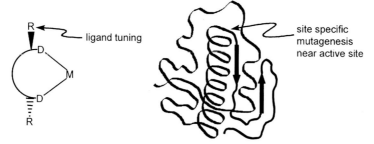

Fig. 11.1. a) Schematic representation of ligand tuning in the design and synthesis of a chiral transition metal (M) catalyst, C_2-symmetry arbitrarily being shown; the arrows symbolize points of potential structural variation and D denote donor atoms.
b) Schematic representation of "*de novo* design" of an enantioselective enzyme, the arrows symbolizing the exchange of amino acids on the basis of site specific mutagenesis.

[6]. However, these catalysts suffer from the disadvantage that for a given synthetic transformation of interest, A → B, enantioselectivity may well be poor. In principle, it should be possible to apply site-directed mutagenesis [7] in order to increase enantioselectivity to an acceptable level (Fig. 11.1b), similar to ligand tuning in transition metal catalysis (Fig. 11.1a). However, this is not a straightforward process.

Recently a radically different approach to the development of enantioselective catalysts has been proposed [8]. It does not rely on any knowledge regarding the structure of the enzyme nor on any speculations concerning enzyme mechanism. Rather, the basic idea is to apply directed evolution.

The combination of proper molecular biological methods for random mutagenesis and gene expression coupled with high-throughput screening systems for the rapid identification of enantioselective variants[*] of the natural (wild-type) enzyme forms the basis of the concept [8 – 10]. The idea is to start with a wild-type enzyme showing an unacceptably low enantiomeric excess (*ee*) or selectivity factor (*E*) value for a given transformation of interest, A → B, to create a library of mutant genes[*], to identify the most enantioselective enzyme variant following expression, and to repeat the process as often as necessary using in each case an improved mutant gene for the next round of mutagenesis. Since the inferior mutant gene and enzyme variants are discarded, the evolutionary character of the overall process becomes apparent (Fig. 11.2) [8 – 10].

[*] In molecular biology a convention has been established according to which the designation "mutant" refers to the altered gene, whereas "variant" denotes the altered protein. In order to avoid confusion we adhere to this nomenclature.

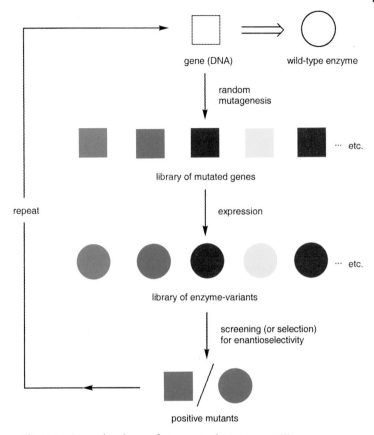

Fig. 11.2. Directed evolution of an enantioselective enzyme [9].

The major challenges in putting this concept into practice involve the development of efficient strategies for exploring protein sequence space with respect to enantioselectivity and the establishment of high-throughput screening or selection methods for assaying enantioselective enzymes. This chapter summarizes the present status of this new and exciting area of research.

11.2
Mutagenesis Methods

In the late 1980s and early 1990s molecular biologists began to develop new and practical techniques for random mutagenesis. One of the landmarks was a report by Leung, Chen and Goedell, who described the technique of error-prone polymerase chain reaction (epPCR) in which the conditions of the classical PCR were varied empirically (e.g. the $MgCl_2$ concentration) so as to attain the desired mutation rate [11a]. Later the method was improved by Cadwell and Joyce [11b]. This procedure of inducing point mutations was followed in 1994 by Stemmer's method of DNA shuffling [12] and in 1998 and 1999 by Arnold's staggered extension process [13] and random priming recombination method [14], respectively, which are all recombinative processes resulting in a high diversity of mutant genes. Since then these and other methods such as saturation mutagenesis (in which the substitution or insertion of codons is performed leading to all possible 20 amino acids at any predetermined position in the gene) or cassette mutagenesis (using DNA-fragments made up of nucleotides encoding one to several hundred amino acids in a defined region of the enzyme) have been applied in the quest to obtain structurally altered enzymes with improved stability and activity [11 – 26].

11.3
Overexpression of Genes and Secretion of Enzymes

The first step in setting up a successful directed evolution protocol is the development of an efficient expression system using an appropriate bacterial host. This is not a trivial task, in particular when overexpression is to be coupled to enzyme secretion. Fortunately, some proteins can easily be overexpressed and secreted by using commercially available systems [27 – 29], a prominent example being subtilisin of *Bacillus subtilis* [30]. However, many enzymes of interest are not amenable to such systems; examples include a variety of different lipases from *Pseudomonas* species.

Genes encoding extracellular enzymes can usually be overexpressed in the standard host strain *Escherichia coli*, but the proteins will form enzymatically inactive inclusion bodies in the bacterial cytoplasm. Attempts have been made to circumvent this problem by construction of overexpression systems which couple gene expression to secretion, thereby directing proteins through the bacterial inner membrane [31]. However, expression levels remain low in these systems as compared to levels which are reached upon overexpression of intracellular proteins in the *E. coli* cytoplasm.

The usefulness of the Gram-negative bacterium *Pseudomonas aeruginosa* has been explored as an expression and secretion host. [32 – 34]. This bacterium produces and secretes an array of different extracellular enzymes including many hydrolases which are of biotechnological interest. For *Pseudomonas* enzymes, it has been demonstrated that folding and secretion are highly specific and regulated processes. At least five different secretion pathways have been identified which are constituted by complex cellular secretion machineries consisting of three up to twenty different proteins [35, 36]. Furthermore, several different enzymes including disulfide oxidoreductases and proteases are involved in efficient folding of enzymes in the bacterial periplasm [37]. These observations explain that the overexpression *and* secretion of extracellular enzymes normally does not function properly in heterologous hosts. Therefore a *P. aeruginosa* strain PABST7.1 was constructed allowing the overexpression of enzyme genes from the strong T7 promoter which is successfully used in *E. coli* [38]. The gene *T7 pol* encoding the RNA polymerase from bacteriophage T7 was cloned behind an inducible promoter which can be repressed by the Lac repressor encoded by the *lacI*^q gene located immediately downstream of *T7 pol* [39]. These genes were stably integrated into the chromosome of the lipase-negative *P. aeruginosa* strain PABS1 (Fig. 11.3) resulting in the overexpression strain *P. aeruginosa* PABST7.1 [40]. Additionally, strain *P. aeruginosa* PAFRT7.7 was constructed which contains the *T7 pol* gene inserted directly into the lipase structural gene *lipA* [41]. Both strains are well suited for overexpression of mutant lipase genes. The technique that was used (Fig. 11.3) can easily be adapted to direct the *T7 pol* gene into any other gene of choice in the bacterial chromosome, thereby allowing the construction of *P. aeruginosa* overexpression strains negative for any enzyme activity to be screened for in subsequent directed evolution experiments.

The second part of the system comprises the plasmid pUCPL6AN in which the lipase operon *lipA/H* was cloned behind the T7-promoter (Fig. 11.3). This plasmid

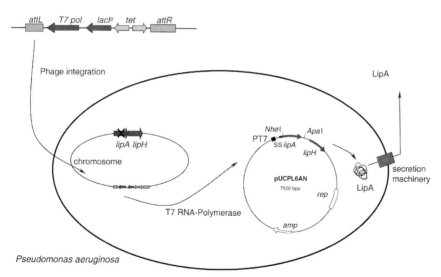

Fig. 11.3. Construction of a system for gene overexpression and enzyme secretion in *Pseudomonas aeruginosa* [32, 37b]. A phagemid containing the *T7 pol* gene [39] was integrated into the chromosome of the lipase-negative strain *P. aeruginosa* PABS1. Plasmid pUCPL6AN was constructed from pUCPKS [45] by cloning of the lipase operon *lipA/H* behind the T7 promoter. Transcription of T7 RNA polymerase is induced by addition of isopropyl-β-D-thiogalactoside. The polymerase then directs the transcription of the lipase operon. The lipase is correctly folded and secreted into the bacterial culture supernatant which can subsequently be used to assay lipase activities.

contains unique *Nhe*I and *Apa*I sites located 5'- and 3' of the lipase structural gene *lipA*. This is particularly important when the construct is used for directed evolution experiments: forced cloning is possible for any mutated lipase gene and the mutations introduced affect neither the signal sequence nor the 3' regulatory regions or the lipase-specific foldase. Thereby, it is possible to avoid the unwanted expression of those mutant genes which are not secreted and can therefore not be screened in a bacterial culture supernatant. This system has been used successfully for overexpression and secretion of large libraries of variant lipases under conditions of growth in microtiter plates thereby demonstrating its efficiency for directed evolution [8 – 10, 42 – 44].

11.4
High-Throughput Screening Systems for Enantioselectivity

One of the major problems in putting the concept described in Fig. 11.2 into practice concerns high-throughput screening for enantioselectivity. The determination of *ee* values is traditionally performed by gas chromatography or HPLC using chiral phases, but only a few dozen samples can be analyzed per day. Indeed, when research was initiated in 1995 concerning the evolution of enantioselective enzymes (see below) [8], high-throughput assays for enantioselectivity were unknown. Since then several methods, including systems based on UV/Vis [8, 46 – 48], IR-thermography [49], MS [50], and capillary array electrophoresis [51] have been developed. A complete review concerning the scope and limitations of these and other methods has appeared recently [10], and only a few highlights as well as new developments are presented here. An extremely useful and practical assay is a method based on electrospray ionization mass spectrometry (ESI-MS) [50]. The (*R*)- and (*S*)-enantiomers of a given chiral product have identical mass spectra and, in the absence of chromatographic separation, cannot be distinguished. However, if one of the enantiomers is deuterium-labeled, the parent peaks appear separately in the mass spectrum of the mixture, and integration then provides the *ee* value. Accordingly, deuterium-labeled substrates in the form of pseudo-enantiomers or pseudo-prochiral compounds are used to test a potentially enantioselective (bio)catalyst. The method is restricted to the kinetic resolution of chiral compounds and to reactions of prochiral compounds having enantiotopic groups (Fig. 11.4). The original form of this system has been automated and allows about 1000 *ee* determinations per day [50a]. A simple kinetic experiment is advisable to exclude possible deuterium isotope effects.

Several specific cases have been described in the literature [10, 50]. A new example concerns the hydrolytic kinetic resolution of the epoxide **1** catalyzed by an epoxide hydrolase (Fig. 11.5) [52]. As in other kinetic resolutions, the reaction is allowed to reach the ideal value of 50 %. Instead of employing a genuine racemate (*R*)-**1**/(*S*)-**1**

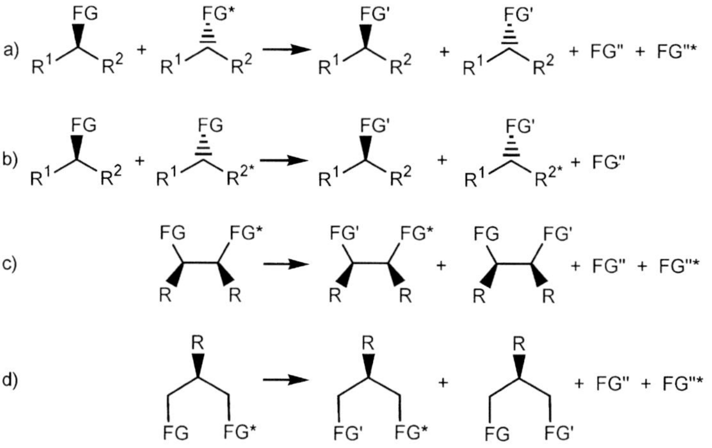

Fig. 11.4. MS-based *ee*-screening of isotopically labeled substrates [50]. a) Asymmetric transformation of a mixture of pseudo-enantiomers involving cleavage of the functional groups FG and labeled functional groups FG*. b) Asymmetric transformation of a mixture of pseudo-enantiomers involving either cleavage or bond formation at the functional group FG; isotopic labeling at R² is indicated by the asterisk. c) Asymmetric transformation of a pseudo-meso substrate involving cleavage of the functional groups FG and labeled functional groups FG*. d) Asymmetric transformation of a pseudo-prochiral substrate involving cleavage of the functional group FG and labeled functional group FG*.

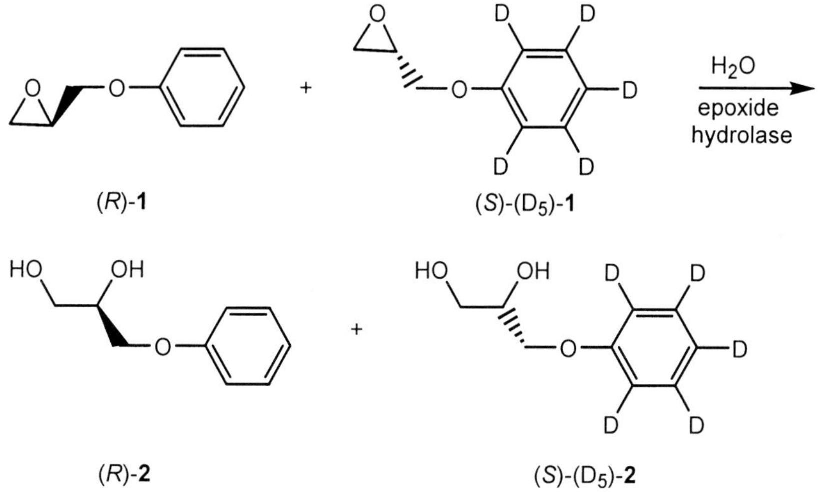

Fig. 11.5. Kinetic resolution of pseudo-enantiomers (*R*)-**1**/(*S*)-(D₅)-**1** [52].

as in a normal lab-scale experiment, a 1/1 mixture of pseudo-enantiomers (*R*)-**1**/(*S*)-(D$_5$)-**1** is used because at any point of the reaction the ratio of (*R*)-**1** : (*S*)-(D$_5$)-**1** and (*R*)-**2** : (*S*)-(D$_5$)-**2** can be determined simply by integrating the appropriate ESI-MS peaks. This provides *ee* as well as *E* values.

In the case of reactions involving desymmetrization of meso-type substrates, kinetic resolution is not involved, which means that the reaction can be run to 100 % conversion [10, 50]. Nevertheless, the ESI-MS-based assay is also applicable. This requires the synthesis and use of pseudo-meso substrates, i.e. meso-substrates that contain deuterium-labeled enantiotropic groups. In summary, although the MS-based assay is restricted to these two symmetry classes (Fig. 11.4), it is highly efficient. Moreover, new MS instruments utilizing an 8-channel multiplex electrospray source allow even higher throughput, which means that about 8000 *ee* determinations can be performed in one day [506]. An alternative MS-based assay requires derivatization using chiral reagents and is also of considerable practical use [53].

In traditional analytical chemistry the determination of enantiomeric purity is sometimes carried out by capillary electrophoresis (CE) in which the electrolyte contains chiral selectors such as cyclodextrin (CD) derivatives [54]. Unfortunately the conventional form of this analytical technique allows only a few dozen *ee* determinations per day. However, as a consequence of the analytical demands arising from the Human Genome Project, CE has been revolutionized in recent years so that efficient techniques for instrumental miniaturization are now available, making ultra-high-throughput analysis of biomolecules possible for the first time [55]. Two different approaches have emerged, namely capillary array electrophoresis (CAE) [55a – e] and CE on microchips (also called CAE on chips) [55f – m]. Both techniques can be used to carry out

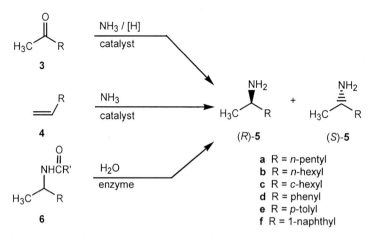

Fig. 11.6. Catalytic formation of chiral amines **5**. Screening was carried out by CAE [51, 57].

DNA sequence analyses and/or to analyze oligonucleotides, DNA-restriction fragments, amino acids or PCR products. Many hundred thousand analytical data points can be accumulated per day using commercially available instruments [55]. Therefore chiral modification of such techniques, if successful, should allow for the first time super-high-throughput analyses of enantiomeric purity (*ee*) [51].

In the case of CAE, instruments have been developed which contain a high number of capillaries in parallel, for example, the commercially available 96-capillary unit MegaBACE which consists of six bundles of 16 capillaries [56], each about 50 cm long. The system can therefore address 96-well microtiter plates. Instruments of this kind have been adapted as a super-high-throughput analytical tool for *ee* determination [51, 57]. In this initial study chiral amines of the type **5**, which are of importance in the synthesis of pharmaceutical and agrochemical products [58], were used as the model substrates. They are potentially accessible by catalytic reductive amination of ketones **3**, Markovnikov addition of ammonia to olefins **4** or enzymatic hydrolysis of acetamides **6** (the reverse reaction also being possible; Fig. 11.6).

In setting up a CAE-based high-throughput *ee*-screening system, the conditions used in the conventional CE assay of the amines **5** were first optimized using various α- and β-CD derivatives as chiral selectors [51, 57]. The amines were derivatized by conventional reaction with fluorescene isothiocyanate (**7**) to give the fluorescence-active compounds **8** (Fig. 11.7) which enables the use of a sensitive detection system, specifically laser-induced fluorescence detection (LIF). Although extensive optimiza-

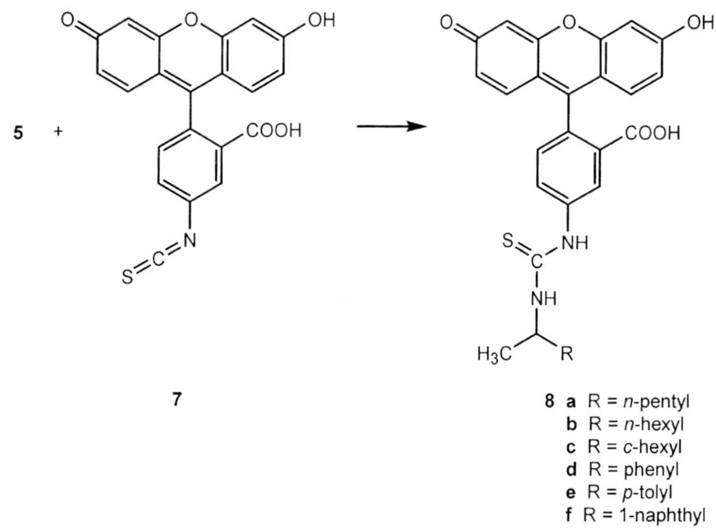

5 +

7

8 a R = *n*-pentyl
b R = *n*-hexyl
c R = *c*-hexyl
d R = phenyl
e R = *p*-tolyl
f R = 1-naphthyl

Fig. 11.7. Derivatization of chiral amines **5** for LIF detection in CAE screening [51, 57].

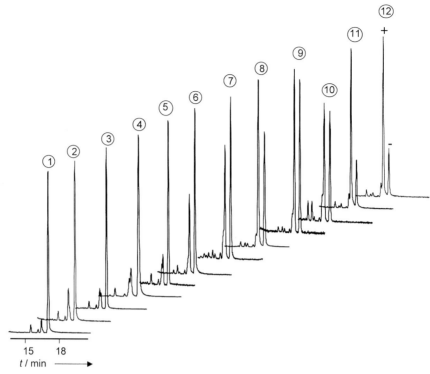

15   18
t / min ⟶

Fig. 11.8. CAE antipode separation of the enantiomers of **8c** in a representative sample [51, 57].

tion was not carried out (only six CD derivatives were tested), satisfactory baseline separation was accomplished in all cases.

The next step involved the use of compounds **5c/8c** as the model substrates for CAE analysis using the MegaBACE system (or similar instruments). Known enantiomeric mixtures of the amine **5c** were transformed into the fluorescence-active derivative **8c**. The latter samples were then analyzed by CAE. Unfortunately, the results of the conventional single capillary system could not be reproduced in the CAE experiments because of unstable electrophoretic runs. The problem was solved by developing a special electrolyte having a higher viscosity. It is composed of 40 mM 2-(N-cyclohexylamino)ethanesulfonic acid (CHES; pH 9.1) and 6.25 mM γ-CD (5/1) diluted with a buffer containing linear polyacrylamide. The instrument was operated at a voltage of -10 kV (at 8 μA) per capillary and a sampling voltage of -2 kV (9 s). Under these conditions baseline separation was excellent (Fig. 11.8). The agreement between *ee* values of mixtures of (*R*)- and (*S*)-**8c** determined by CAE and those of the corresponding mixture of (*R*)- and (*S*)-**5c** as measured by GC turned out to be excellent [51, 57].

The enantiomer separation of $(R)/(S)$-**8c** in this system required about 19 minutes. This means that even though the conditions are far from optimized the automated 96-array system provides more than 7000 *ee* values in a single day [51, 57]. In related cases optimization resulted in shorter analysis times for the separation of the enantiomers, so that a daily throughput of 15 000 to 30 000 *ee* determinations is realistic. Such super-high-throughput screening for enantioselectivity is not possible by any other currently available technology. In view of the possibility of chiral selector optimization and the fact that CAE has many advantages such as the use of extremely small amounts of samples, essentially no solvent consumption, the absence of high pressure pumps and values, as well as high durability of columns, the CAE assay is ideally suited for high-speed *ee* determination in enzymatic reactions (or transition metal catalyzed processes).

The possibility of high-throughput *ee* screening of chiral organic compounds utilizing capillary electrophoresis on microchips has also been considered [51, 59]. The general procedure for CE (or more specifically CAE) on microchips (typically 10 cm × 10 cm) had previously been developed by various groups [55f – m] for the analysis of biomolecules. Traditional photolithographic techniques were used to produce capillary arrays on plastic or glass microchips. However, enantiomer separation of organic molecules on plastic microchips using organic solvents is not generally feasible due to the chemical instability of such chips. The situation is quite different in the case of glass chips [51, 57]. In such a modification the separation of enantiomers, for example, of compound **8c**, was shown to be possible with a detection system based on laser-induced fluorescence (LIF). Automation using robotics is currently being implemented, which means that a second CAE-based assay for super-high-throughput *ee* determination can be expected to emerge in the future. However, this refers to fundamental research [51, 55f – m, 57] and the appropriate instruments need to be developed before commercialization becomes available. Of course, in any CAE-system, derivatization and antipode separation of the relevant substrates need to be efficient, which means that universal generality cannot be claimed.

The two forms of capillary array electrophoresis are emerging as powerful methods for the determination of enantiomeric purity of chiral compounds in a truly high-throughput manner. Various modifications are possible, for example, detection systems based on UV/Vis, MS, or electrical conductivity. Moreover, chiral selectors in the CE electrolyte are not even necessary if the mixture of enantiomers is first converted into diastereomers, for example, using chiral fluorescent-active derivatization agents [51, 57].

Other *ee* assays have been described, although in several cases the actual degree of throughput was not specified [10]. These systems include assays based on color tests [8, 46 – 48, 60], IR-thermography [49, 61], circular dichroism [62], fluorescence [63], and even special forms of gas chromatography [64]. Moreover, optically active compounds capable of enantioselective recognition of chiral substrates can be used as

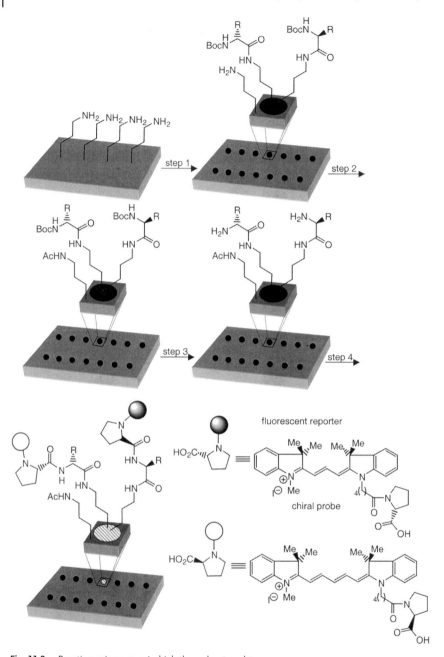

Fig. 11.9. Reaction microarrays in high-throughput *ee*-determination [66]. [a] Reagents and conditions: step 1) BocHNCH(R)-CO$_2$H, PyAOP, *i*Pr$_2$NEt, DMF; step 2) Ac$_2$O, pyridine; step 3) 10 % CF$_3$CO$_2$H and 10 % Et$_3$SiH in CH$_2$Cl$_2$, then 3 % Et$_3$N in CH$_2$Cl$_2$; step 4) pentafluorophenyl diphenylphosphinate, *i*Pr$_2$NEt, 1 : 1 mixture of the two fluorescent proline derivatives, DMF, −20 °C.

chemical sensors, especially if they are fluorescence active, although high-throughput still needs to be developed [65]. In a different approach DNA microarray technology was adapted for the high-throughput determination of enantioselectivity [66]. DNA microarrays had previously been used to determine relative gene expression levels on a genome-wide basis as measured by a ratio of fluorescent reporters [67]. In the newly developed *ee*-assay, amino acids were used as model compounds which were first protected at the amino function (*N*-Boc). Samples were then covalently attached to amine-functionalized glass slides in a spatially arrayed manner (Fig. 11.9) [66]. In a second step the uncoupled surface amines were acylated exhaustively. The third step involved complete deprotection to afford the free amino function of the amino acid. Finally, in a fourth step, two pseudo-enantiomeric fluorescent probes were attached to the free amino groups on the surface of the array. Parallel kinetic resolution in the process of amide coupling provided the basis for determining the *ee* values, these being experimentally accessible by measuring the ratio of the relevant fluorescent intensities. It was reported that 8000 *ee* determinations per day are possible, precision amounting to 10 % *ee* of the actual value [66].

In summary, a variety of different approaches to the high-throughput determination of enantioselectivity have been described [10]. In several cases, as in the ESI-MS and CAE based systems, the appropriate instruments are commercially available. No single system is truly universal, which means that the assays may well be complementary, depending upon the particular problems at hand. Finally, it needs to be pointed out that selection systems, which should be distinguished from screening assays [10], have not yet been developed to include enantioselectivity. Several approaches are possible, including phage display [68] and *in vivo* selection [69].

11.5
Examples of Directed Evolution of Enantioselective Enzymes

11.5.1
Kinetic resolution of a chiral ester catalyzed by mutant lipases

Lipases are the most frequently used enzymes in organic chemistry, catalyzing the hydrolysis of carboxylic acid esters or the reverse reaction in organic solvents [3, 5, 34, 70]. The first example of directed evolution of an enantioselective enzyme according to the principle outlined in Fig. 11.2 concerns the hydrolytic kinetic resolution of the chiral ester **9** catalyzed by the bacterial lipase from *Pseudomonas aeruginosa* [8]. This enzyme is composed of 285 amino acids [32]. It is an active catalyst for the model reaction, but enantioselectivity is poor (*ee* 5 % in favor of the (*S*)-acid **10** at about 50 % conversion) (Fig. 11.10) [71]. The selectivity factor *E*, which reflects the relative rate of the reactions of the (*S*)- and (*R*)-substrates, is only 1.1.

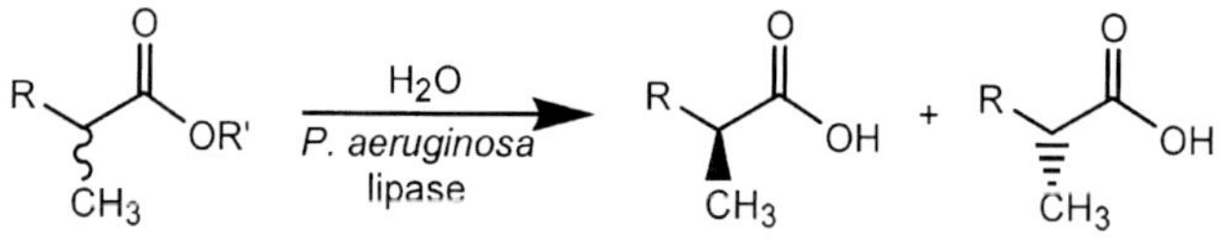

Fig. 11.10. Hydrolytic kinetic resolution of the racemic ester **9** [8, 9].

When performing random mutagenesis, the problem of exploring protein sequence space needs to be considered first. In the present case complete randomization allowing for all possible permutations would theoretically result in 20^{285} different variant enzymes, the masses of which would greatly exceed the mass of the universe, even if only one molecule of each enzyme were to be produced [9]. The other end of the scale entails the minimum amount of structural change, namely the substitution of just one amino acid per molecule of enzyme. On the basis of the algorithm $N = 19^M \times 285!/ [(285 - M)! \times M!]$, in which M = number of amino acid substitutions per enzyme molecule (here M = 1), the library of variants N would theoretically contain 5415 members [8, 9]. However, due to the redundancy of the genetic code, it is impossible to generate by epPCR a library which in fact contains all of the 5415 variants. If the mutation rate is increased to such an extent that an average of two amino acids are exchanged per enzyme molecule (M = 2), then the number of variant *P. aeruginosa* lipases predicted by the above algorithm increases dramatically to about 15 million, which would be very difficult to assay even when applying the best high-throughput screening systems currently available. In the case of M = 3, the number of variants is 52 billion!

The initial strategy was based on a relatively low mutation rate (M = 1), relying on step-wise improvements in enantioselectivity [8, 9, 42]. When creating thousands of lipase variants (or of any other enzyme), the challenge of deconvolution might appear to be an insurmountable task. However, this problem never arises because subsequent to mutagenesis and expression, bacterial colonies are obtained on agar plates, each originating from a single cell. This means that each bacterial colony produces only one variant enzyme (although some may occur more than once on the agar plate). The bacterial colonies are then collected automatically by a colony picker, and are grown in the wells of microtiter plates containing nutrient broth. In this way the supernatant of each well contains just one species of variant, spatially addressable as a catalyst in the model reaction.

Upon generating a library of only 1000 variants in the first generation, about twelve improved variants were identified, the best one resulting in an *ee* value of 31 % ($E = 2.1$) in the test reaction. The process was repeated in the second, third and fourth round of mutagenesis as outlined in Fig. 11.2 with formation of slightly larger libraries (2000 –

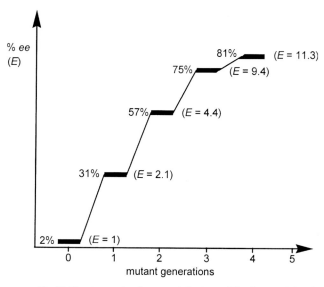

Fig. 11.11. Increasing the *ee* and *E* values of the lipase-catalyzed hydrolysis of the chiral ester **9** [8, 9].

3000 variants), an endeavor that led to an *ee* of 81 % in the fourth generation; this corresponds to a selectivity factor of *E* = 11.3 (Fig. 11.11) [8].

Following these remarkable observations a larger library of variant enzymes was created in the fifth generation, which led to further improvements. Moreover, alternative ways to explore protein sequence space with respect to enantioselectivity in the given test reaction were developed which turned out to be even more efficient [9, 26, 42]. The basic problem relates to the fact that upon passing from one mutant generation to the next, many different "pathways" in protein sequence space are possible. Thus, as in natural evolution itself [72], the analogy with a tree having many branches is useful (Fig. 11.12) [9]. The challenge is to find the shortest possible route in climbing up the "*ee* tree" (or "*E* tree"). Parenthetically, the tree also has roots (not shown in Fig. 11.12), symbolizing the evolution of variants which catalyze the formation of the product having the opposite absolute configuration [9]. Although the cartoon in Fig. 11.12 illustrates considerable complexity, it also suggests that the solution to the problem of creating and finding a highly enantioselective catalyst for a given reaction is not unique. This means that it should be possible to obtain a whole family of different variants, all with high degrees of enantioselectivity for a given reaction. Indeed, this was verified experimentally as delineated below.

A strategy was developed that not only works well in the present situation involving the lipase-catalyzed model reaction **9** → **10** [9, 10, 42], but which may prove to be useful in other cases regarding the evolution of enantioselective enzymes as well. Accord-

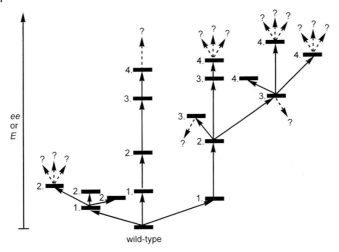

ee
or
E

wild-type

Fig. 11.12. An "evolutionary tree" illustrating the complexity of protein sequence space with respect to enantioselectivity [9]. The simplified scheme is actually much more complicated. The numbers denote positive enzyme-variants obtained from repetitive cycles of mutagenesis and screening in the respective generations. The arrows pointing down symbolize inferior variants which of course out-number the few positive variants.

Variant 01E4 (*E* = 2.1):	S 149G
Variant 08H3 (*E* = 4.4):	S 149G
	S 155L
Variant 13D10 (*E* = 9.4):	S 149G
	S 155L
	V 47G
Variant 04H3 (*E* = 11.3):	S 149G
	S 155L
	V 47L
	F 259L

Fig. 11.13. Data of amino acid exchanges in the best lipase-variants of the first four generations [9, 42].

ingly, DNA sequencing was first performed on the best mutant in each generation in order to define the position and nature of amino acid substitutions responsible for the increase in enantioselectivity. Typical data for the present case are presented in Fig. 11.13 [9].

At this point the actual three-dimensional structure of the wild-type or of the mutant lipases was still of no concern. Indeed, not even the enzyme mechanism was considered! Rather, it was logical to conclude that sensitive positions ("hot spots") in the protein that are instrumental in improving the enantioselectivity had been identified. Moreover, it was reasonable to assume that the observed amino acid exchanges imply the correct position, but owing to the limitations of epPCR not necessarily the optimal amino acid. Thus, it seemed worthwhile to apply saturation mutagenesis [9, 42] at one or more of these "hot spots". In doing so, an excess of 300 – 400 bacterial colonies had to be screened, which is much less than in a normal epPCR-based mutagenesis experiment. For example, position 155 was chosen, saturation mutagenesis being possible in any of the mutant generations or even the wild-type lipase. Saturation mutagenesis was first applied to the mutant gene encoding the best variant with $E = 4.4$ in the second generation in which serine (S) has been substituted by leucine (L), that is, with O8H3. Indeed, the best enzyme in this newly formed library showed a slightly improved selectivity factor ($E = 5.3$) (Fig. 11.14). The new amino acid at position 155 of

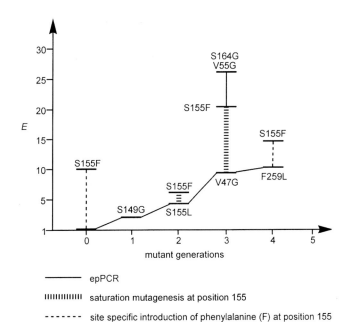

Fig. 11.14. Further improvements in enantioselectivity of the lipase-catalyzed model reaction of ester **9** (S = serine; F = phenylalanine; G = glycine; L = leucine; V = valine). Details are outlined in [42].

this particular variant turned out to be phenylalanine (F). Saturation mutagenesis at position 155 using the gene encoding the most selective enzyme in the third generation led to the identification of an even better variant ($E = 21$). In this case phenylalanine was again identified as the new amino acid at position 155. This seemed to indicate that position 155 is indeed a sensitive spot and that out of all the 20 natural amino acids phenylalanine has the greatest positive influence on enantioselectivity at this particular position (Fig. 11.14). Thus, in order to minimize further screening efforts, phenylalanine was introduced in the best variant of the fourth generation mutant and in the wild-type lipase by site specific mutagenesis. Indeed, significant improvements were observed. It then appeared logical to utilize any one of these mutants as the starting point for further rounds of epPCR. For the sake of clarity and illustration only part of the data which is typical for this type of strategy is shown, namely the result of epPCR in the third generation leading to a variant with an E value of 25 ($ee = 90$ %) [9, 42]. This variant, designated as 94G12, is characterized by five mutations (V47G, V55G, S149G, S155F and S164G).

It is thus clear that the *combination* of epPCR and saturation mutagenesis constitutes an efficient way to explore protein sequence space with respect to enantioselectivity. Indeed, this strategy led to the creation of several other highly (S)-selective variants (*ee* = 88 – 91 %; $E = 20 – 25$) [9, 42], all of them being the descendents of the parent wild-type lipase, which is only slightly (S)-selective.

Although a continuation of epPCR/saturation mutagenesis would constitute a promising search program, alternative strategies were explored. Specifically, questions regarding higher mutation rate in epPCR experiments and the possibility of applying recombinant methods such as DNA shuffling [12] were considered [43]. Unfortunately initial experiments were disappointing. DNA-shuffling using the mutant 94G12 which encodes the most enantioselective variant and the wild-type failed to lead to better or comparable variants [43]. Thus, back-crossing [73] has no positive effects in this case.

As delineated above, if the mutation rate is doubled in order to induce an average of two amino acid substitutions per enzyme molecule, about 15 million different mutants are theoretically possible. A simple calculation shows that if 3000 mutants are then screened, less than 0.02 % of the total protein sequence space would be scanned. In the case of three amino acid exchanges, the statistics are even more drastic. Nevertheless, experiments at relatively "high" mutation rate leading to an average of 2 – 3 amino acids substitutions were performed. About 15 000 variants of the initial library were screened for enantioselectivity in the model reaction. Several (S)- and even (R)-

$$38C8 \quad + \quad 40H4 \quad + \quad 94G12 \quad \xrightarrow{\text{DNA-shuffling}} \quad 8D1$$
$$(E = 6.5) \quad\quad (E = 3.0) \quad\quad (E = 25) \quad\quad\quad\quad\quad (E = 32)$$

Fig. 11.15. DNA-shuffling of three genes using a DNA fragment size of 30 – 50 bp [43].

selective variants were found. The two most (*S*)-selective enzymes, 38C8 (*E* = 6.5) and 40H4 (*E* = 3.0) were shown to have three mutations (D20N, S161P, and T234S; S53P, C180T, and G272A). DNA-shuffling was then applied using mutants 38C8, 40H4, and 94G12 at a DNA fragment size of 30 – 50 bp. The best variant (8D1) displayed an improved selectivity factor (*E* = 32) (Fig. 11.15) [43]. Lipase 8D1 contains five mutations (V47G, S149G, S155F, S199G, and T234S). Three of them originate from gene 94G12, one of them has its origin in gene 38C8 and the fifth one (S199G) is new.

Although at this point further exploration of protein sequence space with respect to enantioselectivity of variants catalyzing the model reaction would be logical on the basis of additional epPCR, saturation mutagenesis and/or DNA-shuffling, the application of cassette mutagenesis was considered next. Progress in the development of methods appeared more important than the sheer goal of attaining high enantioselectivities. Following the identification of "hot regions" and "hot spots", attention was focused on positions 155 and 162. A cassette was produced by "saturating" these positions using equimolar nucleotide mixtures (cassette size: 69 bp). Following expression and screening the three most enantioselective variants were identified as 1A1 (S155S, L162G), 1A11 (S155F, L162G) and 2E7 (S155V, L162G) displaying selectivity factors *E* of 34, 22 and 30, respectively [43]. It is striking that in all three cases mutation L162G was maintained, thereby supporting earlier conjectures regarding the importance of this amino acid substitution [42].

Although these lipase variants, which have only two amino acid substitutions, show notable degrees of enantioselectivity in the model reaction, they are not better than the best variants of previous experiments. However, the efforts are not in vain because they pave the way to a novel strategy for additional improvements in enantioselectivity as discussed below.

The starting point for further exploration of protein sequence space was the conjecture that Stemmer's method of Combinatorial Multiple Cassette Mutagenesis (CMCM) [74] can be applied in appropriately modified form. It is a special type of DNA-shuffling which can be used to generate mutant gene libraries in which cassettes composed of random or defined sequences and the wild-type are incorporated randomly. CMCM had been developed for use in the area of functional antibodies [74].

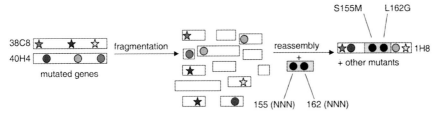

Fig. 11.16. Schematic representation of modified Combinatorial Multiple Cassette Mutagenesis of lipase genes [43].

Tab. 11.1. Best mutant (1H8) and encoded enzyme variant (1H8)
as a result of Combinatorial Multiple Cassette-Mutagenesis [43].

Amino acid exchange	Origin	Base exchange
D20N	gene 38C8	GAC → AAC
S53P	gene 40H4	TCG → CCG
S155M	cassette	TCA → ATG
L162G	cassette	CTG → GGC
T180I	gene 40H4	ACC → ATC
T234S	gene 38C8	ACC → TCC

In the present case of mutant enzymes the goal was to carry out DNA-shuffling in the form of CMCM using the two previously generated mutants 40H4 and 38C8 in combination with an oligocassette representing saturation at positions 155 and 162 (Fig. 11.16). Following expression and screening several enantioselective lipase-variants were found, among them variant 1H8 containing six exchanged amino acids (Tab. 11.1). It shows an unprecedented enantioselectivity of $E > 51$ ($ee > 95\%$) and appears to be particularly active and stable [43].

Interestingly, the result of sequencing (Tab. 11.1) shows that the mutation L → G at position 162 is maintained, whereas S155F transforms into S155M. In summary, the experiments described here constitute the first examples of the application of recombinant methods in the quest to improve the enantioselectivity of enzymes. It is apparent that several forms of DNA shuffling are quite successful.

Another intriguing question concerns the possibility of reversing the sense of enantioselectivity. In order to evolve (R)-selective lipase-variants for the model reaction *rac*-**9** → (R)-**10**, screening needs to be switched to (R)-selectivity. Initial experiments turned out to be promising [9]. Upon carefully scrutinizing the library of variants obtained in the first round of mutagenesis at "high" rate as described above, several (R)-selective enzymes were in fact identified (in addition to the (S)-selective analogs previously mentioned). The best one displayed a selectivity factor of $E = 2$ in favor of the (R)-configured acid **10**. This means that in one and the same library (S)- as well as (R)-selective variants exist. In the second round of "high" rate mutagenesis an (R)-selective variant showing improved enantioselectivity ($E = 3.7$) was found [75]. These experiments provided genes for DNA-shuffling, which led to the formation of a highly improved (R)-selective variant ($E = 30$). At present the overall results concerning (S)- and (R)-selectivity in the model reaction, obtained by screening a total of about 90 000 variants[**], are summarized in Fig. 11.17.

[**] Strictly speaking 90 000 bacterial colonies were picked and scrutinized, some of them occurring more than once.

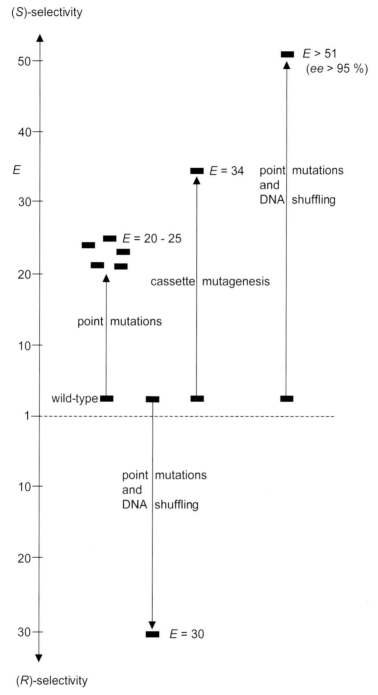

Fig. 11.17. Optional (*S*)- or (*R*)-selectivity in the lipase-catalyzed hydrolysis of ester **9**
[9, 42, 43, 75].

The mutagenesis experiments summarized above not only represent the first example of directed evolution of enantioselective enzymes [8], they also constitute the most comprehensive study concerning the exploration of protein sequence space with respect to enantioselectivity of a given enzyme-catalyzed reaction [9, 42, 43, 75]. Since they do not depend on theoretical predictions based on molecular modeling (which may be imperfect), the approach can be considered to be rational. Indeed, structure or mechanism are not part of the approach. Nevertheless, one of the intriguing aspects of directed evolution of enantioselective enzymes is the fact that once optimal variants of enzymes have been evolved, it is possible to learn some lessons concerning the origin of selectivity. By considering enantioselectivity in the light of structural properties of the wild-type enzyme and of the (*S*)- and (*R*)-selective variants, the foundations are laid for meaningful theoretical interpretations. Although X-ray structural data of the variants are not yet available, an initial interpretation of some of the results in the light of the crystal structure of the wild-type enzyme has been attempted, especially concerning lipase-variant 94G12 [42]. The interpretation is summarized below.

Recently, the 3D structure of *P. aeruginosa* lipase was solved at 2.54 Å resolution as a complex with a covalently bound inhibitor [76]. This lipase shows an α/β hydrolase fold with a six-stranded parallel β-sheet surrounded by α-helices on both sides. The catalytic triad is composed of residues Ser82, Asp229, and His251. The loop containing His251 is stabilized by an octahedrally coordinated calcium ion. The enzyme can adopt a closed or an open conformation depending on the absence or presence of a substrate. The structure was solved in the open conformation with the catalytic site being accessible for a substrate-analogous inhibitor which was bound to the active site serine. In this conformation, the position of the oxyanion hole as well as three substrate binding pockets which can accommodate fatty acid chains could be identified. It is interesting to note that the main forces keeping the substrates' acyl groups in position were identified as van der Waals interactions. A more detailed analysis of the binding mode of the inhibitor further suggested that the size and interactions of the acyl pockets are the predominant determinants of the enzyme's enantiopreference.

The availability of the 3D structure of the wild-type *P. aeruginosa* lipase allowed the positions of the substitutions in the enzyme variants to be located and the effects of the mutations to be rationalized [42] (Fig. 11.18). As delineated above, one of the early lipase variants (94G12) showing excellent enantioselectivity towards the (*S*)-enantiomer of the substrate (**9**) is characterized by five amino acid substitutions, namely V47G, V55G, S149G, S155F and S164G. It is interesting to note that four of these substitutions lead to the introduction of a glycine residue which may increase the overall conformational flexibility of this lipase. Flexibility has been identified before as an important parameter which determines the substrate selectivity of wild-type fungal lipases from *Geotrichum candidum* [77] and *Rhizomucor miehei* [78]. However, the core of these enzymes seems to be markedly rigid whereas the flexible do-

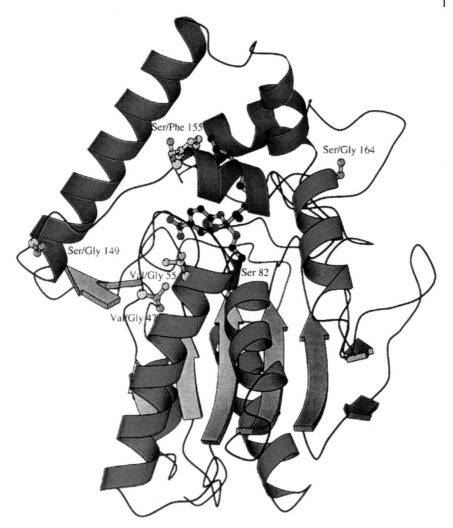

Fig. 11.18. X-ray crystal structure of the wild-type lipase from
Pseudomonas aeruginosa [76], showing the modeled in positions of
the "hot spots" of the enantioselective variant 94G12 (V47G, V55G,
S149G, S155F, and S164G) and the bound substrate (*S*)-**9** [42].

mains are restricted to a few surface-exposed loops as deduced from molecular dynamics simulations [78a, 79]. Such a scaffold comprising a structurally conserved
and rigid α/β core domain and a few surface-exposed and flexible loops has also
been observed in lipases from *Burkholderia cepacia* [80].

Surprisingly, in the case of the evolved variant 94G12, none of the five amino acid
substitutions is in direct contact with the bound substrate at the active center, thus
excluding a direct spatial effect on the enantioselectivity of the reaction [9, 42]. In-

stead, the substitutions are located directly or in close vicinity to loops which are involved in the enzyme's transition from the closed to the open conformation. A similar pattern was found in the case of many of the other enantioselective variants. Remote effects based on amino acid substitutions spatially removed from the active site are known to influence activity and stability of enzymes, as shown by recent studies [81]. However, this is the first observation of remote amino acid substitutions affecting *stereoselectivity*. How can these results be understood?

The amino acids Val47 and Val55 are directly or indirectly involved in a hydrogen bonding network which anchors His14 to helix α2 and to the loop between strand β4 and helix α2. His14 itself is part of the loop containing residue Met16 which is involved in the formation of the oxyanion hole, a structural prerequisite for the formation of the transition state which occurs during substrate hydrolysis. Therefore, the substitutions V47G and V55G can be expected to increase the flexibility of the region between strand β4 and helix α2 and, at the same time, may also affect the stable conformation of the catalytically essential oxyanion hole [42].

The amino acids Ser149, Ser155, and Ser164 are located in the region between helices α5 and α7. Their substitution may affect the secondary structure of the region following the helix α5, not only because of an increased flexibility, but also because of the loss of hydrogen bonds between the side chains of S149 and Ser164. Ultimately, this may even result in a somewhat different relative orientation of helices α5 to α7 in the enzyme-variant 94G12 [42]. This region contains residues which line part of the enzyme's substrate binding site; therefore, the repositioning of these helices could also affect the substrate binding properties of the enzyme. In particular, repositioning of the Leu162 side chain would have a large influence on the enantioselectivity of the enzyme as its side chain would directly interact with the methyl group of the (S)-substrate. Additionally, residues Leu50, Thr158, Leu159 and Leu62 of the substrate binding pocket exert van der Waals interactions with the side chain of Met16 which is involved in oxyanion hole formation. Therefore, variations of the position of these residues might affect the architecture of the oxyanion hole and thus indirectly influence the enantioselectivity of the enzyme. The pronounced positive effect of substitution S155F may be explained by the observation that the introduction of Phe155 brings this residue at van der Waals distance to the acyl group of the substrate thereby directly affecting the enantioselectivity of the enzyme [42]. Final certainty concerning the interpretation of the data will be possible if the results of an X-ray structural analysis of the enantioselective variant 94G12 becomes available. The same applies to (R)-selective variants.

In the case of the most (S)-selective variant 1H8 observed so far, which was evolved by Combinatorial Multiple Cassette Mutagenesis [43] (see above), the analysis based on molecular modeling and molecular dynamics calculations has not been completed. However, based on the results of the sequencing experiments and the X-ray structure of the wild-type lipase, it is already clear that the situation is somewhat different. Only three of the six mutations occur in loops, whereas the three others are found in α-helices [43]. Thus, remote effects resulting in a rearrangement of the enzyme as well as traditional steric effects near the active site may be operating in this case. Here again X-ray structural data would be of great help in understanding the structural changes and their consequences regarding enantioselectivity brought about by the evolutionary process.

11.5.2
Evolution of a lipase for the stereoselective hydrolysis of a meso-compound

In a different ongoing study, a *Bacillus subtilis* lipase has been chosen as the catalyst in the asymmetric hydrolysis of the meso-diacetate **11** with formation of enantiomeric alcohols **12** (Fig. 11.19) [82]. This reaction does not constitute kinetic resolution and can thus be carried out to 100 % conversion. Screening is possible on the basis of the ESI-MS system [50] (see above) using the deuterium labeled pseudo-meso substrate **13** (Fig. 11.20). The ratio of the two pseudo-enantiomeric products **14** and **15** can easily be determined by integrating the two appropriate MS peaks.

The wild-type enzyme leads to an *ee* value of only 38 % in favor of (*R,S*)-**12**. Following an initial round of epPCR-based random mutagenesis, a variant showing an *ee* value of 60 % was identified. This was increased to about 70 % *ee* in the second round. Current

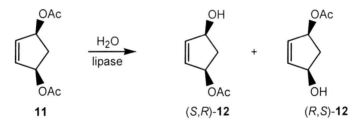

Fig. 11.19. Desymmetrization of a meso-substrate **11** catalyzed by a lipase from *Bacillus subtilis* [82].

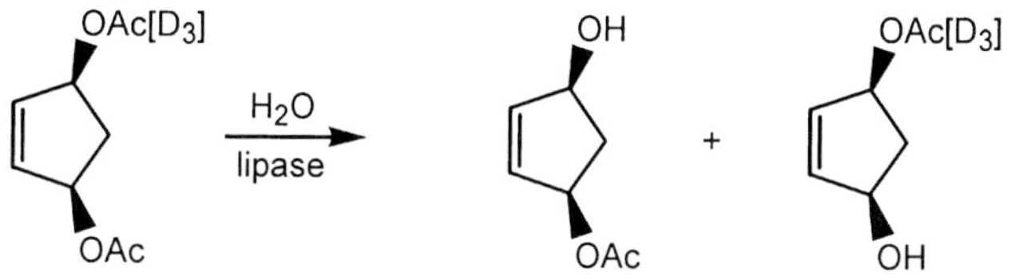

Fig. 11.20. Deuterium labeling in the pseudo-meso compound **13** used in ESI-MS-based *ee*-screening [50].

work includes further rounds of mutagenesis as well as DNA shuffling and the identification of the "hot spots" [82]. Reversal of enantioselectivity (*ee* = 39 %) has also been achieved.

11.5.3
Kinetic resolution of a chiral ester catalyzed by a mutant esterase

A brief report has appeared in which a wild-type esterase from *Pseudomonas fluorescens* (PFE), which shows no activity in the hydrolysis of the ester **16** (Fig. 11.21), was subjected to mutagenesis using the mutator strain *Epicurian coli* XL 1-Red [83]. This resulted in a variant which catalyzes the reaction with an *ee* of 25 %. The absolute configuration of the major product ((*R*)- or (*S*)-**17**) was not determined. Sequencing of the esterase-variant revealed that two point mutations, A 209D and L 181V, had occurred. Since the structure of the enzyme is unknown, a detailed interpretation was not possible, although reasonable speculations were made.

The A 209D/L 181V variant was subjected to a second mutagenesis cycle in hope of increasing enantioselectivity [83]. However, no improvement was observed. Nevertheless, upon isolating the gene of the wild type [84], directed evolution using epPCR and/ or DNA shuffling may lead to success.

11.5.4
Improving the enantioselectivity of a transaminase

Another case of the use of a single round of mutagenesis to improve the enantioselectivity of an enzyme (which therefore is not an evolutionary process) concerns the enantioselectivity of a transaminase. Transaminases are enzymes which catalyze the conversion of ketones to the corresponding primary amines, the *ee* values often being >95 % [85]. However, in one case an *ee* of only 65 % was observed [86]. A single library of variants was produced which contained an enzyme showing 98 % *ee*. Unfortunately

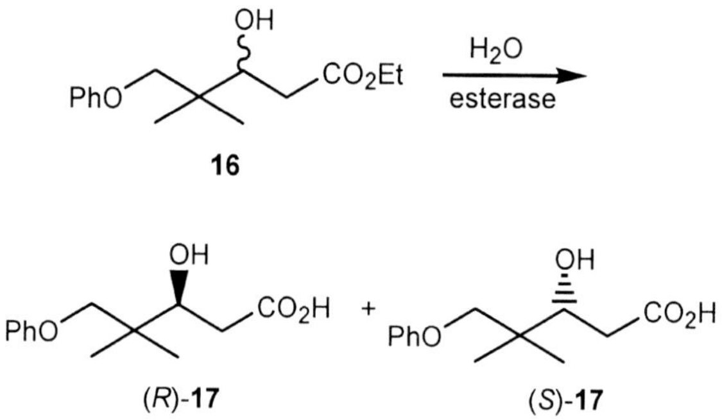

Fig. 11.21. Esterase-catalyzed kinetic resolution of ester **16** [83].

this brief report did not provide any information regarding details of mutagenesis, size of the library or screening system.

11.5.5
Inversion of the enantioselectivity of a hydantoinase

In elegant work the enantioselectivity of a hydantoinase from *Arthrobacter species* for the production of L-methionine in *Escherichia coli* has been inverted [87]. The approach is similar to the one used in the evolution of (S)- and (R)-selective lipases (see above). All known hydantoinases are selective for D-5-(2-methylthioethyl)hydantoin (D-**18**) which leads to the accumulation of N-carbamoyl-D-methionine (D-**19**), conversion being complete if the conditions of dynamic kinetic resolution are upheld [88], in this case by the use of a racemase or pH >8 (Fig. 11.22).

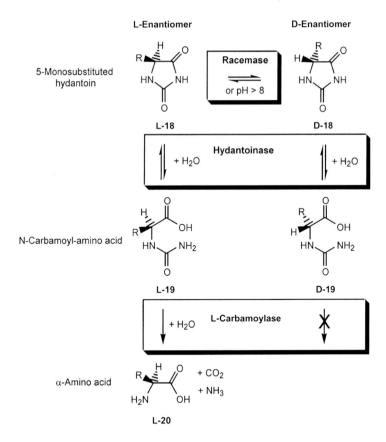

Fig. 11.22. Reactions and enzymes involved in the production of L-amino acids from racemic hydantoins by the three-enzyme hydantoinase process [87].

The wild-type hydantoinase gene was subjected to two successive rounds of random mutagenesis by epPCR followed by saturation mutagenesis and the resulting libraries were screened for altered enantioselectivity by a pH indicator assay [87]. One variant was identified carrying three amino acid exchanges which was 1.5-fold more active than the wild-type enzyme and produced N-carbamoyl-L-methionine (L-**19**) with an *ee* = 20 % at 30 % conversion. When the corresponding gene was co-expressed with a hydantoin racemase and the L-N-carbamoylase in *E. coli*, this led to the conversion of 90 % of the precursor compound D,L-5-(2-methylthioethyl)hydantoin in less than 2 h as compared to 10 h for the wild-type enzyme. Although enantioselectivity needs to be improved, this result clearly demonstrates the possibility of optimizing multi-enzyme pathways by directed evolution of single enzyme genes, thereby improving whole-cell catalysts for the production of chiral compounds.

11.5.6
Evolving aldolases which accept both D- and L-glyceraldehydes

The stereo-controlled formation of carbon – carbon bonds can be performed biocatalytically using aldolases [3, 89]. For example, the 2-keto-3-deoxy-6-phosphogluconate (KDPG) aldolase from *E. coli* catalyzes the reversible addition of pyruvate to a number of aldehydes to form 4-substituted-4-hydroxy-2-ketobutyrates. This enzyme is highly specific for aldehydes with the D-configuration at the C2 position and the formation of the new stereo center occurs with complete selectivity. Recently, directed evolution has been applied to evolve novel aldolases capable of accepting *both* D- and L-glyceraldehydes as substrates [90] (Fig. 11.23).

A library was generated by epPCR with a "low" mutation rate (average of one amino acid exchange per gene and generation) and screened for variants which accept the non-phosphorylated D-2-keto-3-deoxygluconate substrates or the L-glyceraldehyde (L-**22**) instead of the D-enantiomer [90]. The screening method consisted of a coupled enzyme assay on microtiter plates with spectrophotometric determination of the time-dependent decrease of NADH absorbance at 340 nm. The genes of four first generation variants with improved properties were subjected to DNA shuffling and the best gene was again randomly mutated by epPCR yielding several third generation variants which showed a 10-fold improved activity towards the non-phosphorylated substrate D-2-keto-3-deoxygluconate and a more than 5-fold improved rate for the addition of the non-natural L-**22** to pyruvate. This study demonstrates that novel aldolases can successfully be created by directed evolution and that this goal can be reached by screening only a relatively small number of variants [90]. A total of four to five amino acid exchanges introduced into the wild-type enzyme was sufficient here to evolve aldolases which, in contrast to the wild-type enzymes, catalyze the efficient synthesis

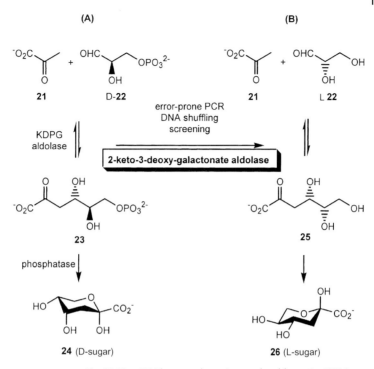

Fig. 11.23. (A) The natural reaction catalyzed by native KDPG aldolase, followed by removal of the phosphate catalyzed by phosphatase. (B) One of the reactions that has been improved by mutagenesis and screening [90].

of intermediates leading to both D- and L-sugars (**24** and **26**, respectively) from non-phosphorylated aldehydes and pyruvate.

11.6
Conclusions

Directed evolution has previously been used to generate enzyme-variants displaying improved properties such as higher activity and stability [11 – 17, 19, 81, 91, 92]. The work summarized in this chapter shows that the difficult goal of controlling enantios-electivity of enzymes can also be reached by directed evolution [8 – 10]. In doing so, two major challenges had to be dealt with:

– development of high-throughput screening systems for assaying enantioselectivity
– development of efficient strategies for exploring protein sequence space with respect to enantioselectivity.

Progress in this new area of asymmetric catalysis has been rapid, but more efforts are necessary including:

– implementation of *ee*-assays based on selection (rather than screening), for example, using phage display or *in vitro* selection
– development of further methods for exploring protein space
– use of novel substrates including those of industrial relevance
– use of other enzymes such as oxidases, reductases, epoxide hydrolases, etc.
– transforming a given enzyme into a novel enzyme type, e.g. a lipase into a Diels-Alderase.

Moreover, it will be interesting to see if the application of genetic algorithms [93] and/or neural networks [94] is of any help in exploring protein sequence space in the quest to create and find enantioselective enzymes [10]. In any case, data management and data analysis will become more important in the future [10]. It is also important to emphasize that directed evolution of enantioselective enzymes offers a unique opportunity to increase our knowledge of the way enzymes function. This is due to the fact that enantioselectivity is a very sensitive parameter [9]. Thus, it is likely that the crystallographic determination of the 3D structures of the (*S*)- and (*R*)-selective lipase variants (see above) in combination with molecular modeling and molecular dynamics calculations will uncover the structural basis of enantioselectivity. It can also be foreseen that directed evolution will be used successfully to create novel enzymes which are highly enantioselective and active as well as stable enough to allow for a variety of biotechnological applications. Comparison with other types of biocatalysts such as catalytic antibodies [4] or ribozymes [95] will then be meaningful.

References

[1] a) A. N. Collins, G. N. Sheldrake, J. Crosby, (Eds.), *Chirality in Industry: The Commercial Manufacture and Applications of Optically Active Compounds*, Wiley, Chichester, 1992; b) A. N. Collins, G. N. Sheldrake, J. Crosby, (Eds.), *Chirality in Industry II: Developments in the Commercial Manufacture and Applications of Optically Active Compounds*, Wiley, Chichester, 1997; c) R. A. Sheldon, *Chirotechnology: Industrial Synthesis of Optically Active Compounds*, Dekker, New York, 1993; d) S. C. Stinson, *Chem. Eng. News* **1999**, *77*(41), 101–120; e) S. C. Stinson, *Chem. Eng. News* **2000**, *78*(43), 55–78.

[2] a) E. N. Jacobsen, A. Pfaltz, H. Yamamoto, (Eds.), *Comprehensive Asymmetric Catalysis*, Vol. I-III, Springer, Berlin, 1999; b) H. Brunner, W. Zettlmeier, *Handbook of Enantioselective Catalysis with Transition Metal Compounds, Vol. I-II*, VCH, Weinheim, 1993; c) R. Noyori, *Asymmetric Catalysis in Organic Synthesis*, Wiley, New York, 1994; d) I. Ojima, (Ed.), *Catalytic Asymmetric Synthesis*, VCH, Weinheim, 1993; e) D. J. Berrisford, C. Bolm, K. B. Sharpless, *Angew. Chem.* **1995**, *107*, 1159–1171; *Angew. Chem. Int. Ed. Engl.* **1995**, *34*, 1059–1070.

[3] a) H. G. Davies, R. H. Green, D. R. Kelly, S. M. Roberts, *Biotransformations in Preparative Organic Chemistry: The Use of Isolated Enzymes and Whole Cell Systems in*

Synthesis, Academic Press, London, 1989;
b) C. H. Wong, G. M. Whitesides, *Enzymes in Synthetic Organic Chemistry* (Tetrahedron Organic Chemistry Series, Vol. 12), Pergamon, Oxford, 1994; c) K. Drauz, H. Waldmann (Eds.), *Enzyme Catalysis in Organic Synthesis: A Comprehensive Handbook, Vol. I-II*, VCH, Weinheim, 1995; d) K. Faber, *Biotransformations in Organic Chemistry*, 3rd ed., Springer, Berlin, 1997; e) G. Carrea, S. Riva, *Angew. Chem.* **2000**, *112*, 2312–2341; *Angew. Chem. Int. Ed.* **2000**, *39*, 2226–2254; f) K. M. Koeller, C.-H. Wong, *Nature (London)* **2001**, *409*, 232–240.

[4] a) P. G. Schultz, R. A. Lerner, *Science (Washington, D.C.)* **1995**, *269*, 1835–1842: b) J.-L. Reymond, *Top. Curr. Chem.* **1999**, *200*, 59–93; c) D. Hilvert, *Top. Stereochem.* **1999**, *22*, 85–135.

[5] R. D. Schmid, R. Verger, *Angew. Chem.* **1998**, *110*, 1694–1720; *Angew. Chem. Int. Ed.* **1998**, *37*, 1608–1633.

[6] a) M. McCoy, *Chem. Eng. News* **1999**, *77*(1), 10–14; b) R. S. Rogers, *Chem. Eng. News* **1999**, *77*(29), 87–91; c) B. Schulze, M. G. Wubbolts, *Curr. Opin. Biotechnol.* **1999**, *10*, 609–615; d) A. Schmid, J. S. Dordick, B. Hauer, A. Kiener, M. Wubbolts, B. Witholt, *Nature (London)* **2001**, *409*, 258–268.

[7] a) M. M. Altamirano, J. M. Blackburn, C. Aguayo, A. R. Fersht, *Nature (London)* **2000**, *403*, 617–622; b) P. N. Bryan, *Biotechnol. Adv.* **1987**, *5*, 221–234; c) J. A. Gerlt, *Chem. Rev.* **1987**, *87*, 1079–1105; d) J. R. Knowles, *Science (Washington, D.C.)* **1987**, *236*, 1252–1258; e) S. J. Benkovic, C. A. Fierke, A. M. Naylor, *Science (Washington, D.C.)* **1988**, *239*, 1105–1110; f) J. A. Wells, D. A. Estell, *Trends Biochem. Sci.* **1988**, *13*, 291–297; g) Y. Hirose, K. Kariya, Y. Nakanishi, Y. Kurono, K. Achiwa, *Tetrahedron Lett.* **1995**, *36*, 1063–1066; h) Z. Shao, F. H. Arnold, *Curr. Opin. Struct. Biol.* **1996**, *6*, 513–518; i) C. Heiss, M. Laivenieks, J. G. Zeikus, R. S. Phillips, *J. Am. Chem. Soc.* **2001**, *123*, 345–346; j) S. Patkar, J. Vind, E. Kelstrup, M. W. Christensen, A. Svendsen, K. Borch, O. Kirk, *Chem. Phys. Lipids* **1998**, *93*, 95–101; k) M. Holmquist, P. Berglund, *Org. Lett.* **1999**, *1*, 763–765; l) P. Kast, D. Hilvert, *Curr. Opin. Struct. Biol.* **1997**, *7*

470–479; m) J. P. Schanstra, A. Ridder, J. Kingma, D. B. Janssen, *Protein Eng.* **1997**, *10*, 53–61; n) M. H. J. Cordes, N. P. Walsh, C. J. McKnight, R. T. Sauer, *Science (Washington, D.C.)* **1999**, *284*, 325–327.

[8] M. T. Reetz, A. Zonta, K. Schimossek, K. Liebeton, K.-E. Jaeger, *Angew. Chem.* **1997**, *109*, 2961–2963; *Angew. Chem. Int. Ed. Engl.* **1997**, *36*, 2830–2832.

[9] M. T. Reetz, K.-E. Jaeger, *Chem. Eur. J.* **2000**, *6*, 407–412.

[10] Review of high-throughput *ee*-screening systems: M. T. Reetz, *Angew. Chem.* **2001**, *113*, 292–320; *Angew. Chem. Int. Ed.* **2001**, *40*, 284–310.

[11] a) D. W. Leung, E. Chen, D. V. Goeddel, *Technique (Philadelphia)* **1989**, *1*, 11–15; b) R. C. Cadwell, G. F. Joyce, *PCR Methods Appl.* **1994**, *3*, S136–S140.

[12] a) W. P. C. Stemmer, *Nature (London)* **1994**, *370*, 389–391; b) P. A. Patten, R. J. Howard, W. P. C. Stemmer, *Curr. Opin. Biotechnol.* **1997**, *8*, 724–733.

[13] H. Zhao, L. Giver, Z. Shao, J. A. Affholter, F. H. Arnold, *Nat. Biotechnol.* **1998**, *16*, 258–261.

[14] Z. Shao, H. Zhao, L. Giver, F. Arnold, *Nucleic Acids Res.* **1998**, *26*, 681–683.

[15] R. C. Cadwell, G. F. Joyce, *Mutagenic PCR.* In: PCR Primer: A laboratory manual, C. H. Dieffenbach, G. S. Dveksler (Eds.), CSHL Press, Cold Spring Harbor, 1995, p 583.

[16] M. Kammann, J. Laufs, J. Schell, B. Gronenborn, *Nucleic Acids Res.* **1989**, *17*, 5405.

[17] A. Crameri, S.-A. Raillard, E. Bermudez, W. P. C. Stemmer, *Nature (London)* **1998**, *391*, 288–291.

[18] A. Urban, S. Neukirchen, K.-E. Jaeger, *Nucleic Acids Res.* **1997**, *25*, 2227–2228.

[19] a) F. H. Arnold, *Acc. Chem. Res.* **1998**, *31*, 125–131; b) J. Affholter, F. H. Arnold, *CHEMTECH* **1999**, *29*, 34–39.

[20] S. A. Benner, *Chem. Rev.* **1989**, *89*, 789–806.

[21] G. MacBeath, P. Kast, D. Hilvert, *Science (Washington, D.C.)* **1998**, *279*, 1958–1961.

[22] D. Naki, C. Paech, G. Ganshaw, V. Schellenberger, *Appl. Microbiol. Biotechnol.* **1998**, *49*, 290–294.

[23] L. D. Bogarad, M. W. Deem, *Proc. Natl. Acad. Sci. USA* **1999**, *96*, 2591–2595.

[24] S. Harayama, *TIBTECH* **1998**, *16*, 76–82.

[25] A. Skandalis, L. P. Encell, L. A. Loeb, *Chem. Biol.* **1997**, *4*, 889–898.

[26] M. T. Reetz, K.-E. Jaeger, *Top. Curr. Chem.* **1999**, *200*, 31–57.

[27] S.-L. Wong, *Curr. Opin. Biotechnol.* **1995**, *6*, 517–522.

[28] W. M. de Vos, M. Kleerebezem, O. P. Kuipers, *Curr. Opin. Biotechnol.* **1997**, *8*, 547–553.

[29] F. Baneyx, *Curr. Opin. Biotechnol.* **1999**, *10*, 411–421.

[30] M. B. Rao, A. M. Tanksale, M. S. Ghatge, V. V. Deshpande, *Microbiol. Mol. Biol. Rev.* **1998**, *62*, 597–635.

[31] a) P. Cornelis, *Curr. Opin. Biotechnol.* **2000**, *11*, 450–454; b) O. Pines, M. Inouye, *Mol. Biotechnol.* **1999**, *12*, 25–34.

[32] K.-E. Jaeger, B. Schneidinger, K. Liebeton, D. Haas, M. T. Reetz, S. Philippou, G. Gerritse, S. Ransac, B. W. Dijkstra, *Lipase of Pseudomonas Aeruginosa: Molecular Biology and Biotechnological Application.* In: Molecular Biology of Pseudomonads, T. Nakazawa, K. Furukawa, D. Haas, S. Silver (Eds.), ASM Press, Washington, **1996**, pp. 319–330.

[33] K.-E. Jaeger, M. T. Reetz, *Trends Biotechnol.* **1998**, *16*, 396–403.

[34] K.-E. Jaeger, B. W. Dijkstra, M. T. Reetz, *Annu. Rev. Microbiol.* **1999**, *53*, 315–351.

[35] M. Koster, W. Bitter, J. Tommassen, *Int. J. Med. Microbiol.* **2000**, *290*, 325–331.

[36] F. Rosenau, J.-E. Jaeger, *Biochimie* **2000**, *82*, 1023–1032.

[37] a) A. Urban, M. Leipelt, T. Eggert, K.-E. Jaeger, *J. Bacteriol.* **2001**, *183*, 587–596; b) K. Liebeton, A. Zacharias, K.-E. Jaeger, *J. Bacteriol.* **2001**, *183*, 597–603.

[38] F. W. Studier, A. H. Rosenberg, J. J. Dunn, J. W. Dubendorff, *Methods Enzymol.* **1990**, *185*, 60–89.

[39] E. Brunschwig, A. Darzins, *Gene* **1992**, *111*, 35–41.

[40] K.-E. Jaeger, B. Schneidinger, F. Rosenau, M. Werner, D. Lang, B. W. Dijkstra, K. Schimossek, A. Zonta, M. T. Reetz, *J. Mol. Catal. B: Enzym.* **1997**, *3*, 3–12.

[41] F. Rosenau, K. Liebeton, K.-E. Jaeger, *Biospektrum* **1998**, *4*, 38–41.

[42] K. Liebeton, A. Zonta, K. Schimossek, M. Nardini, D. Lang, B. W. Dijkstra, M. T. Reetz, K.-E. Jaeger, *Chem. Biol.* **2000**, *7*, 709–718.

[43] M. T. Reetz, S. Wilensek, D. Zha, K.-E. Jaeger, *Angew. Chem.* **2001**, *113*, 3701–3703; *Angew. Chem. Int. Ed.* **2001**, *40*, 3589–3591.

[44] K.-E. Jaeger, T. Eggert, A. Eipper, M. T. Reetz, *Appl. Microbiol. Biotechnol.* **2001**, *55*, 519–530.

[45] A. A. Watson, R. A. Alm, J. S. Mattick, *Gene* **1996**, *172*, 163–164.

[46] L. E. Janes, R. J. Kazlauskas, *J. Org. Chem.* **1997**, *62*, 4560–4561.

[47] L. E. Janes, A. C. Löwendahl, R. J. Kazlauskas, *Chem. Eur. J.* **1998**, *4*, 2324–2331.

[48] a) F. Zocher, M. M. Enzelberger, U. T. Bornscheuer, B. Hauer, R. D. Schmid, *Anal. Chim. Acta* **1999**, *391*, 345–351; b) E. Henke, U. T. Bornscheuer, *Biol. Chem.* **1999**, *380*, 1029–1033.

[49] M. T. Reetz, M. H. Becker, K. M. Kühling, A. Holzwarth, *Angew. Chem.* **1998**, *110*, 2792–2795; *Angew. Chem. Int. Ed.* **1998**, *37*, 2647–2650.

[50] a) M. T. Reetz, M. H. Becker, H.-W. Klein, D. Stöckigt, *Angew. Chem.* **1999**, *111*, 1872–1875; *Angew. Chem. Int. Ed.* **1999**, *38*, 1758–1761; b) W. Schrader, A. Eipper, D. J. Puga, M. T. Reetz, *Can. J. Chem.*, submitted.

[51] M. T. Reetz, K. M. Kühling, A. Deege, H. Hinrichs, D. Belder, *Angew. Chem.* **2000**, *112*, 4049–4052; *Angew. Chem. Int. Ed.* **2000**, *39*, 3891–3893.

[52] M. T. Reetz, A. Eipper, C. Torre, Y. Genzel, R. Furstoss, unpublished results.

[53] J. Guo, J. Wu, G. Siuzdak, M. G. Finn, *Angew. Chem.* **1999**, *111*, 1868–1871; *Angew. Chem. Int. Ed.* **1999**, *38*, 1755–1758.

[54] a) B. Chankvetadze, *Capillary Electrophoresis in Chiral Analysis*, Wiley, Chichester, 1997; b) E. Gassmann, J. E. Kuo, R. N. Zare, *Science (Washington, D. C.)* **1985**, *230*, 813–814; c) L. G. Blomberg, H. Wan, *Electrophoresis* **2000**, *21*, 1940–1952; d) H. Nishi, T. Fukuyama, S. Terabe, *J. Chromatogr.* **1991**, *553*, 503–516; e) S. Fanali, *J. Chromatogr.* **1989**, *474*, 441–446; f) A. Guttman, A. Paulus, A. S. Cohen, N. Grinberg, B. L. Karger, *J. Chromatogr.* **1988**, *448*, 41–53; g) D. Belder, G. Schomburg, *J. Chromatogr. A* **1994**, *666*, 351–365; h) D. Wistuba, V. Schurig, *J. Chromatogr. A* **2000**, *875*,

255–276; i) G. Blaschke, B. Chankvetadze, *J. Chromatogr. A* **2000**, *875*, 3–25.

[55] a) X. C. Huang, M. A. Quesada, R. A. Mathies, *Anal. Chem.* **1992**, *64*, 2149–2154; b) H. Kambara, S. Takahashi, *Nature (London)* **1993**, *361*, 565–566; c) N. J. Dovichi, *Electrophoresis* **1997**, *18*, 2393–2399; d) G. Xue, H. Pang, E. S. Yeung, *Anal. Chem.* **1999**, *71*, 2642–2649; e) S. Behr, M. Mätzig, A. Levin, H. Eickhoff, C. Heller, *Electrophoresis* **1999**, *20*, 1492–1507; f) D. J. Harrison, K. Fluri, K. Seiler, Z. Fan, C. S. Effenhauser, A. Manz, *Science (Washington, D. C.)* **1993**, *261*, 895–897; g) S. C. Jacobson, R. Hergenroder, L. B. Koutny, R. J. Warmack, J. M. Ramsey, *Anal. Chem.* **1994**, *66*, 1107–1113; h) L. D. Hutt, D. P. Glavin, J. L. Bada, R. A. Mathies, *Anal. Chem.* **1999**, *71*, 4000–4006; i) D. Schmalzing, L. Koutny, A. Adourian, P. Belgrader, P. Matsudaira, D. Ehrlich, *Proc. Natl. Acad. Sci. USA* **1997**, *94*, 10273–10278; j) S. C. Jacobson, C. T. Culbertson, J. E. Daler, J. M. Ramsey, *Anal. Chem.* **1998**, *70*, 3476–3480; k) S. Liu, H. Ren, Q. Gao, D. J. Roach, R. T. Loder, Jr., T. M. Armstrong, Q. Mao, I. Blaga, D. L. Barker, S. B. Jovanovich, *Proc. Natl. Acad. Sci. USA* **2000**, *97*, 5369–5374; l) S. R. Wallenborg, C. G. Bailey, *Anal. Chem.* **2000**, *72*, 1872–1878; m) I. Rodriguez, L. J. Jin, S. F. Y. Li, *Electrophoresis* **2000**, *21*, 211–219.

[56] MegaBACE is commercially available from Amersham Pharmacia Biotech (Freiburg, Germany).

[57] M. T. Reetz, K. M. Kühling, A. Deege, H. Hinrichs, D. Belder (Studiengesellschaft Kohle mbH), patent application, 2000.

[58] F. Balkenhohl, K. Ditrich, B. Hauer, W. Ladner, *J. Prakt. Chem./Chem.-Ztg.* **1997**, *339*, 381–384.

[59] M. T. Reetz, A. Zonta, K. Schimossek, K. Liebeton, K.-E. Jaeger (Studiengesellschaft Kohle mbH), DE-A 197 31 990.4, 1997 [*Chem. Abst.* **1999**, *130*, 149528].

[60] D. C. Demirjian, P. C. Shah, F. Mors-Varas, *Top. Curr. Chem.* **1999**, *200*, 1–29.

[61] a) M. T. Reetz, M. H. Becker, M. Liebl, A. Fürstner, *Angew. Chem.* **2000**, *112*, 1294–1298; *Angew. Chem. Int. Ed.* **2000**, *39*, 1236–1239; b) M. T. Reetz,

H. Hermes, M. H. Becker, *Appl. Microbiol. Biotechnol.* **2001**, *55*, 531–536.

[62] a) K. Ding, A. Ishii, K. Mikami, *Angew. Chem.* **1999**, *111*, 519–523; *Angew. Chem. Int. Ed.* **1999**, *38*, 497–501; b) K. Mikami, R. Angelaud, K. Ding, A. Ishii, A. Tanaka, N. Sawada, K. Kudo, M. Senda, *Chem. Eur. J.* **2001**, *7*, 730–737; c) M. T. Reetz, K. M. Kühling, H. Hinrichs, A. Deege, *Chirality*, **2000**, *12*, 479–482.

[63] a) G. Klein, J.-L. Reymond, *Helv. Chim. Acta* **1999**, *82*, 400–406; b) G. T. Copeland, S. J. Miller, *J. Am. Chem. Soc.* **1999**, *121*, 4306–4307; c) Short review of fluorescent bead signaling in combinatorial catalysis: A. H. Hoveyda, *Chem. Biol.* **1999**, *6*, R305–R308; d) P. Geymayer, N. Bahr, J.-L. Reymond, *Chem. Eur. J.* **1999**, *5*, 1006–1012.

[64] M. T. Reetz, K. M. Kühling, S. Wilensek, H. Husmann, U. W. Häusig, M. Hermes, *Catal. Today* **2001**, in press.

[65] M. T. Reetz, S. Sostmann, *Tetrahedron* **2001**, *57*, 2515–2520.

[66] G. A. Korbel, G. Lalic, M. D. Shair, *J. Am. Chem. Soc.* **2001**, *123*, 361–362.

[67] B. Phimister, *Nat. Genet.* **1999**, *21*, supplement, 1.

[68] a) P. Forrer, S. Jung, A. Plückthun, *Curr. Opin. Struct. Biol.* **1999**, *9*, 514–520; b) J. Fastrez, *Mol. Biotechnol.* **1997**, *7*, 37–55; c) M. Olsen, B. Iversen, G. Georgiou, *Curr. Opin. Biotechnol.* **2000**, *11*, 331–337; d) K. Johnsson, L. Ge, in *Current Topics in Microbiology and Immunology, Vol. 243*: Combinatorial Chemistry in Biology, M. Famulok, E.-L. Winnacker, C.-H. Wong (Eds.), Springer, Berlin, 1999, pp. 87–105; e) J. Hanes, A. Plückthun, in *Current Topics in Microbiology and Immunology, Vol. 243*: Combinatorial Chemistry in Biology, M. Famulok, E.-L. Winnacker, C.-H. Wong (Eds.), Springer, Berlin, 1999, pp. 107–122; f) U. Hoffmüller, J. Schneider-Mergener, *Angew. Chem.* **1998**, *110*, 3431–3434; *Angew. Chem. Int. Ed.* **1998**, *37*, 3241–3243.

[69] G. MacBeath, P. Kast, D. Hilvert, *Protein Sci.* **1998**, *7*, 1757–1767.

[70] a) M. T. Reetz, R. Wenkel, D. Avnir, *Synthesis* **2000**, 781–783; b) K.-E. Jaeger, K. Liebeton, A. Zonta, K. Schimossek, M. T. Reetz, *Appl. Microbiol. Biotechnol.*

1996, 46, 99–105; c) M. T. Reetz, K.-E. Jaeger, *Chem. Phys. Lipids* **1998**, 93, 3–14.

[71] In some samples of the wild-type lipase from *Pseudomonas aeruginosa* we have observed small variations in the *ee* (S. Wilensek, projected Dissertation, Ruhr-Universität Bochum, 2001).

[72] a) M. Eigen, R. Winkler-Oswatitisch, *Steps towards Life: A Perspective on Evolution*, Oxford University Press, Oxford, 1996; b) J. M. Smith, E. Szathmáry, *The Major Transitions in Evolution*, Freeman/Spektrum, Oxford, 1995.

[73] M. D. van Kampen, M. R. Egmond, *Eur. J. Lipid Sci. Technol.* **2000**, 102, 717–726.

[74] a) A. Crameri, W. P. C. Stemmer, *Bio-Techniques* **1995**, 18, 194–196; see also previous work on combinatorial cassette mutagenesis: b) J. F. Reidhaar-Olson, R. T. Sauer, *Science (Washington, D.C.)*, **1988**, 241, 53–57; c) J. D. Hermes, S. C. Blackow, J. R. Knowles, *Proc. Natl. Acad. Sci. USA* **1990**, 87, 696–700; d) A. P. Arkin, D. C. Youvan, *Prod. Natl. Acad. Sci. USA* **1992**, 89, 7811–7815.

[75] D. Zha, S. Wilensek, M. Hermes, K.-E. Jaeger, M. T. Reetz, *Chem. Commun. (Cambridge)*, in press.

[76] M. Nardini, D. A. Lang, K. Liebeton, K.-E. Jaeger, B. W. Dijkstra, *J. Biol. Chem.* **2000**, 275, 31219–31225.

[77] a) M. Holmquist, D. C. Tessier, M. Cygler, *Biochemistry* **1997**, 36, 15019–15025; b) M. Holmquist, *Chem. Phys. Lipids* **1998**, 93, 57–65.

[78] a) G. H. Peters, S. Toxvaerd, O. H. Olsen, A. Svendsen, *Protein Eng.* **1997**, 10, 137–147; b) G. H. Peters, R. P. Bywater, *Protein Eng.* **1999**, 12, 747–754.

[79] a) M. Norin, O. Olsen, A. Svendsen, O. Edholm, K. Hult, *Protein Eng.* **1993**, 6, 855–863; b) M. Norin, F. Haeffner, A. Achour, T. Norin, K. Hult, *Protein Sci.* **1994**, 3, 1493–1503; c) G. H. Peters, D. M. F. van Aalten, O. Edholm, S. Toxvaerd, J. Bywater, *Biophys. J.* **1996**, 71, 2245–2255; d) G. H. Peters, D. M. F. van Aalten, A. Svendsen, R. Bywater, *Protein Eng.* **1997**, 10, 149–158.

[80] a) J. D. Schrag, Y. Li, M. Cygler, D. Lang, T. Burgdorf, H.-J. Hecht, R. Schmid, D. Schomburg, T. J. Rydel, J. D. Oliver, L. C. Strickland, C. M. Dunaway, S. B. Larson, J. Day, A. McPherson, *Structure (London)* **1997**, 5, 187–202; b) K. K. Kim, H. K. Song, D. H. Shin, K. Y. Hwang, S. W. Suh, *Structure (London)* **1997**, 5, 173–185.

[81] a) A. Iffland, P. Tafelmeyer, C. Saudan, K. Johnsson, *Biochemistry* **2000**, 39, 10790–10798; b) S. Oue, A. Okamoto, T. Yano, H. Kagamiyama, *Biol. Chem.* **1999**, 274, 2344–2349; c) F. H. Arnold, *Nature (London)* **2001**, 409, 253–257.

[82] M. T. Reetz, A. Eipper, T. Eggert, K.-E. Jaeger, unpublished results.

[83] U. T. Bornscheuer, J. Altenbuchner, H. H. Meyer, *Biotechnol. Bioeng.* **1998**, 58, 554–559.

[84] a) U. T. Bornscheuer, J. Altenbuchner, H. H. Meyer, *Bioorg. Med. Chem.* **1999**, 7, 2169–2173; b) U. T. Bornscheuer, M. Pohl, *Curr. Opin. Chem. Biol.* **2001**, 5, 137–143.

[85] G. W. Matcham, A. R. S. Bowen, *Chim. Oggi* **1996**, 14, 20–24.

[86] X. Zhu, C. M. Lewis, M. C. Haley, M. B. Bhatia, S. Pannuri, S. Kamat, W. Wu, A. R. S. Bowen, IBC's 2nd Annual Symposium on Exploiting Enzyme Technology for Industrial Applications, Feb. 20–21, 1997, San Diego, USA.

[87] O. May, P. T. Nguyen, F. H. Arnold, *Nat. Biotechnol.* **2000**, 18, 317–320.

[88] H. Stecher, K. Faber, *Synthesis* **1997**, 1–16.

[89] W. D. Fessner, *Curr. Opin. Chem. Biol.* **1998**, 2, 85–97.

[90] S. Fong, T. D. Machajewski, C. C. Mak, C.-H. Wong, *Chem. Biol.* **2000**, 7, 873–883.

[91] B. Steipe, in *Current Topics in Microbiology and Immunology, Vol. 243*: Combinatorial Chemistry in Biology, M. Famulok, E.-L. Winnacker, C.-H. Wong (Eds.), Springer, Berlin, 1999, pp. 55–86.

[92] U. Kettling, A. Koltermann, M. Eigen, in *Current Topics in Microbiology and Immunology, Vol. 243*: Combinatorial Chemistry in Biology, M. Famulok, E.-L. Winnacker, C.-H. Wong (Eds.), Springer, Berlin, 1999, pp. 173–186.

[93] See for example: a) D. E. Goldberg, *Genetic Algorithms in Search Optimization and Machine Learning*, Addison Wesley, Reading, **1989**; b) L. Weber, S. Wallbaum, C. Broger, K. Gubernator, *Angew. Chem.*

1995, *107*, 2452–2454; *Angew. Chem. Int. Ed. Engl.* **1995**, *34*, 2280–2282; c) D. Weuster-Botz, C. Wandrey, *Process Biochem.(Oxford)* **1995**, *30*, 563–571; d) M. Bradley, L. Weber, *Curr. Opin. Chem. Biol.* **2000**, *4*, 255–256; e) D. E. Clark (Ed.), *Evolutionary Algorithms in Molecular Design*, Wiley-VCH, Weinheim, 2000.

[94] a) J. Sadowski, *Curr. Opin. Chem. Biol.* **2000**, *4*, 280–282; b) S. Anzali, J. Gastei-ger, U. Holzgrabe, J. Polanski, J. Sadowski, A. Teckentrup, M. Wagener, *Perspect. Drug Discovery Des.* **1998**, *9/10/11*, 273–299.

[95] a) C. Wilson, J. W. Szostak, *Nature (London)* **1995**, *374*, 777–782; b) M. Fa-mulok, A. Jenne, *Top. Curr. Chem.* **1999**, *202*, 101–131; c) B. Seelig, S. Keiper, F. Stuhlmann, A. Jaeschke, *Angew. Chem.* **2000**, *112*, 4764–4768; *Angew. Chem. Int. Ed.* **2000**, *39*, 4576–4579.

12
Applied Molecular Evolution of Enzymes Involved in Synthesis and Repair of DNA

John F. Davidson, Jon Anderson, Haiwei Guo, Daniel Landis, and Lawrence A. Loeb

12.1
Introduction

In natural evolution mutations are often deleterious. While they provide the variability that drives natural selection, they frequently result in loss of function and genetic instability. As a result, a variety of enzymes have evolved to repair DNA damages that can cause mutations and to synthesize DNA during cell replication with exquisite fidelity. The enzymes responsible for DNA repair and replication are in general highly conserved across diverse domains of life as befitting their unique roles in the cell, i.e. the maintenance of genetic stability.

Recently, methods have become available to create large libraries of mutant enzymes *in vitro*. We can now address the tantalizing question as to whether DNA synthesizing enzymes that are central to the maintenance of genetic stability and are highly conserved during evolution, can tolerate changes in amino acid sequence. Moreover, can we create new mutant enzymes that are altered in the fidelity of catalysis and can we exploit these new mutant enzymes for medical and industrial needs? In this chapter we will summarize the status of current methodologies for the creation of mutant enzymes and then consider the application of these technologies to the creation of mutant enzymes that function in different aspects of DNA metabolism.

Our focus will be on studies carried out in our laboratory aimed at both understanding mechanisms of catalysis and on producing enzymes for gene therapy of cancer.

In order to determine the relationship between protein structure and function and to create mutant enzymes with altered properties useful for biotechnology and cancer therapy, a directed evolution approach has been explored and novel proteins developed for Pol I DNA polymerase enzymes: thymidylate synthase, thymidine kinase and O^6-alkylguanine-DNA alkyltransferase. In every case the creation of a large variety of altered proteins has been achieved, and the emerging picture is that even highly conserved proteins can tolerate wide-spread amino acid changes at the active site without substantial loss of activity. In the case of Pol I enzymes, the conservation of some amino acid motifs is absolute across organisms separated by at least a billion years. This poses an interesting question as to what the selection pressures are that seek to maintain a conserved active site in a strict configuration given that considerable plasticity that can be tolerated *in vitro*.

We describe the various approaches that one might use to generate novel enzymes and the application of one such technique, Random Oligonucleotide Mutagenesis, to the DNA synthesis and repair enzymes: DNA polymerase I from *E. coli* and *Thermus aquaticus*, as well as human thymidylate synthase, thymidine kinase and O^6-alkylguanine-DNA alkyltransferase.

12.2
Directed Evolution of Enzymes

Life has been evolving for nearly a billion years, providing nature the opportunity to progressively select for enzymes that function to enhance survival of the organism. The process of natural evolution, however, is not ideally suited to develop enzymes that perform their designated tasks under non-physiological conditions or to display novel functions. We might desire to modify the functionality of a particular enzyme to perform tasks that nature has not intended. It may be thought that nature, with its billions of years of experience, can not be outdone. But even nature has not had the opportunity to explore the vastness of sequence space and test all possible amino acid combinations. Sequence space is the connected network of all protein sequences, with each point in space representing a unique amino acid sequence [1]. The sequence space for a protein of 350 amino acids would contain approximately 20^{350} points. This number is so large that it vastly exceeds even the number of atoms in the visible universe, thought to be approximately 10^{79}. In other words, one would run out of atoms in the universe before one could even get close to synthesizing every possible 350 amino acid protein.

The size of sequence space is also a major concern when modifying existing enzymes. For a protein of 350 amino acids, a total of 6650 substitutions would have to be made in order to fully examine all possible single amino acid changes and 2.2×10^7 substitutions would be needed just to look at all double changes. This problem is particularly acute since recent evidence suggests that the most efficacious mutant enzymes frequently require multiple amino acid substitutions. Furthermore, it is likely that the evolution of enzymes with enhanced stability will require large numbers of amino acid substitutions throughout the protein. Thus, in order to make our way through sequence space and identify mutant enzymes that exhibit interesting properties, it is clear that we need to formulate new rules to predict the effects of multiple substitutions on protein stability and catalysis.

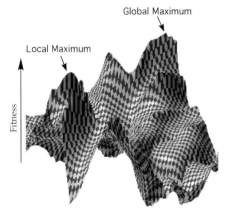

Fig. 12.1. Schematic representation of sequence space. This landscape shows possible
amino acid sequences for an enzyme with the
height of the peaks representing the fitness of the
enzyme under some set conditions. Mutational
studies that try to increase fitness by creating
single amino acid changes have the potential of
becoming trapped at local maximums.

Under the conditions that we are interested in, each individual protein sequence will
maintain a fitness. When the fitness of each protein is mapped in sequence space, the
three-dimensional landscape forms peaks and valleys of fitness with the highest peak
representing the most fit protein (Fig. 12.1). Therefore, the successful enhancement of
an enzyme can be thought of as a hill-climbing exercise with the goal of reaching the
highest peak by mutating the most effective combination of amino acid changes.

Several techniques can be used to mutate genes and thus search sequence space.
Mutations can be created over the entire gene, over a selected region of the gene, or can
be targeted to specific nucleotides within the gene. In this section, we will discuss
several of the techniques that are currently being used for mutating enzymes in
an attempt to enhance or alter their function. These techniques include: site-directed
mutagenesis, random genetic damage, polymerase chain reaction (PCR) mutagenesis,
DNA shuffling, and random oligonucleotide mutagenesis.

12.2.1
Site-directed mutagenesis

As the name implies, site-directed mutagenesis involves making predetermined nucleotide substitutions at specified positions within a gene. *In vitro* techniques of site-
directed mutagenesis generally either exchange a mutant DNA fragment with the wild-
type DNA or use oligonucleotides harboring designated substitutions by incorporating
them into newly synthesized DNA using PCR. This technology directly yields information on structure-function relationships. The technique can also be used to determine if certain amino acid substitutions can be tolerated by the enzyme without loss of
activity. It is frequently assumed that loss of enzyme activity by specific substitutions
indicates the involvement of a specific residue in catalysis. This need not be the case,

since substitutions at a distance can also result in structural collapse and loss of enzyme activity. It should be noted that site-specific mutagenesis is dependent on detailed knowledge of structure and enzyme mechanism and does not require screening or selection because the effects of each substitution can be individually analyzed.

This single step mutation procedure also has several important limitations. Using site-directed mutagenesis as a method to search through sequence space is logistically ineffectual. This method relies on the fact that an informed decision can be made on which amino acid should be mutated and to what new amino acid it should be changed. Information on the enzyme may be gathered through crystal structure analysis, homology searches, and biochemical measurements, however, this information often does not provide sufficient knowledge to know how a protein will respond to even a single amino acid substitution. A single step mutation procedure may also be a disadvantage in cases where single amino acid changes cannot overcome a local maximum in sequence space and only a group of simultaneous changes can move the search to the global maximum (Fig. 12.1).

As we better understand protein folding, the technique of site-directed mutagenesis should become more useful in rationally designing enzymes with new properties. Currently, the most powerful application for site-directed mutagenesis is its use in conjunction with other mutagenesis techniques. Following selection of an enzyme containing several amino acid substitutions, site-directed mutagenesis can be used to individually test each of the substituted amino acids and thus yield information on mechanisms of catalysis.

12.2.2
Directed evolution

Just as nature uses evolution to naturally select enzyme variations that provide an advantage to the host, directed evolution of the amino acid substitution utilize techniques to screen or select for mutant enzymes that perform better than wild-type enzymes. Unlike site-directed mutagenesis, directed evolution techniques do not require a detailed understanding of the enzyme in order to identify useful variants. The techniques of applied molecular evolution however, do require a screening or selection step to identify the individual mutants of interest (Fig. 12.2).

The power of the screening or selection method ultimately delineates the extent to which the sequence space can be explored. Screening usually involves the physical separation of mutants identified on some phenotypic change such as colony size or color. It is essentially a brute force approach that is amenable to amplification by robotics and is limited to identification of mutants in a population of transfected bacteria or cells totaling at most one million and more with typically only a few thousand clones each harboring a different mutant gene. Selection takes advantage

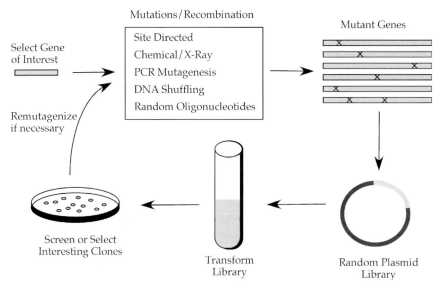

Fig. 12.2. Mutagenesis experiment overview. After selecting a gene of interest, the gene is mutagenized by one of several methods and cloned into an expression vector, creating a mutant gene library. Genes from the library are inserted into bacteria, producing the enzyme variants. The bacteria expressing the mutant enzymes are selected or screened for interesting characteristics, which can be further mutated if necessary.

of the power of genetics to identify genes harboring mutant genes that can substitute for enzymes that are either deleted or conditionally inactivated in recipient host cells. The power of selection is the identification of active mutants and the vast numbers of mutations that can be surveyed. If the difference between inactive variants and active variants in complementing the missing required enzyme in the host cells is sufficient, one can routinely screen 10 million transfected bacteria to identify a few mutants of interest. We will cover several different techniques for applied molecular evolution that generate variants through the incorporation of random mutations.

12.2.3
Genetic damage

DNA can be damaged by chemicals and physical agents and then copied by DNA polymerases to produce mutations opposite the altered sites in DNA. In principle one could damage whole genes or operons, insert them into plasmids and then transform bacteria of other host cells. However, most agents that damage DNA do so randomly and many sites of damage inactivate key genes that are required for plasmid replication. An alternative approach is to incorporate nucleotide analogs into oligonu-cleotides either site-specifically or over specified regions of a gene. Chemicals and X-rays were originally used in classical genetic studies to generate chromosomal DNA damage [2]. Later, the manipulation of DNA *in vitro* allowed researchers to use muta-gens to create libraries of altered genes that could be screened for interesting variants [3, 4]. These chemicals can cause transitions, transversions, and even frameshifts through such actions as alkylation, deamination, and intercalation. The use of chemi-cals to cause DNA damage and ultimately mutations, may not produce a random spec-trum of mutations. These biases in types of mutations can cause regions of sequence space to remain untestable and may affect one's ability to identify interesting variants.

12.2.4
PCR mutagenesis

The simple method of PCR using enzymes such as *Taq* polymerase, that do not con-tain a proofreading mechanism, can produce a library of mutated genes. As amplifica-tion of a gene or gene segment proceeds via PCR, mutations are formed in the pool of amplified products, which can later be cloned and analyzed for altered activity. *Taq* polymerase produces approximately one mutation in every 9000 bases synthesized [5]. The error rate can be increased by several techniques, making this method an effective approach to introduce a spectrum of mutations within a delineated seg-ment. The region being mutated can be demarcated by using primers that flank a specific segment, while the number of mutations within the amplified segment can be controlled by altering the PCR reaction conditions. The use of manganese in-stead of magnesium as the divalent cation has been shown to greatly increase the mutation frequency of a variety of DNA polymerases [6 – 9]. The error rate of a DNA polymerase is proportional to the biases in the deoxynucleoside triphosphates in the reaction mixture [10]. Altering the pool of natural deoxynucleoside triphosphates (dNTP) has also been shown to increase the number of misincorporations and inser-tion/deletion events [11, 12]. Further enhancement in mutagenesis *in vitro* can be obtained by the use of PCR reactions containing nucleoside analogs. The oxidized guanosine triphosphate (8-oxo-dGTP) and the base analog, dPTP, have been success-fully used to create libraries of mutant genes [13, 14]. Finally, the production of mod-ified *Taq* polymerases that are more mutagenic than wild-type *Taq* may lead to more efficient ways of creating mutant libraries [15, 16].

PCR mutagenesis, however, has several disadvantages compared to other methods. Even though the mutation rate of PCR mutagenesis can be increased by altering the reaction conditions, the ability of this technique to create multiple mutations within a narrow region of the gene (for example, the catalytic region) is severely limited. Furthermore, the mutations created with this method are not completely random and may cause biases in the mutant library, thus limiting the ability to search effectively regions of sequence space. Even with these limitations, PCR mutagenesis is an efficient and straightforward technique to produce a sparse array of mutations throughout the length of a protein.

12.2.5
DNA shuffling

The process of DNA shuffling (also known as sexual PCR and family shuffling) entails the disassembly and eventual random reassembly of a gene, yielding a library of reconstituted chimeric genes. The starting material for DNA shuffling is made up of either homologous genes from several different species or several mutagenized forms of a single gene. The pool of genes (homologs or mutants) are amplified and then fragmented by DNAse, sonication, or restriction enzymes to create a new pool of short random fragments. The fragments are allowed to recombine at random and the new mutant genes are reassembled using primerless PCR (Fig. 12.3). The technique of

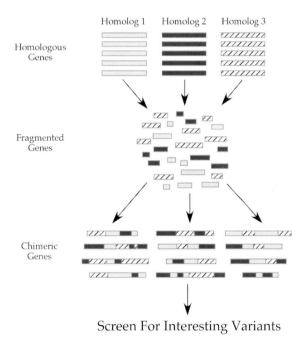

Homolog 1   Homolog 2   Homolog 3

Homologous
Genes

Fragmented
Genes

Chimeric
Genes

Screen For Interesting Variants

Fig. 12.3. DNA Shuffling. Several homologous genes (or a mutated library of genes) are fragmented using DNAse, sonication, or restriction enzymes to create a pool of short DNA fragments. The fragments are allowed to recombine creating chimeric genes that are made up of combinations of the original genes.

DNA shuffling can also be combined with other forms of directed evolution to en-hance the diversity of the recombinants. PCR mutagenesis can be used in either the amplification of the original gene pool or in the reassembly of the shuffled gene, creating a final mutant library containing rearranged segments that harbor ran-dom mutations.

The logic behind DNA shuffling involves the idea that we cannot predict the tertiary structure and function of an enzyme and therefore, the incorporation of mutations throughout a protein will often produce nonfunctional products. However, when seg-ments of the original gene are conserved but recombined in an alternative order, the conserved segments should produce structures seen in the original protein. The final gene product that results from these reordered segments should have a higher prob-ability of coding for a functional enzyme [17, 18].

The use of homologous genes from many separate organisms (family shuffling) allows millions of years of evolution to aid in the production of new enzyme var-iants. Since the homologs all derived from a common ancestor, the changes that have been made over time should mainly not affect the major structure and function of the enzyme, but merely alter the specific properties. By combining the genetic ma-terial from several homologs to create a new enzyme variant, DNA shuffling has the potential to make large moves in sequence space while still preserving the function of the enzyme [19]. The approach is limited in that it focuses on the scaffoldings that Nature has evolved and does not thoroughly survey other structures that Nature may not have chosen and yet could be more effective.

12.2.6
Substitution by oligonucleotides containing random mutations (random mutagenesis)

Whereas many of the directed evolution methods produce mutations over an entire gene and site-directed mutagenesis often alters only a single base within a gene, ran-dom mutagenesis focuses somewhere between these extremes by creating mutations within limited regions of the gene. This technique utilizes degenerate oligonucleotides to replace one or several regions of the gene. This provides a library of mutant genes that contain a random assortment of amino acid substitutions over the region(s) of interest. Proteins produced by this method can then be screened or selected for en-zymatic functionality.

The randomness of the library can be precisely controlled by adjusting the fraction of wild-type nucleotides at each position compared to the other three degenerate nucleo-tides [20]. The advantages of this procedure over other directed evolutionary proce-dures include the ability to target a specific region of the gene and to introduce a predetermined percentage of nucleotide substitutions at each position. By cleverly assembling the oligonucleotide inserts, a spectrum of noncontiguous nucleotides can be substituted and thus mimic the three-dimensional array of amino acids that contact the substrate at the active site. However, one major disadvantage of this technique is that only a small region of the gene can be mutated at a given time, limiting the amount of sequence space that can be searched.

Overall, several techniques have been developed to systematically or randomly mutate genes that encode enzymes and other proteins. Each method presents advantages and disadvantages, without any particular method perfectly suited for every situation. If only a single amino acid is substituted, then a selection step can be avoided but no information is gained about other mutations that may be more advantageous to the change made. If one saturates even a small segment of a gene with 100 % random substitutions at each position, then the majority of mutants will contain multiple substitutions. Assuming one can utilize a powerful positive genetic selection technique to identify mutants, then many of the mutants will contain multiple substitutions. Mutants with multiple substitutions have proven to exhibit interesting variant properties but seldom yield mechanistic information; we lack sufficient knowledge to predict the effects of more than one substitution on catalytic rates or substrate specificities of enzymes. If an entire gene is randomly mutated, then there are no limits imposed on the location of possible mutations and little information is needed about the structure of the enzyme prior to carrying out mutagenesis. However, the number of possible mutations made over an entire gene will greatly exceed any feasible selection system and only a portion of the mutants will be observed.

Random oligonucleotide mutagenesis was first applied to promotor sequences that regulate the production of enzymes in cells [21] and was the first method used to alter systematically the functions of enzymes by directed evolution [22]. Based on our experience, we will focus on this approach and emphasize recent applications of this methodology to enzymes involved in DNA repair and synthesis, including DNA polymerase enzymes, thymidylate synthase, thymidine kinase, and O^6-alkylguanine-DNA alkyltransferase.

12.3
Directed Evolution of DNA polymerases

A major goal of directed evolution of DNA polymerases has been to elucidate the structural elements that confer high fidelity during DNA replication. If DNA polymerases were to rely solely on the stability of nucleotides that aligned with template for discrimination of correct template-directed polymerization, the error frequency would be in the order of one mispaired nucleotide per 100 incorporated [23]. The measured error rate for incorporation and extension of a mismatched nucleotide attributable to DNA polymerases lacking an error correcting exonucleolytic activity range from 10^{-4} – 10^{-5}, highlighting the large input that the protein structure imparts on polymerase fidelity. Factors such as exonucleolytic proofreading and mismatch repair further increase the fidelity of DNA replication [24, 25].

Site-directed approaches have yielded polymerases with altered properties including loss of 3'-5' exonucleolytic proofreading, increased rates of dideoxynucleotide incorporation and loss of 5'-3' exonuclease activity [26 – 28]. It is usually implied that amino acid substitutions at specified positions that result in loss of activity indicate the involvement of that amino acid in the chemical step of catalysis or in substrate recognition but this need not be the case. Even single substitutions can cause collapse of the three-dimensional structure of the catalytic site on the enzyme. Other interesting properties have been evolved from DNA polymerases including enzymes that have the ability to bypass replication blocking lesions and the ability to synthesize RNA from a DNA template [29, 30]. These studies have important practical implications for biotechnology, for example, DNA sequencing and incorporation of fluorescent or modified bases for DNA labeling, as well as providing information for the structural determinants and mechanisms involved.

The nucleotide sequence of DNA polymerases is highly conserved within families. Amongst families, DNA polymerases demonstrate a remarkable conservation of structure, even in particularly divergent organisms. The overall shape of DNA polymerases is reminiscent of a partially closed right hand, with distinct finger, palm and thumb subdomains [31]. There are currently six known families of DNA polymerase classed according to their amino acid sequence identity [32]. The families are A, B, C, X, reverse transcriptase and UmuC/DinB. Family A, B, and C are classed according to homologies with *E. coli* Pol I, II, and III.

Amino acid alignments of the C-terminal polymerase domain from diverse Pol I enzymes from family A, have enabled identification of six conserved regions. Region 1 amino acids are located at the tip of the thumb subdomain and form a helix-loop that interacts with the minor groove of the double stranded DNA during nucleic acid binding. Region 2 amino acids are located within the palm subdomain and interact with the template strand along the minor groove, gripping the template. Regions 3 – 5 resemble the motifs A, B and C from DNA Pol α and are highly conserved across divergent prokaryotes and eubacteria (Tab. 12.1). Region six contains amino acids that interact with the first template base.

Evidence based on site-directed mutagenesis studies and Pol I crystal structures show motifs A, B and C to be located near the incoming nucleotide triphosphate and constitute the polymerase active site [33 – 35]. Motif A structurally consists of an antiparallel β-strand containing primarily hydrophobic residues continuing into a α-helix. The essential catalytic aspartic acid (Asp-705 of *E. coli* Pol I) is located within motif A, and forms the floor of the active site in the "Palm". Motif B consists of the α-helix O, and comprises one wall of the active site. Motif B has been demonstrated to be

Tab. 12.1.

	MOTIF A	MOTIF B	MOTIF C
Thermus aquaticus	LLVALDYSQIELR	RRAAKTINFGVLY	LLQVHDELVLE
Thermus thermophilus	ALVALDYSQIELR	RRAAKTVNFGVLY	LLQVHDELLLE
Thermus filiformis	LLLAADYSQIELR	RRAAKTVNFGVLY	LLQVHDELVLE
Deinococcus radiodurans	TLIAADYSQIELR	RRAAKTVNFGVLY	LLQVHDELLIE
Escherichia coli	VIVSADYSQIELR	RRSAKAINFGLIY	IMQVHDELVFE
Haemophilus influenzae	SIVAADYSQIELR	RRNAKAINFGLIY	IMQVHDELVFE
Streptococus pneumoniae	VLLSSDYSQIELR	RRNAKAVNFGVVY	LLQVHDEIVLE
Mycobacterium tuberculosis	ELMTADYSQIEMR	RRRVKAMSYGLAY	LLQVHDELLFE
Mycobacterium laprae	ELMTADYSQIEMR	RRRVKAMSYGLAY	LLQVIIDELLFE
Treponema pallidum	ELISADYTQIELV	RRIAKTINFGIVY	LLQVHDELIFE
Chlamydia trachomatis	YFLAADYSQIELR	RYQAKAVNFGLVY	LLQIHDELLFE
Boriela burgdorferi	IFISADYSQIELA	RRIAKSINFGIIY	LLQVHDEMLIE
Helicobacter pyroli	CLLGVDYSQIELR	RSIAKSINFGLVY	LLQVHDELIFE
Lactococcus lactis	LLLSSDYSQIELR	RRNAKAVNFGVVY	LLQVHDEIILD
Mythelobacterium	KLISADYSQIELR	RRRAKTINFGIIY	LLQVHDELVFE
Rhodothermus obamensis	KLLSADYVQIELR	RRRAKMVNYGIPY	LLQVHDELVFE
Rickettsia prowazekii	KLISADYSQIELR	RRKAKAINFGIIY	ILQIHDELLFE
Streptomyces coelicolor	SLMTADYSQIELR	RRKIKAMSYGLAY	LLQVHDEIVLE
Bacillus stearothermophilus	LIFAADYSQIELR	RRQAKAVNFGIVY	LLQVHDELILE
Synechocystis sp	LLVSADYSQIELR	RNLGKTINFGVIY	LLQVHDELIFE
Aquifex aeolicus	TFVISDFSQIELR	RQLAKAINFGLIY	VNLVHDEIVVE
Apse-1 DNA polymerase	KLVISDLSNIEGR	RQIGKVMELGLGY	IVTVHDEIISE
T7 DNA polymerase	VQAGIDASGLELR	RDNAKTFIYGFLY	MAWVHDEIQVQ
T5 DNA polymerase	RVIAWDLTTAEVY	RQAAKAITFQILY	VMLVHDSVVAI

involved in binding both template primer and incoming dNTP and is located in the "fingers" subdomain [33].

A major breakthrough that has facilitated the identification of active mutant DNA polymerases has been the use of *E. coli recA718polA12*, a bacteria that encodes a temperature-sensitive mutant DNA polymerase I. At elevated temperatures the mutant *E. coli* fails to form colonies unless complemented by a DNA polymerase that can effectively substitute for DNA polymerase I [34]. The *E. coli recA718 polA12* strain was first utilized in identification of active mutants of rat DNA Pol-β and then in analyzing mutations in HIV reverse transcriptase [35]. In both situations, complementation required very active mutant enzymes and it was only feasible to screen libraries that contained thousands of mutant genes.

12.3.1
Random mutagenesis of *Thermus aquaticus* DNA Pol I

The most extensive studies on directed evolution have been carried out with *Thermus aquaticus* DNA Pol I (*Taq* Pol I). Motifs A and B have both been targeted for random mutagenesis by complementing the *recA718polA12* temperature-sensitive Pol I mutation in *E. coli*. Once mutants are identified, initial screening for specific properties can be carried out by boiling crude bacterial extracts, because the host cell polymerases are inactivated.

Two random libraries were generated for *Thermus aquaticus* DNA polymerase I (*Taq* Pol I) in which nucleotides coding for amino acids 659 – 671 of motif B containing the consensus sequence RRxhKhhNFGhhY (where x is any amino acid and h is hydrophilic amino acids), or amino acids 605 – 617 of motif A, including the consensus sequence DYSQIELR were randomized [29, 36]. For the motif A random library, replacement of the segment with hybridized random oligonucleotide sequences having approximately 88 % wild type and 4 % for each of the other three nucleotides was achieved. For motif B, the level of randomization was 91 % wild type and 3 % of each of the other nucleotides at each position. In both cases a large library was created containing on average three to four substitutions at the amino level. Selection *in vivo* for active polymerases was achieved by transformation of the library into *E. coli* containing the *recA718 polA1* temperature-sensitive mutation. Growth at 37 °C was restored fully by complementation with wild-type *Taq* polymerase. Mutant *Taq* polymerases having >10 % activity have the capacity for complementation and the frequency of false positives is <1 %, thus enabling a high degree of stringency in the selection. Active mutant enzymes were then individually sequenced to determine the spectrum of permissible amino acid changes, and crude extracts of *E. coli* harboring the mutant proteins were studied for alterations in DNA synthetic activities prior to extensive purification of *Taq* Pol I for more detailed characterization.

12.3.1.1 Determination of structural components for *Taq* DNA polymerase fidelity

The fidelity of polymerization depends on the ability of the polymerase active site to discriminate complementary from non-complementary incoming nucleotide triphosphates as well as to extend matched from unmatched primer termini. *Taq* DNA polymerase does not contain a 3'-5' exonuclease or "proofreading" activity. This makes it advantageous for studying the intrinsic fidelity of the polymerase active site for discrimination between complementary versus non-complementary base pairing.

Polymerases are proposed to undergo at least four significant conformational changes during catalysis. The first occurs upon DNA binding and is followed by a second shift during the incoming dNTP binding step. Immediately preceding chemical catalysis is the third conformational shift subsequent to phosphodiester bond formation and PPi release. The final change occurs during translocation towards the next primer 3'-OH terminus. Theoretical studies indicate multiple conformational changes contribute to the fidelity of catalysis by DNA polymerases [37].

A library consisting of 5×10^4 transformants was generated with random mutations in the 13 amino acids of motif B. From this library, 67 active mutants were screened for their ability to elongate primers annealed to a template, in the absence of one of the four dNTPs [16]. Of 67 mutants screened, 13 displayed extended products indicating an increase in incorporation and/or extension of non-complementary nucleotides. Two mutants (A664R and A661E) were purified and showed 7 – 25-fold increases in forward mutation frequencies using gapped M13mp2 template DNA. Both of these residues are located near the distal portion of the "fingers" subdomain and may stabilize the closed conformation following dNTP binding, thus contributing to the reduction in fidelity of DNA synthesis. High fidelity mutants were also obtained from the screen [15]. Notably, a triple mutant (A661E, I665T, F667L) displayed an overall threefold decrease in mutation frequency compared to wild-type enzyme, principally via 10-fold reductions in G → T and A → T transversions. In addition this mutant has an increased discrimination against synthesis of A:A mispairs and reduced ability to extend mismatches. As described above, A661E is a strong mutator when present singly, thus the I665 and/or F667L mutations act as intragenic suppressors.

Using a similar primer misextension screen for mutants obtained from the randomization of motif A, both high and low fidelity mutants were identified [29]. All low fidelity mutants were found to have hydrophilic substitutions at I614. This position is able to tolerate a variety of substitutions and still maintain wild-type activity levels. Some of the mutants exhibit a 10-fold higher efficiency of misinserting nucleotides, as well as at least 10-fold higher efficiency in extending the resulting mismatches. When used in PCR reactions, *Taq* Pol I mutants containing hydrophilic residues at position 614 exhibit up to 20-fold higher error rates relative to the WT enzyme and efficiently catalyze both transition and transversion errors. In addition to conferring low base pairing fidelity, hydrophilic substitutions for I614 also allow mutant *Taq* Pol I to bypass blocking template lesions such as abasic sites and the vinyl chloride alkylation product, ethenoA. Hydrophilic substitutions for I614 presumably confer a low fidelity by "widening" the dNTP-binding pocket to accommodate unusual template base and incoming nucleotide structures. These mutants or similar others should facilitate the use of mutagenic PCR methods to rapidly create a variety of mutated genes.

12.3.1.2 Directed evolution of a RNA polymerase from *Taq* DNA polymerase

The intracellular concentration of ribonucleotides is much higher than that of deoxynucleotides and accordingly DNA polymerases specifically discriminate against incorporation of ribonucleotides. Based on the crystal structure of the active site of *Taq* Pol I bound to dNTP, the discrimination involves the amino acid side chain of Glu615 which provides a steric hindrance to the 2'-OH of an incoming rNTP [38]. Site-directed mutagenesis of the homologous residue to alanine in *E. coli* Pol I resulted in enhanced incorporation of single ribonucleoside triphosphates but did not result in a polymerase that could progressively incorporate ribonucleotides [39], presumably because the overall enzymatic activity was impaired by introduction of a neutral residue.

By selecting for active polymerases following random mutagenesis of motif A, and then screening for the ability to incorporate rNTPs, *Taq* polymerase mutants were isolated that could efficiently synthesize long stretches of RNA from a DNA template. Of the 23 ribonucleotide-incorporating mutants obtained, many contained multiple substitutions but could be divided into two major classes: 1) those encoding a hydrophilic substitution at Ile-614, and 2) those that encode a Glu615Asp substitution. Kinetic analyses of some of the mutants show that each incorporates ribonucleotides at an efficiency (k_{cat}/K_m) 1000-times higher relative to the wild-type enzyme.

12.3.1.3 Mutability of the *Taq* polymerase active site

For both *Taq* Pol I motif A and motif B randomized libraries, a spectrum of substitutions was permitted without loss in catalytic activity. In motif A, only one of the thirteen residues was absolutely immutable (Asp610) and likewise only two of the thirteen motif B residues were found to be immutable (Arg659 and Lys663). These residues were absolutely required for activity and concur with the crystal structure prediction for coordination of the metal-mediated catalysis reaction and contacts with the incoming dNTP. Four additional amino acids allowing conservative substitutions were found in motif A and two amino acids allowed conservative substitutions in motif B. Overall, the data from the random mutagenesis show that only 3 out of 26 active site (motif A and B) residues analyzed are absolutely required for function *in vivo*, six amino acids can tolerate only conservative substitutions, and all other residues are highly substitutable. This was a surprising find considering that motifs A and B are so highly conserved across diverse prokaryotes (Tab. 12.1). In order to determine if the considerable plasticity observed for the Taq DNA polymerase active site was a general phenomena for Pol I like enzymes, the motif A of *Escherichia coli* DNA polymerase (Pol I) was similarly randomized.

12.3.2
Random oligonucleotide mutagenesis of *Escherichia coli* Pol I

A random library of E. coli Pol I motif A mutants was constructed as for *Taq* Pol I. All of the motif A residues were mutable and the spectrum of mutations was remarkably similar to that obtained for the *Taq* Pol I studies. In particular, substitutions for D705, the catalytically essential amino acid, did not appear amongst the active mutants obtained, as expected, and residue E710 was substituted solely by aspartic acid, highlighting the importance of a negative charge at this position. Single mutations in the N- and C-terminal portions of motif A tended to have opposite effects on DNA polymerase activity. Those mutations in the N- terminal end of motif A yielded enzymatic activities equal to or better than that of the wild type. This region forms part

Tab. 12.2.

E												C
F2	A4	D12										F
G	E11	E2										G5
H2	F12	F	G11				C13					I2
I21	I	G3	I	F2			D		A			K18
K2	M5	I2	K	I6			G10		K12			L
M	Q2	K2	N2	M8			I		L7			M13
P2	R3	L29	P	P3			L	H17	M18	A		Q
Q8	S3	M21	S6	Q9			N20	L15	N2	F10		S15
R8	T	P	T14	R		F12	R30	N	Q2	I6		T12
T	V4	S	V22	S2		H2	T3	R	T16	D15	P4	V
V12	W16	T	Y3	V21		W	Y	S	V13		V7	W5
L	L	V	A	L	D	Y	S	Q	I	E	L	R
605												617

Taq Pol I

			A9									
			E1									
A5			F1									
D1	F3		G1	E4								
E9	H1		I2	G11								
F2	L6		K1	L3			A5		A1			
G6	M6	F8	L2	P2		C6	G4	E1	F10			A1
I2	N4		P2	Q1		F3	L7	H10	L2			C6
L12	S5	G4	R1	S7		G1	R3	K4	M11			G8
M8	T1	I13	T10	T8		H7	T4	L3	N8		M8	H3
R1	V8	L7	W1	V14		I1	V1	M1	S6		P2	L2
W2	Y1	M1	Y1	W1		S1	W8	R4	T4		Q2	M1
						V1	Y1	V1	V10	D6	V11	S11
V	I	V	S	A	D	Y	S	Q	I	E	L	R
700												712

E.coli Pol I

of the anti-parallel β-sheet structure that accommodates the triphosphate moiety of the substrate dNTP. Mutation in the C-terminal region tended to render the enzyme less active than wild-type. The Overall plasticity of the active site was remarkable given the high conservation of motif A and paralleled the findings from the *Taq* Pol I studies (Tab. 12.2). The main differences in mutability between the two enzymes are the greater number of positively charged residue substitutions in *Taq* compared to *E. coli* mutants and the restriction to planar ringed amino acids in *Taq* Pol I at Y611. These differences may be due to the different folding and packing constraints necessary to confer the thermostability of *Taq* Pol I.

12.4
Directed Evolution of Thymidine Kinase

A unifying approach in cancer therapy is to exploit biochemical and cellular differences between tumor and normal cells and use methods that selectively kill tumor cells. Along these lines, the toxicity of certain drugs toward cancerous cells can be enhanced by the specific introduction of selected genes into these malignant cells using gene therapy. One of these genes, Herpes Simplex Virus type 1 (HSV-1) thymidine kinase (TK), phosphorylates a variety of nucleoside analogs, including ganciclovir (GCV) and acyclovir (ACV). These analogs lack 3' hydroxyl termini and thus terminate DNA synthesis upon incorporation into host and/or viral DNA [40]. In contrast, human TK exhibits a high stringency in substrate selection, as it catalyzes the phosphorylation of thymidine and only a limited number of related analogs. One possible approach to treat cancers is by specifically introducing HSV TK into tumors and then administering GCV or ACV systemically. This approach has been tested in animal models, and tumors regress following treatment [40 – 43]. Now in clinical trials, the efficacy of this method is limited by the side effects of the drugs required for tumor regression. The power of this approach can be further enhanced by identifying HSV-1 TK mutants that confer heightened GCV and ACV sensitivity when expressed in tumor cells.

We have carried out random oligonucleotide mutagenesis of the HSV-1 TK gene with the goal of isolating mutants with altered substrate specificities [22, 44 – 48]. A series of libraries containing amino acid substitutions at the nucleoside binding domain were created, and active mutant HSV-1 TKs were selected in a thymidine kinase deficient *E. coli* strain [45]. These studies identified key residues that affect substrate specificity. Using the information provided by these studies a library of 1.1×10^6 *E. coli* transformants were created from a pool of HSV-1 TK that was 100 % randomized at 18 nucleotides in the nucleoside binding domain; 426 (0.039 %) of these mutants were active [48]. *E. coli* harboring active mutants were then screened for preferential killing by low doses of GCV and ACV. Among these 426 candidates, 26 showed enhanced sensitivity to GCV, 54 were more sensitive to ACV, and 6 to both analogs. Sequencing the ten most promising mutants showed that each contains three to six amino acid substitutions. Some of these variants were stably transfected into baby hamster kidney cells. A particular mutant (number 75) harboring four substitutions (I160L, F161L, A168V, L169M) caused the cells to become as much as 43-fold more sensitive to GCV and 20-fold more sensitive to ACV, compared to cells expressing wild-type HSV-1 TK. Kinetic analysis revealed that the K_m of mutant 75 for GCV is $10 \mu M$ and is one-fifth of the wild-type enzyme K_m. The K_m ratio of GCV to thymidine for mutant 75 is 11, while that for wild-type HSV-1 TK is 10-fold higher at 120. The turnover rate or k_{cat} is the same between wild-type TK and mutant 75 for thymidine (0.23 and 0.21 sec^{-1}), GCV (0.05 and 0.05 sec^{-1}), and ACV (0.008 and 0.01 sec^{-1}). Another mutant demonstrates an 8.5-fold reduction in ACV IC$_{50}$ compared to wild-type TK [49]. Thus, these results suggest that the enhanced killing results from increased enzyme affinity for GCV and ACV. The use of mutant 75 or similar mutant genes in gene therapy could enable more effective killing of cancer cells at less systemically toxic doses of GCV and ACV.

HSV TK mutants with altered activity toward azidothymidine (AZT) have also been developed using directed evolution. Herpes simplex virus type-1 and type-2 TK genes were randomly combined using DNA shuffling. Mutants with multiple amino acid substitutions were discovered that sensitize *E. coli* to 32-fold less AZT compared with HSV-1 TK and 16,000-fold less than HSV-2 TK [50]. The increased sensitivity has been attributed to decreased mutant K_m for AZT as well as decreased specificity for thymidine. These mutants present the possibility of using the chain terminator AZT in cancer gene therapy.

The success of cancer gene therapy is likely to require the introduction of genes into all or the majority of malignant cells within a tumor and thus even if efficacious mutant genes are created we will still require vectors that can introduce these genes specifically into malignant cells. A more modest approach to gene therapy is to protect bone marrow cells against commonly used chemotherapeutic agents since bone marrow toxicity frequently limits therapy by many effective agents. We have utilized directed evolution to target two enzymes that offer potential in protection of human bone marrow: thymidylate synthase and DNA methyltransferase. It is important that transductions can be carried out on isolated bone marrow cells and these can be reintroduced into the same individual prior to chemotherapy. Successfully transfected cells harboring mutant enzymes resistant to subsequent chemotherapy would have a selective advantage. In addition, transduction need not be 100 % efficient.

12.5
Directed Evolution of Thymidylate Synthase

Mutant thymidylate synthases have the potential to render cells either sensitive or resistant to chemotherapeutic agents. Thymidylate synthase (TS) is a 72 kDa homodimer responsible for the methylation of dUMP to dTMP using methylenetetrahydrofolate both as a methyl donor and reductant [51]. After subsequent phosphorylations to dTTP, residues of dTMP are incorporated into the newly synthesized DNA strand. The production of dTMP by this *de novo* pathway is absolutely required for DNA synthesis and cell survival, and not surprisingly, TS activity has been demonstrated to be highest in rapidly dividing cells [52, 53]. In the absence of TS, cells undergo a characteristic demise known as thymineless death [54]. For these reasons, human TS has been an important target for the design of chemotherapeutic agents.

The most widely known chemotherapeutic agent directed against TS is 5-fluorouracil (5-FU). 5-FU was first used clinically almost 50 years ago, yet still remains a mainstay for the treatment of carcinoma of the breast and gastrointestinal tract. In cells, 5-FU is metabolized to 5-FdUMP, which forms a stable inhibitory ternary complex with the co-substrate N^5N^{10}-methylene-5,6,7,8-tetrahydrofolate (CH_2H_4-folate) and TS. In this complex, a covalent bond links the thiol of cysteine 195 of human TS to C6 of deoxyuracil monophosphate (dUMP) and the methylene carbon of the co-substrate is joined to C5 of the nucleotide [55]. The fluorine at C5, unlike the proton, cannot be abstracted, leading to a stable covalent complex that inactivates TS [56]. In addition to the pyrimidine-based inhibitors, more recent analysis of the catalytic mechanism and structural interactions of TS has resulted in the creation of stable and specific folate-based inhibitors of TS. One compound, Tomudex (Raltitrexed, ZD 1694; Zeneca Pharmaceuticals, UK), has been recently approved for treatment of advanced colorectal cancer in the UK. Several other novel folate-based TS inhibitors are in various stages of development and clinical evaluation at this time.

The study of drug-resistant variants of TS began over 10 years ago with the discovery of a naturally occurring drug-resistant mutant variant of TS, Y33H. This mutant, expressed heterozygously in a human HCT116 colonic cell line, was found to confer a fourfold resistance to 5-FdUR, a prodrug of FdUMP, in cell culture [57 – 59]. In the last two years, there has been a virtual explosion of new studies identifying drug-resistant TSs that contain amino acid alterations in several different regions of the protein (Fig. 12.4).

The extensive studies on catalysis by TS and the existence of protein crystallography data, has facilitated studies using site-specific mutagenesis. By subjecting highly conserved residues of TS that are important in cofactor binding to site-directed mutagenesis, two drug-resistant mutants of TS were discovered. I108A was found to be resistant to the antifolates Raltitrexed and Thymitaq (AG337), and F225W was found resistant to the antifolate BW1843U89 and 5-FdUR [60]. Recently, a 5-FdUR resistant cell line was determined to encode a P303L mutant that, although metabolically unstable, was nonetheless able to confer resistance to transfected cells against FdUR, Raltitrexed, Thymitaq, and BW1843U89. An engineered P303D mutant was also resistant to both 5-FdUR and BW1843U89 [61].

Contemporaneously with these recent discoveries, our laboratory has used directed evolution by random sequence mutagenesis of a region of human TS near the catalytic cysteine (position 195) to produce novel 5-FU resistant TS mutants. A plasmid-encoded library of TS mutants was constructed using an oligonucleotide-based random mutagenesis technique. The end product of this process was a family of over 2 million

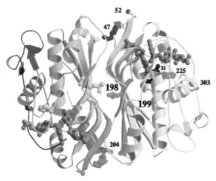

Fig. 12.4. The human TS dimer. Residues reported by others to be associated with drug resistance include Y33, F225, I108 (not depicted), P303, K47, D49 (not depicted), and G52 [57, 60 – 64, 82]. We have found several drug-resistant TS mutants through mutation of both the Arg^{50}-loop (defined by residues 47 – 52, shown), and a loop near the catalytic C195 encompassing A197 – V204) [62, 64, 82]. Residues 198, 199, and 204, contained within this randomization scheme and located on the dimer interface, are shown for reference.

TS enzymes with a random peptide sequence of 13 amino acid surrounding the active site. This library of mutants was transfected into *E. coli* lacking the *ThyA* (TS) gene. Clones that were able to survive in the presence of increasing doses of 5-FdUR were sequenced, and it was discovered that altering residues 195 – 199 is sufficient to provide increased 5-FdUR. Kinetic analysis of a triple mutant (A197V, L198I, C199F) demonstrated a 20-fold increase in K_d for 5-FdUMP, with retention of near wild-type catalytic efficiency and affinities for dUMP and CH_2H_4-folate [62].

Studies that used chemical mutagenesis followed by selection in human HT1080 cells with the antifolate Thymitaq have identified three mutants that demonstrate resistance to Thymitaq and 5-FdUR. All three of these identified mutants contain substitutions in the essential and conserved Arg^{50}-loop. This loop is believed to form an important bridge linking the substrate, cofactor, and enzyme C-terminus. These mutants, all of which harbor a single amino acid alteration (K47E, D49G, G52S), confer a high degree of drug resistance to mammalian cells in culture [63]. The mutant G52S conferred almost a 100-fold increase in IC_{50} for Thymitaq while still retaining resistance to 5-FdUR. Inhibition studies with 5-FdUMP and three antifolate inhibitors indicated that D49G and G52S demonstrated an increase in the K_i for FdUMP of 5.4 to 20-fold, respectively, while retaining a K_i for Thymitaq approximately six times that of wild type. Interestingly, despite a 5-fold increase in IC_{50} for Thymitaq and 5-FdUR, K47E did not demonstrate variant kinetics for any of the inhibitors tested. The power of this chemical mutagenesis technique employed by Tong *et al.* lies in the ability to screen residues of the entire TS polypeptide sequence for critical mutations. However, chemical mutagenesis using EMS is statistically unlikely to test synergistic effects of multiple amino acid substitutions in one polypeptide nor can it test substitution of each of the 20 amino acids at each position of interest. Therefore, we view the mutations identified by Tong *et al.* as prototype, or first-generation, drug-resistant mutants that can guide the discovery of mutants with yet greater resistance using directed evolution.

Having developed a robust *E. coli* selection scheme for identifying human 5-FdUR resistant TS mutants, we employed targeted random mutagenesis to exhaustively examine all mutations within the Arg^{50}-loop (residues 47-52) [64]. Approximately 1.5 million mutants of the Arg^{50}-loop were created and selected to identify those mutants with the ability to complement a thymidylate synthase deficient *E. coli* strain and form colonies in the presence of increasing amounts of 5-FdUR. *E. coli* harboring plasmids that were encoding TS with single, double, and triple amino acid substitutions were identified that survived at dosages of 5-FdUR lethal to *E. coli* harboring either wild-type TS or constructs encoding previously characterized drug-resistant mutants. Four 5-FdUR resistant mutants were purified to apparent homogeneity and kinetic studies indicate that these enzymes are highly efficient, with no mutant displaying more than a 50 % decrease in k_{cat}. Inhibition constants (K_i) for the double mutant K47Q, D48E and the triple mutant D48E, T51S, G52C in the presence of 5-FdUMP were determined to be 75 to 100 times higher, respectively, than that of the wild-type enzyme. In addition, this library reselected two of the mutants identified through EMS mutagenesis (G52S and K47E). As these mutants were originally selected in mammalian cells using a folate based inhibitor, this finding validates the use of an *E. coli*-based selection assay and indicates that the identical mutant can be selected via either 5-FdUR or a folate-based inhibitor. This likely can be explained by the central role of the Arg^{50}-loop in coordination with both of the corresponding substrates. Work is currently underway to test these new mutants in human cells and to further understand the structural consequences of the mutations in the Arg^{50}-loop.

Because 5-FdUMP is structurally similar to the natural substrate, dUMP, it is difficult to predict how single amino acid substitutions or multiple substitutions could restrict the binding of 5-FdUMP without affecting binding of dUMP. Random oligonucleotide mutagenesis provides a combinatorial alternative that can examine a large amount of sequence space and create altered enzymes without requiring detailed knowledge about amino acid interactions or effects of specific alterations.

The decrease in affinity to 5-FdUMP in mutants with near normal catalytic activity suggests that these mutants may be suitable for use as drug-resistant genes in gene therapy applications. Although effective, the use of 5-FU as a chemotherapeutic agent has been limited by toxicity to bone marrow, gastrointestinal, and other tissues [65]. The introduction and expression of mutants of human TS that can function in the presence of systemic 5-FU could protect normal cells from cytotoxicity or allow augmentation of the maximally tolerated dose of 5-FU (For a recent review please see [66]).

12.6
O^6-Alkylguanine-DNA Alkyltransferase

Another approach in cancer gene therapy is to protect normal host tissues (hematopoietic stem cells, for example) against the dose-limiting toxicity of many commonly used alkylating drugs [67]. Many alkylating agents, including *N*-methyl-*N'*-nitro-*N*-nitrosoguanidine (MNNG) and the chemotherapeutic 1,3-bis(2-chloroethyl)-1-nitrosourea (BCNU), produce cytotoxic and mutagenic alkyl adducts at the O^6 position of guanine. The human O^6-alkylguanine-DNA alkyltransferase (AGT) removes alkyl groups at the O^6 position of guanine, and to a lesser degree, adducts at the O^4 position of thymine, by transferring them to a cysteine residue on the protein. Since this transfer is irreversible, any alkyltransferase molecule can react only once, and as such, can be classified as a suicide protein [68]. Wild-type human AGT has been over-expressed in bone marrow of mice using retrovirus mediated gene transfer, and shown to protect against the cytotoxicity of BCNU [69].

In order to find mutant AGT that confer greater resistance against alkylating agents, we have generated a library in which twelve codons surrounding, but not including, the active cysteine were replaced by a random nucleotide sequence [70]. After three rounds of stringent selection with MNNG in an alkyltransferase deficient *E. coli* background, different mutants containing a valine to phenyalanine change at codon 139 (V139F) accounted for more than 70 % of survivors. Mutants containing this particular substitution were selected from the library at a greater frequency than the wild-type sequence. The single V139F mutant was shown to provide greater protection than wild-type AGT, increasing the MNNG resistance over fourfold and reducing the mutagenesis rate 2.7 – 5.5 fold. In a different study, human AGT was shuffled with other mammalian alkyltransferases, and active mutants were identified in bacteria by selecting for MNNG resistance. The most resistant mutant discovered contains seven amino acid substitutions, all at positions distant from the active site, again providing better protection than wild-type AGT (Christians, F.C. and Stemmer, W.P.C., personal communication).

These studies have been expanded to obtain mutants that are more resistant to competitive inhibition. Currently in clinical trials, O^6-benzylguanine (BG) is a competitive inhibitor of AGT that promises to improve cancer chemotherapy by depleting tumor cells of AGT [71]. However, BG also reduces the already low AGT activity in bone marrow and further exacerbates dose-limiting myelosuppression in the patient [72, 73]. Therefore, AGT variants that both remove alkylation damage and that are resistant to inhibition by BG would be highly advantageous for protecting bone marrow. Mutants have been identified by site-directed mutagenesis, the presumptive mechanism of resistance being a restructuring of the active site to exclude BG's larger benzyl group. However, these mutants, most of which contain single amino acid substitutions, are only partially resistant to BG *in vitro* (for example G156A) [68].

Instead of examining a limited number of amino acid replacements by site-directed mutagenesis, we have selected mutants from over 8×10^6 AGT variants from two libraries that were created by random oligonucleotide mutagenesis. Active mutants were isolated after challenge with combined BG and MNNG. The first library contains 6.5×10^6 members, and is 100 % randomized at 6 codons over a 13 codon stretch. In this library, V139 was fixed as phenylalanine, the active site region between 143 to 147 (IPCHR) were unchanged, and all other codons were randomized [74]. Another library was generated with 10 % randomness over a 23 codon stretch downstream of the active site cysteine (amino acids 150 to 172), and is composed of greater than 1.5×10^6 members [75]. These studies have yielded a large number of active mutants, many of which show enhanced resistance to BG. In addition to producing novel gene therapy candidates, the above libraries have also produced AGT variants that show altered substrate specificity. One mutant, 56-8, containing eight substitutions near the active site (C150Y, S152R, A154S, V155G, N157T, V164M, E166Q, and A170T), shows enhanced repair of O^4 methylated thymine (O^4mT) *in vitro*. The second-order rate constant for repair of O^4mT was up to 11.5-fold greater than that of wild-type AGT. The relative O^4mT specificity, $k(O^4mT)/k(O^6mG)$, was 75-fold greater [76]. Further, this mutant shows enhanced protection against O^4mT induced mutations in *E. coli* when compared to wild-type AGT [77].

   In order to select the fittest mutant for further evaluation we made use of Darwinian evolution. Seventeen AGT mutants were introduced into human hematopoietic cells by retrovirus transduction, and selected for resistance to killing by the combination of BG and BCNU [78]. Competitive analysis among the variants revealed one mutant (MGMT-2) with remarkable resistance to the combination of BG and BCNU. Starting with equal numbers of transfected bone marrow cells harboring each of the 17 mutants and the wild type, one mutant was recovered in 54 % of the clones isolated. A repeat of the competition experiment recovered predominantly the same mutant. The resistance of the cells harboring the mutant was analyzed. With a fixed dose of 20 μM BCNU, IC_{90} of cells transduced with MGMT-2 was 410 – 500 μM BG, compared with less than 20 μM BG with wt-MGMT expressing cells. With BG fixed at 25 μM, MGMT-2 transduced cells had an IC_{90} of 60 μM, approximately fourfold higher than the 15 μM IC_{90} of the wt-MGMT expressors. Increased BG and BCNU resistance by MGMT-2 was not due to increased expression or increased protein stability, but due to greater resistance to inhibition by BG. *In vitro* analysis showed that MGMT-2 protein retaining full activity at up to 2 mM BG inhibition, compared with loss of activity at less than 0.2 μM with wild-type AGT. MGMT-2 harbors five mutations (S152H, A154G, Y158H, G160S, L162V). Although all five substitutions are possibly not required for the change in substrate specificity, it is noteworthy that a mutant with these particular substitutions would likely not have been rationally designed. This mutant displays a level of BG resistance in human hematopoietic cells greater than any previously reported mutant, and is a prime candidate for gene therapy to protect bone marrow of cancer patients undergoing BCNU chemotherapy.

   In the case of AGT, we have demonstrated the isolation of novel mutants that show enhanced function, variants that are more resistant to inhibitors, and mutants that have altered substrate specificity. Mutants in each category have important applications in structure/function studies, gene therapy, and mutation analysis.

12.7
Discussion

Mutants chosen by natural selection may not be those that are most desirable for biotechnology, industry, medicine nor for probing the catalytic function of an enzyme. Having a more active enzyme or one with altered substrate specificities may not be advantageous to cells or organisms. Recently, a variety of techniques, under the rubric of applied molecular evolution, have been established that allow scientists to explore large regions of sequence space and to identify new mutant enzymes with unique properties. Spiegelman and Eigen were pioneers in directing evolution *in vitro* using Qβ bacteriophage [79, 80]. Since then, the concept of extensively mutating DNA *in vitro* and identifying specific mutants by genetic selection has been applied to a variety of genes that encode different enzymes. A spectrum of techniques has evolved, many of which can be utilized sequentially or in combination to evolve mutant enzymes *in vivo*. As a result, mutant proteins are being created for both industrial and medical utility and have provided powerful tools to explore biochemical mechanisms of catalysis. In contrast to the millennia required by natural evolution, we can now evolve mutant enzymes with desired specific properties in weeks.

In this chapter we focused on random sequence mutagenesis and on enzymes involved in DNA synthesis, DNA repair and nucleotide metabolism. These enzymes have a central role in DNA metabolic processes. They are highly conserved in nature and mutations in these enzymes can result in a mutator phenotype, that is, an enhanced rate of mutagenesis throughout the genome. The general protocol has been to replace a portion of a plasmid encoded gene with a chemically synthesized oligonucleotide containing random sequences in place of a portion of the active site of an enzyme, transfect the plasmid into *E. coli,* and identify active mutants by genetic complementation.

The exploration of sequence space by any of the techniques enumerated is restricted by the number of mutants that one can effectively select or screen. The most mutants that we contemplated screening is 10^{10}, and this requires a particularly strong selection assay, β-lactamase resistance (unpublished results). In routine experiments, we can seldom analyze more than 10^7 mutants. A similar limitation is also apparent using other procedures in which an oligonucleotide is inserted into a gene, or in which genes are shuffled amongst homologous sequences. Thus, one needs to develop technologies for progressive selection for desired traits, or one needs to carry out successive rounds of selection. In the latter case, one needs to establish, in an early selection step, the most favorable sites for mutagenesis and then use subsequent rounds of re-randomization to further enrich for the desired changes.

The limitation on the number of mutants that can be analyzed results from at least several variables. First is the transfection efficiency. Transfection can be carried out very effectively in *E. coli* (1 transfectant per 100 *E. coli*) and to a similar efficiency with mammalian cells. With mammalian cells, one can essentially transform 100 % of the cells using retroviral vectors, as illustrated in the case of methyltransferase. Yeast offers powerful systems for potential complementation by mutant foreign genes, but the transfection efficiency is lower. Secondly, one is limited by the ligation efficiency of inserting the oligonucleotide into the plasmid. The efficiency of ligation is governed by a second-order reaction and is frequently sequence dependent. Thirdly, one is limited by the number of host cells that can be grown and transfected economically. Bacteria offer an enormous advantage due to their small size and the population density to which they can be cultured. Lastly, and most importantly, we are limited by the boundaries of our imaginations in devising robust genetic selection strategies to identify specific mutations amongst populations of cells harboring billions of other mutations.

It is conceivable that selection for mutations within the active site of enzymes may not be the most efficacious approach to obtain mutations with useful altered properties. Most amino acid substitutions within the active site are likely to reduce catalytic activity. Considering the vastness of sequence space within a protein, it seems reasonable that substitutions at a distance may be equally or more likely to altered substrate specificity without reducing catalytic efficiency. Some of these substitutions could alter the binding pocket for the docking of the substrate without affecting the chemical or rate limiting step in catalysis. The major disadvantage of sequence space alterations at a distance is our lack of knowledge about how alterations in the overall three-dimensional structure of an enzyme effects the structure of the amino acid residues that form the catalytic site of enzymes. Without this knowledge we are at a loss to determine specific residues to target for mutagenesis.

If one reviews the achievements of random sequence mutagenesis, one is struck by two general and perhaps unexpected conclusions. First is the exceptional plasticity of the active sites of enzymes, even those that are highly conserved during natural evolution. For example, random sequence mutagenesis of motif A in both *Taq* and *E. coli* DNA polymerase I resulted in multiple substitutions for every amino acid residue except the absolutely required aspartic acid. Many of these substitutions did not result in a reduction of size of colonies formed nor resulted in a reduction of catalytic efficiency. Many of these substitutions altered the fidelity of DNA synthesis, which could conceivably be advantageous during periods of environmental stress. In contrast, only a few natural substitutions have been observed within the DYSQIELR sequence within motif A. More strikingly are the results of sequence alignments within the *Rickettsia* genus which show conservation of sequence identity even at the nucleotide level [81]. The lack of mutability in nature, even at the level of third position codon variation, suggests that prokaryotes have evolved a mechanism to maintain sequence homogeneity of key genes or selected segments. Based on the results obtained with random sequence mutagenesis, we have proposed that genes encoding many proteins are indeed highly variable and can tolerate a high mutation burden, facilitating the emergence of beneficial mutations during periods of environmental stress. However, cells have also evolved mechanisms of recombination involving lateral transfer that restore the wild-type sequence [82].

Secondly, from the results so far obtained utilizing applied molecular evolution, it can be surmised that natural evolution of enzymes has been finely tuned to select for high catalytic efficiency using natural substrates. A major achievement of applied molecular evolution has been to tailor enzymes to utilize non-canonical substrates at exceptionally high efficiencies or to become resistant to potent inhibitors. For example, mutant DNA polymerases have been identified that exhibit 1000-fold enhancement in catalytic efficiency using ribonucleoside triphosphates as substrates, and mutant Herpes thymidine kinases can exhibit a 100-increase in the rate of incorporation of ganciclovir triphosphate, a chain terminator guanosine analog. Moreover, mutant thymidine synthases have been obtained that exhibit a K_i for fluorouracil that is 75-greater than that of the wild-type and mutant methyl transferases are more than 500-fold resistant to the inhibitor, benzyl guanine. In sharp contrast, we have not succeeded in identifying mutant "DNA enzymes" that exhibit more than a fivefold increase in utilizing natural substrates. Future experiments will determine whether repetitive selection or continuous selection under conditions with increasing stringency will result in enzymes with enhanced activities with natural substrates. Nevertheless it is important to note that for industrial purposes, only small increases in catalytic efficiencies or stability may be advantageous.

Acknowledgments
Studies from our laboratory have been supported by grants from the NIH (CA-78885) and by the NCI Molecular Training and Cancer research grant (5T32CA09437) to JD and HG and the NIHS environmental pathology/toxicology training program (532ES07032) to JA.

References

[1] Smith, J. M. (1970) *Nature 225*, 563–4.

[2] Muller, H. J. (1927) *Science 66*, 84-87.

[3] Myers, R. M., Lerman, L. S., and Maniatis, T. (1985) *Science 229*, 242–7.

[4] Diaz, J. J., Rhoads, D. D., and Roufa, D. J. (1991) *Biotechniques 11*, 204–6, 208, 210–1.

[5] Tindall, K. R., and Kunkel, T. A. (1988) *Biochemistry 27*, 6008–13.

[6] El-Deiry, W. S., Downey, K. M., and So, A. G. (1984) *Proc. Natl. Acad. Sci. USA 81*, 7378–82.

[7] Goodman, M. F., Keener, S., Guidotti, S., and Branscomb, E. W. (1983) *J. Biol. Chem. 258*, 3469–75.

[8] Fromant, M., Blanquet, S., and Plateau, P. (1995) *Anal. Biochem. 224*, 347–53.

[9] Beckman, R. A., Mildvan, A. S., and Loeb, L. A. (1985) *Biochemistry 24*, 5810–7.

[10] Kunkel, T. A., and Loeb, L. A. (1980) *J. Biol. Chem. 255*, 9961–6.

[11] Bebenek, K., Roberts, J. D., and Kunkel, T. A. (1992) *J. Biol. Chem. 267*, 3589–96.

[12] Bebenek, K., and Kunkel, T. A. (1990) *Proc. Natl. Acad. Sci. USA. 87*, 4946–50.

[13] Pavlov, Y. I., Minnick, D. T., Izuta, S., and Kunkel, T. A. (1994) *Biochemistry 33*, 4695–701.

[14] Zaccolo, M., Williams, D. M., Brown, D. M., and Gherardi, E. (1996) *J. Mol. Biol. 255*, 589–603.

[15] Suzuki, M., Yoshida, S., Adman, E. T., Blank, A., and Loeb, L. A. (2000) *J. Biol. Chem. 275*, 32728–35.

[16] Suzuki, M., Avicola, A. K., Hood, L., and Loeb, L. A. (1997) *J. Biol. Chem. 272,* 11228–11235.

[17] Harayama, S. (1998) *Trends Biotechnol. 16,* 76–82.

[18] Crameri, A., Raillard, S.-A., Bermudez, E., and Stemmer, W. P. C. (1998) *Nature 391,* 288–291.

[19] Arnold, F. H. (1998) *Nat. Biotechnol. 16,* 617 8.

[20] Black, M. E., and Loeb, L. A. (1996) *Meth. Mol. Biol. 57,* 335–349.

[21] Horwitz, M. S., and Loeb, L. A. (1986) *Proc. Natl. Acad. Sci. USA 83,* 7405–9.

[22] Dube, D. K., Parker, J. D., French, D. C., Cahill, D. S., Dube, S., Horwitz, M. S., Munir, K. M., and Loeb, L. A. (1991) *Biochemistry 30,* 11760–11767.

[23] Loeb, L. A., and Kunkel, T. A. (1982) *Annu. Rev. Biochem. 51,* 429–57.

[24] Echols, H., Lu, C., and Burgers, P. M. (1983) *Proc. Natl. Acad. Sci. USA 80,* 2189–92.

[25] Modrich, P. (1987) *Annu. Rev. Biochem. 56,* 435–466.

[26] Tabor, S., and Richardson, C. C. (1995) *Proc. Natl. Acad. Sci. USA 92,* 6339–6343.

[27] Tabor, S., and Richardson, C. C. (1989) *J. Biol. Chem. 264,* 6447–58.

[28] Gutman, P. D., and Minton, K. W. (1993) *Nucleic Acids Res. 21,* 4406–7.

[29] Patel, P. H., Kawate, H., Adman, E., Ashbach, M., and Loeb, L. A. (2000) *J. Biol. Chem. 7,* 7.

[30] Patel, P. H., and Loeb, L. A. (2000) *J. Biol. Chem. 275,* 40266–72.

[31] Ollis, D. L., Brick, P., Hamlin, R., Xuong, N. G., and Steitz, T. A. (1985) *Nature 313,* 762–766.

[32] Braithwaite, D. K., and Ito, J. (1993) *Nucleic Acids Res. 21,* 787–802.

[33] Joyce, C. M., and Steitz, T. A. (1994) *Annu. Rev. Biochem. 63,* 777–822.

[34] Sweasy, J. B., and Loeb, L. A. (1992) *J. Biol. Chem. 267,* 1407–1410.

[35] Kim, B., and Loeb, L. A. (1995) *Proc. Natl. Acad. Sci. USA 92,* 684–688.

[36] Suzuki, M., Baskin, D., Hood, L. E., and Loeb, L. A. (1996) *Proc. Natl. Acad. Sci. USA 93,* 9670–75.

[37] Beckman, R. A., and Loeb, L. A. (1993) *Quart. Rev. Biophys. 26,* 225–331.

[38] Li, Y., Korolev, S., and Waksman, G. (1998) *Embo. J. 17,* 7514–25.

[39] Astatke, M., Ng, K., Grindley, N. D., and Joyce, C. M. (1998) *Proc. Natl. Acad. Sci. USA 95,* 3402–7.

[40] Elion, G. B. (1980) *Advan. Enzy. Regul. 18,* 53–66.

[41] Ram, Z., Culver, K. W., Walbridge, S., Frank, J. A., Blaese, R. M., and Oldfield, E. H. (1993) *J. Neurosurg. 79,* 400–407.

[42] Caruso, M., Panis, Y., Gagandeep, S., Houssin, D., Salzmann, J.-L., and Klatzmann, D. (1993) *Proc. Natl. Acad. Sci. USA 90,* 7024–28.

[43] Vile, R. G., and Hart, I. R. (1993) *Cancer Res. 53,* 3860–3864.

[44] Munir, K. M., French, D. C., Dube, D. K., and Loeb, L. A. (1992) *J. Biol. Chem. 267,* 6584–89.

[45] Black, M. E., and Loeb, L. A. (1993) *Biochemistry 32,* 11618–626.

[46] Munir, K. M., French, D. C., and Loeb, L. A. (1993) *Proc. Natl. Acad. Sci. USA 90,* 4012–4016.

[47] Munir, K. M., French, D. C., Dube, D. K., and Loeb, L. A. (1994) *Protein Eng. 7,* 83–89.

[48] Black, M. E., Newcomb, T. G., Wilson, H.-M. P., and Loeb, L. A. (1996) *Proc. Natl. Acad. Sci. USA 93,* 3525–3529.

[49] Kokoris, M. S., Sabo, P., and Black, M. E. (2000) *Anticancer Res 20,* 959–63.

[50] Christians, F. C., Scapozza, L., Crameri, A., Folkers, G., and Stemmer, W. P. (1999) *Nat. Biotechnol. 17,* 259–64.

[51] Davisson, V. J., Sirawarapora, W., and Santi, D. V. (1989) *J. Biol. Chem. 264,* 9145–9148.

[52] Rode, W., Scanlon, K. J., Moroson, B. A., and Bertino, J. R. (1980) *J. Biol. Chem. 255,* 1305–1311.

[53] Conrad, A. H., and Ruddle, F. H. (1972) *J. Cell Sci. 10,* 471–486.

[54] Seno, T., Ayusawa, D., Shimizu, K., Koyama, H., Takeishi, K., and Hori, T. (1985) *Basic Life Sci. 31,* 241–63.

[55] Montfort, W. R., Perry, K. M., Fauman, E. B., Finer-Moore, J. S., Maley, G. F., Hardy, L., Maley, F., and Stroud, R. M. (1990) *Biochemistry 29,* 6964–6977.

[56] Pogolotti, A. L., Jr., Danenberg, P. V., and Santi, D. V. (1986) *J. Med. Chem. 29,* 478–82.

[57] Berger, S. H., Barbour, K. W., and Berger, F. G. (1988) *Molec. Pharm. 34*, 480–484.

[58] Barbour, K. W., Hoganson, D. K., Berger, S. H., and Berger, F. G. (1993) *Molec. Pharm. 42*, 242–248.

[59] Barbour, K. W., Berger, S. H., and Berger, F. G. (1990) *Mol. Pharmacol. 37*, 515–18.

[60] Tong, Y., Liu-Chen, X., Ercikan-Abali, E. A., Zhao, S., Banerjee, D., Maley, F., and Bertino, J. R. (1998) *J. Biol. Chem. 273*, 31209–14.

[61] Kitchens, M. E., Forsthoefel, A. M., Barbour, K. W., Spencer, H. T., and Berger, F. G. (1999) *Mol Pharmacol 56*, 1063–70.

[62] Landis, D. M., and Loeb, L. A. (1998) *J. Biol. Chem. 273*, 25809–17.

[63] Tong, Y., Liu-Chen, X., Erickan-Abali, E. A., Capiaux, G. M., Zhao, S., Banerjee, D., and Bertino, J. R. (1998) *J. Biol. Chem. 273*, 11611–18.

[64] Landis, D. M., Heindel, C. C., and Loeb, L. A. (2001) *Cancer Res. 61*, 666–72.

[65] Grem, J. L. (1990) in *Cancer Chemotherapy: Principles and Practice* (Chabner, B. A., and Collins, J. M., Eds.) pp 180–224, J.B. Lippincott, Philadelphia.

[66] Banerjee, D., Tong, Y., Liu-Chen, X., Capiaux, G., Ercikan-Abali, E. A., Takebe, N., O'Connor, O. A., and Bertino, J. R. (1999) *Prog. Exp. Tumor Res. 36*, 107–14.

[67] Encell, L. P., Landis, D. M., and Loeb, L. A. (1999) *Nat. Biotechnol. 17*, 143–147.

[68] Pegg, A. E., Dolan, M. E., and Moschel, R. C. (1995) *Prog. Nucleic Acid Res. 51*, 167–223.

[69] Maze, R., Carney, J. P., Kelley, M. R., Glassner, B. J., Williams, D. A., and Samson, L. (1996) *Proc. Natl. Acad. Sci. USA 93*, 206–210.

[70] Christians, F. C., and Loeb, L. A. (1996) *Proc. Natl. Acad. Sci. USA 93*, 6124–28.

[71] Spiro, T. P., Gerson, S. L., Liu, L., Majka, S., Haaga, J., Hoppel, C. L., Ingalls, S. T., Pluda, J. M., and Willson, J. K. (1999) *Cancer Res. 59*, 2402–10.

[72] Gerson, S. L., and Willson, J. K. (1995) *Hematol. Oncol. Clin. North Am. 9*, 431–50.

[73] Gerson, S. L., Phillips, W., Kastan, M., Dumenco, L. L., and Donovan, C. (1996) *Blood 88*, 1649–55.

[74] Christians, F. C., Dawson, B. J., Coates, M. M., and Loeb, L. A. (1997) *Cancer Res. 57*, 2007–2012.

[75] Encell, L. P., Coates, M. M., and Loeb, L. A. (1998) *Cancer Res. 58*, 1013–1020.

[76] Encell, L. P., and Loeb, L. A. (1999) *Biochemistry 38*, 12097–12103.

[77] Encell, L. P., and Loeb, L. A. (2000) *Carcinogenesis 21*, 1397–402.

[78] Davis, B. M., Encell, L.P., Zielske, S.P., Christians, F.C., Lie, L., Friebert, S.E., Loeb, L.A., and Gerson. S.L. *Proc. Natl. Acad. Sci. USA*, in press.

[79] Mills, D. R., Peterson, R. L., and Spiegelman, S. (1967) *Proc. Natl. Acad. Sci. USA 58*, 217–224.

[80] Biebricher, C. K., Eigen, M., and Gardiner, W. C. J. (1983) *Biochemistry 22*, 2544–59.

[81] Patel, P. H., and Loeb, L. A. (2000) *Proc. Natl. Acad. Sci. USA 97*, 5095–100.

[82] Landis, D. M., Gerlach, J. L., Adman, E. T., and Loeb, L. A. (1999) *Nucleic Acids Res. 27*, 3702–11.

13
Evolutionary Generation versus Rational Design of Restriction Endonucleases with Novel Specificity

Thomas Lanio, Albert Jeltsch, and Alfred Pingoud

13.1
Introduction

13.1.1
Biology of restriction/modification systems

Restriction/modification systems (RM systems) can be regarded as an "immune" system to protect the host against foreign DNA, in particular bacteriophage DNA [1]. RM systems comprise two different enzyme activities directed against the same short, often palindromic, DNA sequence – (i) a methyltransferase activity that protects the host DNA by methylation of the recognition sequences and (ii) a restriction endonuclease activity, that cuts unmethylated recognition sequences usually present in foreign DNA (Fig. 13.1). Since erroneous cleavage of the host DNA would be highly toxic for the cell, restriction enzymes have evolved a remarkable specificity for their canonical recognition site. They are known to cleave "star" sites, which are DNA sequences differing only in one base pair from the canonical sequence, and also methylated recognition sites, by at least 5 – 6 orders of magnitude more slowly

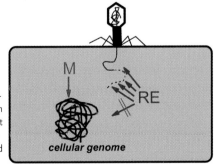

Fig. 13.1. Biological function of restriction/modification systems: RM systems recognize and act on short palindromic DNA sequences. While the host genome is protected by the methyltransferase activity of the system, invading phage DNA is cleaved by the endonuclease activity.

than their canonical recognition sequence [2 – 6]. With this specificity, restriction endonucleases are among the most specific enzymes known that do not employ an energy-dependent proofreading step.

RM systems are very effective against bacteriophages and are ubiquitously distributed over the eubacterial and archaeal kingdoms. The observation that genes encoding RM systems may comprise up to 4 % of the genome of a prokaryotic organism emphasizes the importance of these enzymes for their hosts [7]. In addition to their protective functions, the involvement of RM systems in recombination and transposition is still being discussed [8 – 10], as well as the concept that genes encoding these systems could be regarded as selfish genetic elements [11].

13.1.2
Biochemical properties of type II restriction endonucleases

RM systems are usually divided into three different types (I, II and III), and then into subtypes, according to certain properties like cofactor requirement, subunit composition and mode of DNA cleavage. Among these three types, the type II restriction enzymes are the most important ones for molecular biology and biotechnology. Today, more than 3300 restriction endonucleases with more than 200 different specificities are known and their number is steadily increasing [12]. These enzymes have been studied over the last 30 years as model systems for specific protein-DNA interactions, in general, and enzymes acting on DNA, in particular (for reviews see [13, 14]). Besides many biochemical studies that were carried out to analyze the mechanism of DNA recognition and cleavage by these enzymes, crystal structures of 12 different type II restriction enzymes have been determined, in part in complex with DNA allowing for a detailed insight into the structure and function of this class of enzyme (for a review see [15]).

The orthodox type II restriction endonuclease is a homodimer of around $2 \times 30\,\mathrm{kDa}$. It recognizes a palindromic sequence of 4 – 8 base pairs and cleaves both strands of its DNA substrate in the presence of Mg^{2+} ions within or directly adjacent to the recognition site, leading to 5' phosphate and 3' OH ends. Depending on the enzyme employed, the cleavage reaction results either in *blunt ends* (for example *Eco*RV), or *sticky ends* with a 5' overhang (for example *Eco*RI) or a 3' overhang (for example *Bgl*I, Fig. 13.2).

An interesting subgroup is represented by the type IIS restriction endonucleases. These enzymes recognize an asymmetric site and cleave this sequence at a defined distance after dimerizing on the substrate (for example *Fok*I).

Restriction enzymes constitute one of the largest families of related enzymes known so far and therefore exemplify how natural evolution has generated many different specificities from one (or a few) ancestral protein(s). A detailed comparison of the

EcoRV

GATATC   GAT   pATC
CTATAG   CTAp   TAG

EcoRI

GAATTC   G   pAATTC
CTTAAG   CTTAAp   G

BglI

GCCNNNNNGGC   GCCNNNN   pNGGC
CGGNNNNNCCG   CGGNp   NNNNCCG

Fig. 13.2. Cleavage of the specific recognition sites by the type II restriction endonucleases *EcoRV*, *EcoRI* and *BglI*: The cleavage reaction, which requires Mg^{2+} as cofactor, leads to 5' phosphate and 3' OH ends. While *EcoRV* cleavage results in blunt ends, *EcoRI* and *BglI* generate sticky ends with a 5' and 3' overhang, respectively.

structures and functions of these enzymes will not only improve our understanding of natural evolution but also guide design approaches.

Information regarding all known RM systems are available at REBASE (http://www.rebase.neb.com).

13.1.3
Applications for type II restriction endonucleases

The discovery of restriction endonucleases was one of the key events leading to and being a prerequisite for the development of modern gene technology. These enzymes are indispensable tools for DNA analysis and cloning as well as for medical applications where the study of restriction fragment length polymorphisms (RFLPs) provided first insights into the genetic diversity of man as well as into the genetic basis of some diseases. Recently the use of restriction enzymes has been partly replaced by the PCR technique for certain applications but restriction enzymes still are the work horses in the molecular biology laboratory with a worldwide market of approximately 100 – 200 million US$. Today, an almost full set of enzymes with different specificities is commercially available for the 4 base pair and 6 base pair cutters (16 or 64, respectively), but only 9 out of 256 possible 8 base pair cutters were found. This under-representation of restriction enzymes with long recognition sites can be understood, because palindromic sites comprising 8 base pairs statistically occur only once every 65 536 base pairs and phage genomes usually are short ($10^4 - 10^5$ base pairs). Therefore, a restriction enzyme with an 8 or even 10 base pair recognition site could not fulfill its biological role efficiently. Restriction endonucleases having a recognition sequence of 8 or more base pairs, therefore, in general provide only a limited selective advantage for the bacterial host and in consequence to date only a few 8 base pair cutters were isolated from natural sources and most likely not many more will be found by screening bacterial strains. On the other hand these "rare cutters" with recognition sites longer than 8 base pairs would be very helpful for the manipulation of large DNA fragments such as human genes which are often larger than 5 kb. The availability of such "rare cutters" would facilitate the generation of artificial chromosomes and perhaps make the manipulation of such constructs as easy as plasmid cloning is today. Moreover, such enzymes could be extremely useful for gene targeting and gene replacement, not only for experimental genetics but also for biomedical applications [16 – 19].

Currently, the only strategy available to introduce single cuts in large DNA fragments is the Achilles' heel cleavage procedure. In this method, an enzyme is converted into a "rare cutter" by allowing specific cleavage only at one recognition site within a large substrate by protecting all other specific sites via methylation. This is achieved by incubating the DNA substrate with an oligodeoxyribonucleotide, which forms a triple-helix with the recognition site to be cleaved, thereby protecting the site. In the next step the DNA is incubated with the methyltransferase corresponding to the restriction endonuclease to be employed. The methyltransferase methylates all specific recognition sites of the plasmid except the single site protected by triple-helix formation. After inactivation of the methyltransferase the triple-helix is disrupted and the substrate is washed extensively. The single non-methylated site is then cut by the restriction endonuclease. More recently instead of oligodeoxyribonucleotides, PNAs (**p**eptide **n**ucleic **a**cids) have been used to improve triple-helix formation by sequence-specific strand invasion even at shorter recognition sequences. With PNA-assisted rare cleavage [20] a restriction enzyme was demonstrated to cut genomic DNA into a relatively small number of fragments in the range of Mbp. Instead of triplex-forming oligonucleotides or PNAs also other DNA binding protein (like transcription factors) may be used to protect one site from methylation. Employing this method it was for example, possible to cleave a single site in the yeast genome [21]. A major disadvantage of this method is the requirement for complete methylation of the substrate and, moreover, to prevent methylation of the single site of interest. In consequence the Achilles' heel approach, in the absence of better methods, might be quite useful for DNA analysis, but not for the manipulation of whole genomes.

13.1.4
Setting the stage for protein engineering of type II restriction endonucleases

From what has been said above it is obvious that there is a great demand for restriction enzymes with long recognition sequences of 8 base pairs and more. In the last few years several projects were initiated towards achieving this goal by protein engineering, taking advantage of the fact that restriction endonucleases have not only been studied over the last 30 years but also introduced as model systems for protein engineering in the late 1980s [22]. Whereas initial efforts were devoted to rational protein design, more recently the focus has shifted to directed evolution to change (or extend) the specificity of restriction endonucleases. In the present review we will summarize what has been achieved so far using these different approaches and present an outlook on future developments.

13.2
Design of Restriction Endonucleases with New Specificities

13.2.1
Rational design

13.2.1.1 **Attempts to employ rational design to change the specificity of restriction enzymes**

A prerequisite for the rational design of proteins is detailed knowledge about the structure, or, even better, about the apoenzyme, the enzyme in the specific complex with its cognate substrate, and the enzyme in complex with competitive inhibitors (transition state analogs) (Fig. 13.3). Until now 12 different type II restriction endonucleases have been studied crystallographically (for a review see [15]). Of special interest are those studies which reveal the differences between the structures of the enzyme bound to its specific DNA, the enzyme bound to non-specific DNA, and the free enzyme. Even more valuable are crystallographic studies of enzymes complexed with slightly different (with respect to sequences flanking the recognition site) cognate and near-cognate (with respect to the recognition site) substrates. The restriction endonuclease *Eco*RV (recognition site: GATATC) is one of the best characterized type II restriction enzymes. For this enzyme the structure of the free enzyme [23, 24], the structure of the enzyme bound non-specifically to DNA [23], and the structure of the enzyme bound to various cognate substrates [23, 25 – 27] as well as the structure of a product complex [25] are known. In addition, structures of *Eco*RV variants bound to specific substrates [28] and of *Eco*RV bound to modified substrates [29] have been determined. All these different structures, and those of other restriction endonucleases, show that the recognition of the specific site and the discrimination between the specific and non-specific sites is achieved by multiple contacts between enzyme and substrates. These contacts are not only formed between the enzyme and the bases of the recognition sequence (direct readout) [30] but also between the enzyme and the phosphodiester backbone of the DNA (indirect readout) [31]. With detailed structural knowledge available it was suggested that it should be possible to produce enzymes with new specificities by altering some of these contacts. However, all mutagenesis studies carried out in particular with *Eco*RI and *Eco*RV revealed that none of the mutants generated by rational design displayed the desired change in specificity (for a review see [32]). Instead, the substitution of amino acid residues involved in substrate recognition usually resulted in a severe loss of activity. Obviously, a complex interwoven network in the protein-DNA interface is formed cooperatively, leading to the accurate positioning of the catalytic amino acid residues of both enzyme subunits, which explains the very high level of specificity and accuracy. Any manipulation in this network seems to lead inevitably to a strong reduction in the catalytic rate [33, 34] and/or a relaxation of specificity [35], as shown for *Eco*RI.

The observation that it was not possible to change the substrate specificity of any restriction enzyme might be attributed to our limited understanding of the process of DNA recognition by enzymes for two reasons: (i) we do not fully understand the dynamic structures of the specific enzyme-DNA complexes, which means that we have no reasonable estimates for the energies associated with any of the observed contacts and, more importantly, we have no quantitative information regarding the influence of

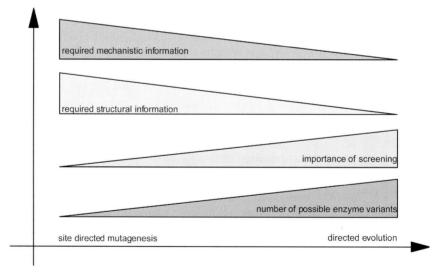

Fig. 13.3. Correlation of required mechanistic information, required structural information, importance of screening and number of possible enzyme variants in protein engineering by rational protein design and directed evolution.

one contact on all other contacts, i.e. the cooperativity of the interactions. (ii) We do not have sufficient knowledge of alternative conformations of the enzyme-DNA complexes that might be adopted if one or more positions of the target sequence have been exchanged. Only a detailed understanding of these alternative conformations, which do not support catalytic activity, will allow the elucidation of the enthalpic and entropic origins of the extreme specificity of restriction endonucleases. However, the lack of success of rational design might also in part be due to the biology of these systems: restriction endonucleases, like all proteins, are synthesized by protein biosynthesis on the ribosomes in the cell. This process has a limited accuracy: about one wrong amino acid residue is incorporated per $10^3 - 10^4$ residues [36 – 38]. Assuming that there are 10 copies of a restriction enzyme in the cell (2×300 amino acid residues) more than 99 % of all cells will contain at least one restriction enzyme variant with at least one amino acid exchange. Given the possibility that even one copy of an enzyme with an altered or relaxed specificity will cause severe problems, there must have been a strong evolutionary pressure to make restriction enzymes not only accurate but also secure in the sense that variants produced by errors in protein biosynthesis do not cleave an altered DNA sequence, i.e. a sequence unprotected by the methyltransferase of the given RM system.

Rational design was applied with notable success only to the generation of variants which recognize and cleave modified substrates. For example, an *Eco*RI variant was obtained that had lost the ability to discriminate between thymine and uracil at the forth position of the *Eco*RI recognition sequence (GAAUTC and GAATTC) [39]. An *Eco*RV variant was produced in which a single enzyme-substrate contact was altered by site-directed mutagenesis leading to a 220-fold higher selectivity for GATAUC over GATATC modified substrates [40]. The largest alteration of the specificity of a restriction enzyme achieved to date was obtained with another *Eco*RV variant. This variant cleaves a modified substrate in which the GATATpC phosphate group is replaced by a methylphosphonate, with the same rate as the wild-type enzyme, which cleaves modified and non-modified substrate at almost the same rate. The variant, however, prefers the modified over the non-modified substrates by more than three orders of magnitude [41]. As none of the modifications targeted by these variants occurs in regular DNA, the described variants can only serve as the proof of principle that new enzyme-substrate contacts can be generated by rational design. Changing the target specificity of a restriction enzyme, however, would require the reengineering of 3 – 4 contacts of the enzyme to the DNA (co-crystal structure analyses of specific restriction enzyme-DNA complexes tell us that on average each base pair of the recognition sequence is involved in four specificity-determining contacts) against the evolutionary pressure that exists to prevent the generation of variants with new specificities by chance!

14.2.1.1 **Changing the substrate specificity of type IIs restriction enzymes by domain fusion**

Type IIs restriction endonucleses are composed of a DNA recognition domain and a DNA cleavage domain. Whereas the DNA cleavage domain of *Fok*I, a type IIs enzyme, has a typical restriction endonuclease fold [42], the structure of the DNA recognition domain is completely unrelated and resembles that of the catabolite activator protein [43]. The separation of recognition and cleavage in two different subunits can be employed to generate a chimeric nuclease with altered specificity by genetically fusing other non-related DNA binding domains to the *Fok*I catalytic domain [44]. The most successful class of chimeric nucleases is based on the linkage of the cleavage domain from *Fok*I and a DNA binding domain consisting of three zinc fingers [45]. Since each of the zinc fingers binds to three base pairs, the length and sequence of the DNA to be recognized can be addressed by modulation of the number and nature of the amino acids residues in contact with the DNA. Similar fusions have been prepared with transcription factors of the homeodomain [46] and Gal4-type [47]. Efficient double strand cleavage is only observed if two copies of the recognition sequence are in close proximity and in the correct orientation, reflecting the requirement for dimerization of the cleavage domain of *Fok*I [48]. While native *Fok*I cleaves 9 and 13 base pairs away from its recognition site the dimerized cleavage domains sometimes produce cuts with overhangs other than four base pairs, which could be explained by the uncoupling of substrate recognition and catalysis. A potential application of these chimeric nucleases could be the stimulation of homologous recombination *in vivo* by introducing a double strand break and thereby producing recombinogenic ends [49].

Similar approaches have been explored for other non-specific nucleases, like DNaseI. However, none of the alternative chimeric nucleases showed a similar degree of specificity as the *Fok*I chimera although it was possible to change the cleavage preferences of DNaseI significantly [50]. The reason for this difference might be that DNaseI is a very active nuclease whereas the catalytic domain of *Fok*I only has a low catalytic activity. Therefore, with DNaseI DNA cleavage always occurs also at non-targeted sites, which does not happen as easily with *Fok*I.

13.2.1.3 **Rational design to extend specificities of type II restriction enzymes**

A much more promising approach than to employ rational design for the change of a specificity is the *extension of existing specificities*. As we know from the published co-crystal structures and biochemical studies, restriction endonucleases contact not only their recognition sequence but also some additional base pairs. In the case of *Eco*RV it was shown that, besides the contacts to the nucleotides of the recognition sequence, the protein interacts non-specifically with the backbone and the bases upstream and downstream of the GATATC site [23]. These contacts are largely water-mediated and are considered to be responsible for the dependence of the cleavage rate on the sequences flanking the recognition sequence. This flanking sequence effect suggested a promising starting point for a rational protein design project. It seemed to be possible to engineer new specific protein-DNA contacts outside of the GATATC site enabling the nuclease to recognize an extended site of up to 10 base pairs. This idea was put forward and tested by Schöttler *et al.* [51]: they exchanged an amino acid residue involved in a water-mediated contact to the base on the 5'-side of the GATATC recognition sequence against all naturally occurring amino acid residues. Some variants, having amino acid residues with long or bulky side chains at that position show altered preferences, namely an extended specificity for **PuGATATCPy** (Fig. 13.4). More recently, on the basis of crystal structures obtained with *EcoRV* and oligonucleotides with a canonical site but different flanking sequences, amino acid residues which are involved in contacts to the flanking sequences were identified, and substitutions suggested which could possibly discriminate between different sequences [28]. However, the detailed biochemical characterization of the mutants led to disappointing results, because only one had a considerably altered selectivity, which, however, was not the predicted one [52].

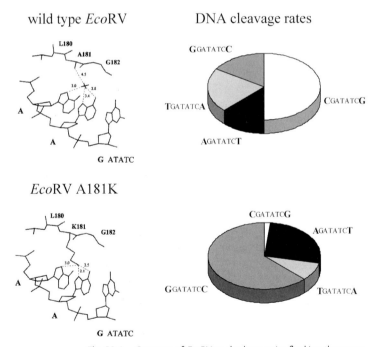

Fig. 13.4. Contacts of *EcoRV* to the base pairs flanking the recognition sequence. Exchange of Ala 181 (top) to Lys (bottom) leads to a variant with a strong preference for purine residues 5' to the recognition sequence [51].

Obviously it is very difficult to predict amino acid substitutions that would lead to an extended specificity of a restriction enzyme. This does not necessarily mean that additional specific contacts cannot be designed, but more likely that the starting point of the design is not well-defined. In most of the co-crystals obtained so far, the enzymes cannot be activated by Mg^{2+} ions soaked into the crystal. Thus, the structures do not really represent the specific complex as it is formed before DNA cleavage. Moreover, no structural information on the transition state is available with respect to the specificity determining step and in consequence, predictions regarding the effects of amino acid residue substitutions are unreliable. The lack of our understanding of the key steps of the different conformational changes that are required to couple recognition to catalysis, has so far prevented successful rational design, not only of enzyme variants with an altered specificity but also of those with an extended specificity.

13.2.2
Evolutionary design of extended specificities

A possible way to overcome these limitations might be a "non-rational" approach by repertoire selection methods which have been shown to be extremely useful for developing new enzyme specificities. With a combination of random mutagenesis and selection it was possible to change enzyme specificities for a wide range of different purposes, even in the absence of detailed structural information (Fig. 13.3). One of the most successful techniques proved to be the shuffling method which was introduced by Stemmer in 1994 [53] leading to the molecular breeding approach, which was published a few years later by the same group [54]. In the last few years a variety of enzymes was improved regarding stability, solubility, substrate specificity and other properties using this and other methods of directed evolution [55], which all make use of large libraries of up to $10^9 - 10^{10}$ different variants. To find the variants of interest, namely those with the desired improved or altered specificity, a reliable high-throughput assay or a stringent *in vivo* assay must be available (Fig. 13.3). Even more important is the coupling of phenotype to genotype in order to facilitate recovery of the genes of interest. This can be achieved *in vivo*, because in the cell the synthesis of each protein is instructed by the cellular DNA, or *in vitro* by using phage display [56], ribosome display [57] or coupled transcription/translation in water-oil emulsions [58].

The application of directed evolution approaches for the change or extension of the specificity of a restriction enzyme is hampered by the fact that an *in vivo* selection assay is not available and that examination of endonuclease activity *in vitro* usually requires purification of the enzymes to avoid background activity by other nucleases prevalent in all cells. This means that an altered or extended specificity can only be observed with sufficiently purified protein preparations, thereby unfortunately separating genotype and phenotype. As neither a reliable *in vivo* test nor the secretion of restriction endonucleases followed by a colony assay could be established in the past, experiments with larger libraries of these enzymes require a solid clone management, a fast and un-complicated, nevertheless specific, purification scheme and a fast and robust assay method.

Such an approach was chosen to extend the recognition sequence of *Eco*RV, for example from GATATC to GGATATCC or to AGATATCT. For this enzyme, the ex-pression of His$_6$-tagged fusion proteins is established, enabling protein purification via affinity chromatography on a Ni^{2+}-NTA column or Ni^{2+}-NTA covered magnetic beads, and assays have been worked out to test for extended specificities using sub-strates with two differently flanked GATATC sites. As randomization of a whole gene results in libraries far too large to be screened *in vitro*, for *Eco*RV a structure guided semi-rational approach was employed. The cocrystal structure of *Eco*RV reveals differ-ent regions of the enzyme which are in close proximity (<0.5 mm) to the nucleotide directly flanking the recognition site on the 5' end. The three regions identified are between four and eight amino acid residues in length and located in different parts of the enzyme (Fig. 13.5). To limit the number of possible variants and to make an evolu-tionary approach possible, the amino acid residues in the three regions were rando-mized and screened for the desired specificity in three subsequent rounds of muta-genesis and selection, employing the genes of the selected variants of the first cycle as template for the second mutagenesis step and so forth (Fig. 13.6). For that purpose, a mutagenesis protocol was developed to ensure that each mutation cycle resulted in additional mutations instead of only accumulating variants that originated from the template pool [59]. Employing this method and screening more than 500 different variants, generated in three subsequent cycles, several variants were found which cleave differently flanked *Eco*RV recognition sites with more than 10-fold different rates. Compared to the wild-type enzyme, the selectivity for sites with differently flanked sequences has been increased by two orders of magnitude [60]. During the experiment not only the desired preferences increased, but also the number of var-iants displaying this preference. To further improve these results, the genes of the most successful variants were mixed and cleaved with two restriction endonu-cleases, thereby separating the three mutated regions. Religation of the fragments yielded recombinants with new combinations of beneficial mutations. Interest-ingly, some of the resulting variants had lost their extended specificity, indicating that individual beneficial substitutions can neutralize each other. However, this last shuffling step also resulted in *Eco*RV variants with further improved proper-ties, namely variants which differ in their activity towards differently flanked GA-TATC sites by more than two orders of magnitude: the non-preferred site is only at-tacked after complete cleavage of the preferred AGATATCT site (Fig. 13.7). A careful look at the wild-type enzyme cocrystal structure did not reveal an obvious explanation for the effects of the different amino acid substitutions, neither for a single substitu-

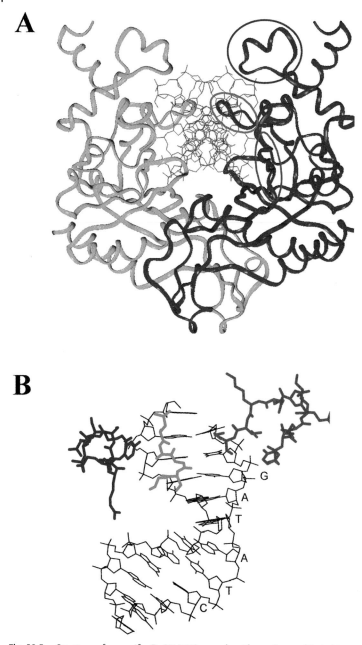

Fig. 13.5. Structure of a specific *Eco*RV-DNA complex. The regions subjected to mutagenesis as described in Lanio *et al* (1998) are highlighted in red (amino acid residues 95 to 104), green (amino acid residues 180 to 184) and blue (amino acid residues 219 to 226). In B only the regions subjected to mutagenesis are displayed, together with the DNA.

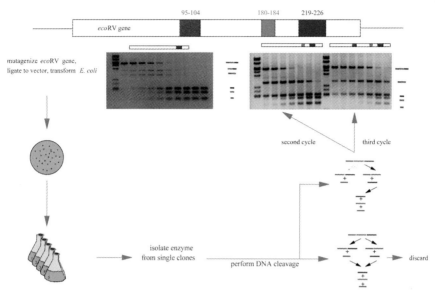

Fig. 13.6. Schematic overview of the *in vitro* evolution cycle employed. The locations of the regions subjected to mutagenesis are shown in Fig. 13.5. The substrate employed to assay the activity and preference of the variants was a prelinearized plasmid harboring two differently flanked EcoRV recognition sites (CGGGATATCGTC and AAAGATATCTTT). Both sites are cleaved by wild-type EcoRV with identical rates, such that two cleavage intermediates are formed. Mutants that prefer one of the sites produce both intermediates in unequal amounts.

tion itself, nor for the combination of several different substitutions. As these substitutions most probably affect the transition state whose formation is presumed to be the specificity determining step, even a cocrystal structure of *Eco*RV variants with differently flanked substrates might not really improve our ideas about how to extend specificity.

Although in the course of the experiments described only single and double mutants were introduced per cycle, the number of variants screened was several orders of magnitude lower than the number of variants generated. Since the bottleneck of the whole procedure was the isolation and purification of the enzyme variants, a high-throughput process was developed in order to allow parallel protein expression, purification and subsequent characterization of 96 protein variants in an automated protocol in microplate format [61] (Fig. 13.8). To test the reliability of the system three codons of the *ecoRV* gene were randomized using a spiked oligodeoxynucleotide yielding a library of about 1200 different clones. 1056 different clones, corresponding to an almost complete representation of the library, were inoculated and expressed in 96-well blocks. The purification of the proteins after cell lysis is based on the interaction between the His$_6$-tagged *Eco*RV variants and Ni^{2+}-NTA coated microplate allowing for the purifica-

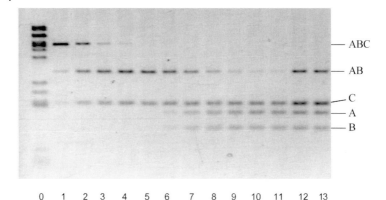

Fig. 13.7. Cleavage experiments with prelinearized substrate and the *EcoRV* variant with the most pronounced preference for AT flanked recognition sites (Y95H; K98E; E99V; N100T; S183A; Q224K). Substrate (4 nM) (see Fig. 13.6) was incubated with enzyme (3 nM). Lane 0 = length marker, 1 – 13 = 1', 3', 5', 6', 10', 30', 60', 90', 120', 150', 170' – addition of new substrate to exclude that the variant lost activity during the extended incubation time – 180', 190'. Note that the substrate ABC is cleaved very rapidly into the two intermediate products AB and C – the GC-flanked recognition site in intermediate AB is only cleaved after complete cleavage of the preferred AT-flanked recognition site, to give the final products A, B and C.

tion of 10 pmol protein per well. The functional characterization of the variants was performed with the assays previously described (Figs. 13.6 and 13.7). Since the whole process is done automatically by a pipetting robot it is possible to characterize 10^3 different clones within a week and $10^4 – 10^5$ clones in a reasonable period of time. In the relatively small library described several variants displayed a more than 10-fold extended specificity against differently flanked *EcoRV* recognition sites [61].

In the future, similar approaches will be made easier by the availability of an assay that is suited for high-throughput sample analysis [62]. Rather than using macromolecular DNA substrates and analyzing their cleavage by electrophoresis, a 96-well microplate-based fluorescence assay has been developed employing two oligodeoxynucleotides with differently flanked recognition sites and labeled with different fluorescence markers. Before cleavage the fluorescence is quenched. Upon cleavage, the fluorescence of the preferentially cleaved substrate increases. Recording the fluorescence of the two differently labeled substrates allows a rapid automated analysis of variants with an extended specificity (Fig. 13.9).

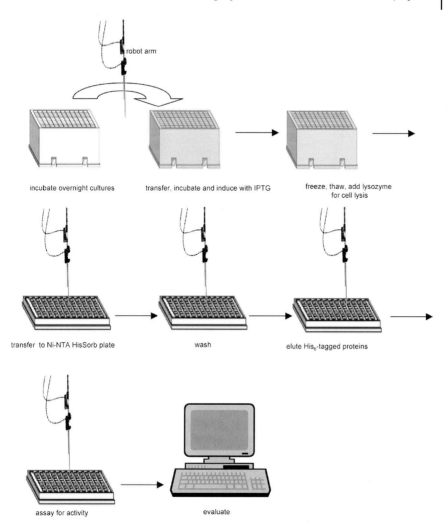

robot arm

incubate overnight cultures

transfer, incubate and induce with IPTG

freeze, thaw, add lysozyme
for cell lysis

transfer to Ni-NTA HisSorb plate

wash

elute His$_6$-tagged proteins

assay for activity

evaluate

Fig. 13.8. Purification of 96 different enzyme variants in a high-throughput approach. Cells are grown, harvested and lysed in microplates. His$_6$-tagged variants are isolated by transferring the lysate to Ni-NTA coated microplates from which they are eluted after washing. Specific enzyme activity can be checked by suitable assays.

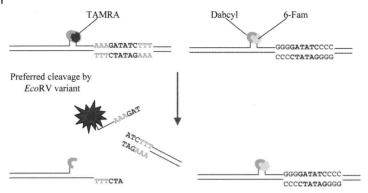

Fig. 13.9. Schematic overview of a fluorescence-based cleavage assay for determination of cleavage preferences of restriction enzymes. The two substrates employed harbor differently flanked recognition sites and carry different labels whose fluorescence is quenched before cleavage. Upon cleavage the fluorescence of the preferentially cleaved substrate increases.

13.3
Summary and Outlook

In the past two decades different strategies have been employed to change or to extend the specificity of restriction enzymes. Since substrate recognition is highly redundant and recognition itself coupled to catalysis via a multitude of intra- and intermolecular interactions, which are not really understood, all rational design projects, by and large, have failed so far. The rare positive results concerned variants with preferences for artificial substrates. Taking these results into account rational design turns out to be ill-suited for the alteration of substrate specificity of restriction enzymes at present.

Promising results were obtained with chimeric nucleases, in which combination of the cleavage domain of the type IIs restriction endonuclease *Fok*I was fused to different DNA binding proteins. This technique was successfully employed to initiate homologous recombination *in vivo*, but since the uncoupling of recognition and catalysis leads to different single strand overhangs these chimeric endonucleases are not suitable for cloning experiments which require a precise double strand break at a defined position with defined overhangs. As this can only be achieved by restriction enzymes, the Achilles' heel approach is the superior technique, especially for cloning experiments. The requirement for complete methylation of the DNA at all sites that should not be cleaved by the restriction enzyme, prevents the usage of this for DNA manipulation *in vivo*.

The extension of recognition sequences of type II restriction endonucleases also turned out to be a more than difficult job. The function of the different contacts within the protein DNA interface is still poorly understood, as is the mechanism of how substrate recognition is coupled to catalysis. Presumably, the most serious deficit for attempts to alter the specificity of restriction enzymes is the lack of detailed knowledge of the transition state of the DNA cleavage reaction. In consequence, one does not really know what the specificity determining step looks like and how the catalytic centers are activated during the recognition process. This lack of information could be bypassed by an evolutionary design approach which does not presuppose this information. Unfortunately, for directed evolution large libraries of clones have to be generated, which demands a stringent selection of the clones coding for the gene product with the desired property. As screening for restriction enzyme variants with extended cleavage specificities is not yet possible *in vivo,* a protocol for expression, purification and characterization of the different variants is required. This limits the size of the library. It was shown recently, however, that by employing nucleoside analogues in an error-prone PCR, even relatively small libraries of variants that carry many amino acid exchanges could yield the same variants that are otherwise only generated by gene shuffling [63]. Libraries comprising up to 10^5 different clones could be expressed and screened with recently established methods [61] and the high-throughput generation and characterization of restriction enzyme variants with extended specificities could be further improved.

Acknowledgements
Research on restriction enzymes in the authors' laboratory was supported by the Deutsche Forschungsgemeinschaft, the Bundesministerium für Bildung, Wissenschaft, Forschung und Technologie, the European Union, and by the Fonds der Chemischen Industrie.

References

[1] T. A. Bickle, D. H. Kruger, *Microbiol. Rev.* **1993**, *57*, 434.

[2] D. R. Lesser, M. R. Kurpiewski, L. Jen-Jacobson, *Science* **1990**, *250*, 776.

[3] V. Thielking, J. Alves, A. Fliess, G. Maass, A. Pingoud, *Biochemistry* **1990**, *29*, 4682.

[4] J. Alves, U. Selent, H. Wolfes, *Biochemistry* **1995**, *34*, 11191.

[5] M. Nelson, E. Raschke, M. McClelland, *Nucleic Acids Res.* **1993**, *21*, 3139.

[6] L. Jen-Jacobson, L. E. Engler, D. R. Lesser, M. R. Kurpiewski, C. Yee, B. McVerry, *Embo J* **1996**, *15*, 2870.

[7] L. F. Lin, J. Posfai, R. J. Roberts, H. Kong, *Proc. Natl. Acad. Sci. U. S. A.* **2001**, *98*, 2740.

[8] J. Heitman, *Genet. Eng.* **1993**, *15*, 57.

[9] M. McKane, R. Milkman, *Genetics* **1995**, *139*, 35.

[10] K. Carlson, L. D. Kosturko, *Mol. Microbiol.* **1998**, *27*, 671.

[11] T. Naito, K. Kusano, I. Kobayashi, *Science* **1995**, *267*, 897.

[12] R. J. Roberts, D. Macelis, *Nucleic Acids Res.* **2001**, *29*, 268.

[13] R. J. Roberts, S. E. Halford, in *Nucleases, 2nd edition* (Eds.: S. M. Linn, R. S. Lloyd, R. J. Roberts), Cold Spring Harbor Laboratory Press, Cold Spring Harbor, **1993**, pp. 35.

[14] A. Pingoud, A. Jeltsch, *Eur. J. Biochem.* **1997**, *246*, 1.

[15] A. Pingoud, A. Jeltsch, *Nucleic. Acids Res.* **2001**, *29*, 3705.

[16] A. Choulika, A. Perrin, B. Dujon, J. F. Nicolas, *Mol. Cell. Biol.* **1995**, *15*, 1968.

[17] B. Elliott, C. Richardson, J. Winderbaum, J. A. Nickoloff, M. Jasin, *Mol. Cell. Biol.* **1998**, *18*, 93.

[18] M. Cohen-Tannoudji, S. Robine, A. Choulika, D. Pinto, F. El Marjou, C. Babinet, D. Louvard, F. Jaisser, *Mol. Cell. Biol.* **1998**, *18*, 1444.

[19] C. Richardson, M. E. Moynahan, M. Jasin, *Genes Dev.* **1998**, *12*, 3831.

[20] K. I. Izvolsky, V. V. Demidov, N. O. Bukanov, M. D. Frank-Kamenetskii, *Nucleic Acids Res.* **1998**, *26*, 5011.

[21] M. Koob, W. Szybalski, *Science* **1990**, *250*, 271.

[22] J. Rosenberg, B. Wang, C. Frederick, N. Reich, P. Greene, J. Grable, J. McClarin, in *Protein Engineering*, (Eds.: D. Oxender, C. F. Fox), A.R. Liss Inc., **1987**, pp. 237.

[23] F. K. Winkler, D. W. Banner, C. Oefner, D. Tsernoglou, R. S. Brown, S. P. Heathman, R. K. Bryan, P. D. Martin, K. Petratos, K. S. Wilson, *EMBO J.* **1993**, *12*, 1781.

[24] J. J. Perona, A. M. Martin, *J. Mol. Biol.* **1997**, *273*, 207.

[25] D. Kostrewa, F. K. Winkler, *Biochemistry* **1995**, *34*, 683.

[26] N. C. Horton, J. J. Perona, *J. Mol. Biol.* **1998**, *277*, 779.

[27] N. C. Horton, J. J. Perona, *Proc. Natl. Acad. Sci. U. S. A.* **2000**, *97*, 5729.

[28] N. C. Horton, J. J. Perona, *J. Biol. Chem.* **1998**, *273*, 21721.

[29] N. C. Horton, M. D. Sam, B. A. Connolly, J. J. Perona, *Biophys. J.* **2000**, *78*, 417A.

[30] N. C. Seeman, J. M. Rosenberg, A. Rich, *Proc. Natl. Acad. Sci. U. S. A.* **1976**, *73*, 804.

[31] Z. Otwinowski, R. W. Schevitz, R. G. Zhang, C. L. Lawson, A. Joachimiak, R. Q. Marmorstein, B. F. Luisi, P. B. Sigler, *Nature* **1988**, *335*, 321.

[32] A. Jeltsch, C. Wenz, W. Wende, U. Selent, A. Pingoud, *Trends Biotechnol.* **1996**, *14*, 235.

[33] J. Alves, T. Rüter, R. Geiger, A. Fliess, G. Maass, A. Pingoud, *Biochemistry* **1989**, *28*, 2678.

[34] R. Geiger, T. Rüter, J. Alves, A. Fliess, H. Wolfes, V. Pingoud, C. Urbanke, G. Maass, A. Pingoud, A. Dusterhoft, M. Kroger, *Biochemistry* **1989**, *28*, 2667.

[35] H. Flores, J. Osuna, J. Heitman, X. Soberon, *Gene* **1995**, *157*, 295.

[36] R. B. Loftfield, D. Vanderjagt, *Biochem. J.* **1972**, *128*, 1353.

[37] P. Edelmann, J. Gallant, *Cell* **1977**, *10*, 131.

[38] J. Parker, T. C. Johnston, P. T. Borgia, G. Holtz, E. Remaut, W. Fiers, *J. Biol. Chem.* **1983**, *258*, 10007.

[39] A. Jeltsch, J. Alves, T. Oelgeschlager, H. Wolfes, G. Maass, A. Pingoud, *J. Mol. Biol.* **1993**, *229*, 221.

[40] C. Wenz, U. Selent, W. Wende, A. Jeltsch, H. Wolfes, A. Pingoud, *Biochim. Biophys. Acta* **1994**, *1219*, 73.

[41] T. Lanio, U. Selent, C. Wenz, W. Wende, A. Schulz, M. Adiraj, S. B. Katti, A. Pingoud, *Protein Eng.* **1996**, *9*, 1005.

[42] D. A. Wah, J. A. Hirsch, L. F. Dorner, I. Schildkraut, A. K. Aggarwal, *Nature* **1997**, *388*, 97.

[43] S. C. Schultz, G. C. Shields, T. A. Steitz, *Science* **1991**, *253*, 1001.

[44] S. Chandrasegaran, J. Smith, *Biol. Chem.* **1999**, *380*, 841.

[45] J. Smith, M. Bibikova, F. G. Whitby, A. R. Reddy, S. Chandrasegaran, D. Carroll, *Nucleic Acids Res.* **2000**, *28*, 3361.

[46] Y. Kim, S. Chandrasegaran, *Proc. Natl. Acad. Sci. U. S. A.* **1994**, *91*, 883.

[47] Y.-G. Kim, J. Smith, M. Durgesha, S. Chandrasegaran, *Biol. Chem.* **1998**, *379*, 489.

[48] J. Bitinaite, D. A. Wah, A. K. Aggarwal, I. Schildkraut, *Proc. Natl. Acad. Sci. U. S. A.* **1998**, *95*, 10570.

[49] M. Bibikova, D. Carroll, D. J. Segal, J. K. Trautman, J. Smith, Y. G. Kim, S. Chandrasegaran, *Mol. Cell. Biol.* **2001**, *21*, 289.

[50] S. Cal, K. L. Tan, A. McGregor, B. A. Connolly, *EMBO J.* **1998**, *17*, 7128.

[51] S. Schöttler, C. Wenz, T. Lanio, A. Jeltsch, A. Pingoud, *Eur. J. Biochem.* **1998**, *258*, 184.

[52] T. Lanio, A. Jeltsch, A. Pingoud, *Protein Eng.* **2000**, *13*, 275.

[53] W. P. Stemmer, *Nature* **1994**, *370*, 389.

[54] A. Crameri, S. A. Raillard, E. Bermudez, W. P. Stemmer, *Nature* **1998**, *391*, 288.

[55] F. H. Arnold, P. L. Wintrode, K. Miyazaki, A. Gershenson, *Trends Biochem. Sci.* **2001**, *26*, 100.

[56] T. Clackson, H. R. Hoogenboom, A. D. Griffiths, G. Winter, *Nature* **1991**, *352*, 624.

[57] J. Hanes, A. Plückthun, *Proc. Natl. Acad. Sci. U. S. A.* **1997**, *94*, 4937.

[58] D. S. Tawfik, A. D. Griffiths, *Nat. Biotechnol.* **1998**, *16*, 652.

[59] T. Lanio, A. Jeltsch, *Biotechniques* **1998**, *25*, 958.

[60] T. Lanio, A. Jeltsch, A. Pingoud, *J. Mol. Biol.* **1998**, *283*, 59.

[61] T. Lanio, A. Jeltsch, A. Pingoud, *Biotechniques* **2000**, *29*, 338.

[62] K. Eisenschmidt, Diploma Thesis, Justus-Liebig Universität (Giessen), **2000**.

[63] M. Zaccolo, E. Gherardi, *J. Mol. Biol.* **1999**, *285*, 775.

14

Evolutionary Generation of Enzymes with Novel Substrate Specificities

Uwe T. Bornscheuer

14.1
Introduction

The application of enzymes is now well documented. They are used in organic chemistry – mainly for the production of optically pure compounds – synthesis of detergents and emulsifiers, lipid modification, laundry formulations, food and beverage production etc. [1 – 6]. Due to their high chemo-, regio- and stereoselectivity, activity at ambient temperatures and pH, biocatalysts are often superior to chemical ones and this might explain that a considerable number of industrial bioprocesses have been established in recent years [2,7].

Traditional methods to identify new enzymes are based on screening of e.g. soil samples or strain collections by enrichment cultures. Once a suitable biocatalyst is identified, strain improvement as well as cloning and expression of the encoding gene(s) enable production on a large scale. Unfortunately, not all microorganisms can be cultured using common fermentation technology and the number of accessible microorganisms from a sample is estimated to 0.001 – 1 % depending on their origin [8]. In addition, not all enzymes found in nature are suitable for a specific synthetic problem and the useful properties mentioned above are not always satisfactory in practice. Until recently, these limitations were overcome by i) screening for alternative biocatalysts, ii) changes of the reaction system (e.g., from hydrolysis to esterification) or iii) by rational protein design followed by site-directed mutagenesis, if the gene encoding the enzyme and its three-dimensional structure are available [9]. All these methods are very time-consuming and in case of rational protein design also information-intensive.

A very promising alternative for the generation of enzymes with improved or altered properties is the strategy of directed evolution (also called evolutive biotechnology or molecular evolution) which came to the fore in the early 90s. In most cases, directed evolution is comprised of a random mutagenesis of the gene encoding the catalyst by

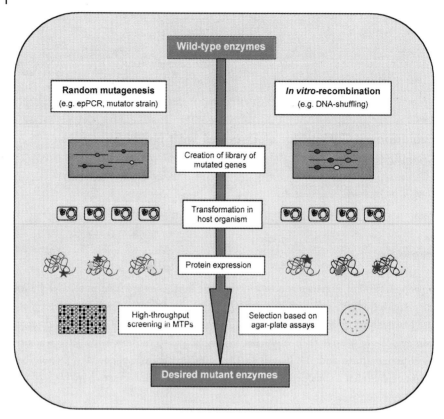

Fig. 14.1. Principle of directed evolution.

error-prone polymerase chain reaction (epPCR) resulting in large libraries of mutants. Another method is based on DNA- (or gene-) shuffling and related techniques. Libraries thus created are then assayed by using high-throughput technologies to identify improved variants (Fig. 14.1). Further details on the principles and applications of directed evolution can be found elsewhere in this book and in a considerable number of recent reviews dealing with various aspects of directed evolution [8 – 21].

This chapter focuses on the creation of novel substrate specificities of enzymes. Although in principle, this would also include altered enantio- or stereoselectivity, this aspect is already covered extensively in Chapter 11.

14.2
General Considerations

During natural evolution, a broad variety of enzymes has been developed, which are classified according to the Nomenclature Committee of the International Union of Biochemistry and Molecular Biology (IUBMB). Thus, for each type of characterized enzyme an EC (Enzyme Commission) number has been provided (see: http://www.expasy.ch/enzyme/). For instance, all hydrolases have EC number 3 and further subdivisions are provided by three additional digits, e.g. all lipases (official name: triacylglycerol lipases) have the EC number 3.1.1.3 and are thus distinguished from esterases (official name: carboxyl esterases) having the EC number 3.1.1.1. This classification is based on the substrate (and cofactor) specificity of an enzyme only, however often very similar amino acid sequences and also related three-dimensional structures can be observed.

This suggests that the creation of novel enzyme specificity has a high probability if a closely related catalytic machinery and/or architecture is present for both types of activities, i.e. both substrates should be converted by the Ser-His-Asp triad present in lipases and amidases if an amidase substrate should be accepted by a lipase or vice versa. The chance for creating activity is usually enhanced if both enzymes share high sequence and/or structural homology. In other words, most readers will agree that it will be rather impossible to convert an esterase into a monooxygenase as these two enzymes differ substantially in their architecture, substrate and cofactor-requirement and consequently mode of action.

With this in mind, it is not surprising that the examples reported in the literature (Table 14.1) for novel substrate specificities by directed evolution can be subdivided into the following three groups (Scheme 14.1):

1) Conversion of substrates, which in principle should be acceptable by the enzyme under investigation, but do not react due to e.g. steric hindrance, electronic effects etc.
2) Conversion of substrates which are usually converted by another (sometimes closely related) enzyme.
3) The compounds are already converted by the enzyme of interest, but at very low activity compared to well accepted (natural) substrates.

It should be noted that a clear-cut distinction between these groups is not always feasible.

Tab. 14.1. Examples for the evolutionary generation of enzymes with novel substrate specificities.

Wild-type activity	Method	Assay	Main Results	Ref.
Pseudomonas fluorescens esterase	MS	growth/ pH-indicator	active double mutant acting on sterically hindered substrate	22, 24, 26
E. coli IGPS	SDM	growth assay	evolved PRAI activity from IGPS scaffold	28
E. coli ProFARI	shuffling	growth assay	evolved PRAI activity from ProFARI	29
E. coli β-galactosidase	DNA-shuffling	*p-/o*-nitro-phenyl glycosides	300 – 1000 fold increased specificity for fucosides	31
Bacillus megaterium P450 MO	epPCR/SM	visual inspection	generation of indigo/ indirubin	33
Pseudomonas SY77 glutaryl acylase	epPCR/SM	growth assay	increased V_{max}/K_m for adipyl-7-ADCA	34
Pseudomonas putida P450 MO	epPCR/StEP	digital image screening	enhanced naphthalene hydroxylation/formation of dihydroxy compounds	35, 37
Staphylococcus aureus lipase	epPCR/ shuffling	chromogenic assay	12-fold increase in phospholipase activity	38
Bacillus thermocatenulatus lipase	epPCR/SDM	visual inspection	17-fold increase in phospholipase activity	39
Saccharomyces cerevisiae peroxidase	shuffling	visual inspection	300-fold increased activity for guaiacol	40
E. coli aldolase	epPCR/ shuffling	UV assay	acceptance of phosphate free substrates	41

Abbreviations: ADCA, adipyl-cephalosporic acid; epPCR, error-prone PCR; IGPS, indole-3-glycerol phosphate synthase; MO, monooxygenase; MS, mutator strain; PRAI, phosphoribosylanthranilate isomerase; ProFARI, *N'*-[5-phosphoribosyl)formimino]-5-aminoimidazole-4-carboxylamide ribonucleotide isomerase; SDM, site-directed mutagenesis, StEP, staggered extension process

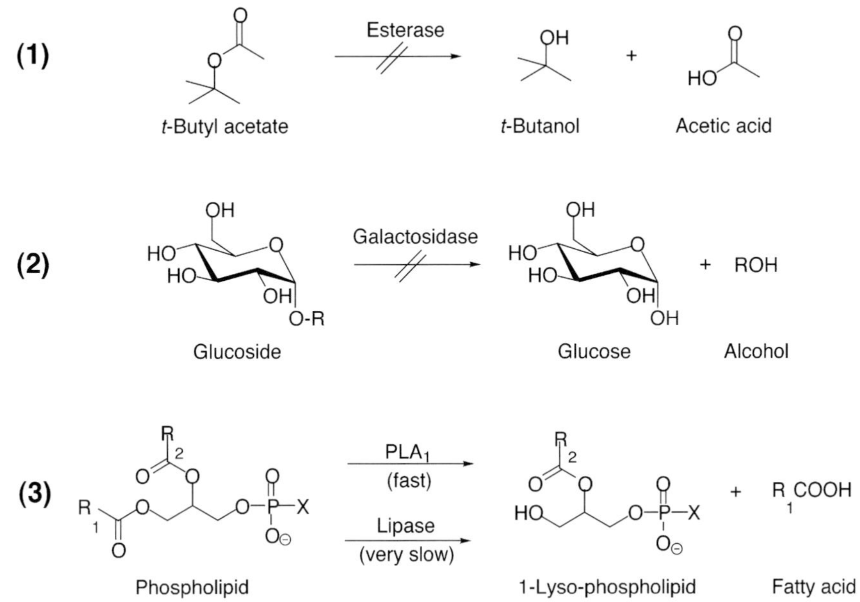

(1)

t-Butyl acetate   Esterase   t-Butanol   Acetic acid

(2)

Glucoside   Galactosidase   Glucose   Alcohol

(3)

Phospholipid

PLA₁ (fast)

Lipase (very slow)

1-Lyso-phospholipid   Fatty acid

Scheme 14.1. Selected examples illustrating possible targets for novel substrate specificities of enzymes. Esterases do not act on esters of tertiary alcohols and this problem thus belongs to group 1. Galactosidases only accept galactosides as substrates, but not glucosides and therefore this example resembles group 2. Some lipases – e.g from *Rhizopus* sp., *Bacillus* sp. or *Staphylococcus* sp. – exhibit low activity towards the sn1-group of phospholipids. However, this activity is significantly lower compared to phospholipase A₁ (PLA₁) and consequently it can be regarded as a group 3 target.

14.3
Examples

14.3.1
Group 1

The first example in which a substrate not accepted by the wild-type enzyme could be converted by a mutant obtained by directed evolution was published by my group [22]. A sterically hindered 3-hydroxy ester (Scheme 14.2, **1a, b**) resembling a building block in the envisaged total synthesis of epothilones was not accepted by 20 wild-type hydrolases (18 lipases, 2 esterases). Although closely related 3-hydroxy esters could be stereoselectively converted by a range of commercial lipases, this specific compound bearing two methyl groups at C4 did not react in hydrolysis of the corresponding ethyl ester as well as in acylation attempts at the 3-hydroxy group using vinyl acetate in toluene [23]. As the basic structure of the 3-hydroxy ester was set by the epothilone scaffold, the

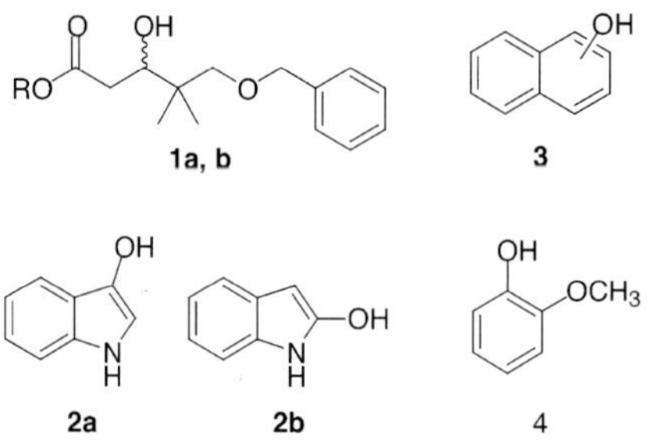

Scheme 14.2. Structures of compounds used (**1a,b, 4**) and obtained (**2, 3**) by directed evolution of biocatalysts aiming at novel substrate specificities.

most promising alternative to obtain the compound in optically pure form was the creation of a hydrolase mutant exhibiting activity towards this substrate, preferentially in a stereoselective manner. Thus, a mutant library of an esterase from *Pseudomonas fluorescens* (PFE) was created using the mutator strain *Epicurian coli* XL1-Red (available from Stratagene). Key to the identification of improved variants from these libraries was an agar plate assay system based on pH indicators, thus leading to a change in color upon hydrolysis of the ethyl ester **1a** (Scheme 14.3). Parallel assaying of replica-plated colonies on agar plates supplemented with the glycerol derivative **1b** was used to refine the identification, because only *E. coli* colonies producing active esterases had access to the carbon source glycerol, thus leading to enhanced growth and in turn larger colonies [22, 24]. By this strategy, a double mutant (Leu181Val, Ala209Asp) was obtained, which indeed hydrolyzed the sterically hindered substrate. This was confirmed by preparative scale reactions followed by gas chromatographic analysis to calculate conversion and enantiomeric excess as well as by determination of the optical rotation by polarimetry. The best mutant gave an enantiomeric excess of 25 % *ee* which corresponds to an enantioselectivity of E ~5.0, which is in accordance

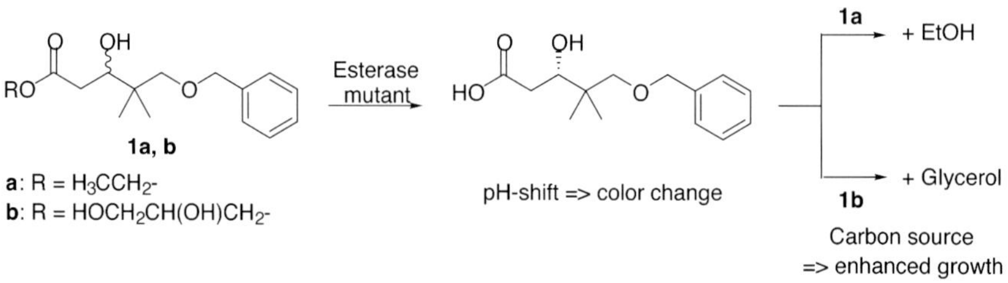

Scheme 14.3. Principle of the assay system used to identify active variants of esterase from *Pseudomonas fluorescens* (PFE) acting on the sterically hindered 3-hydroxy ester **1**. Both substrates **1a,b** yield the free acid leading to a color change; **1b** also releases the carbon source glycerol leading to enhanced growth of esterase mutant producing *E. coli* clones.

with the, in general, rather low selectivity of PFE. Several other mutants were identified, but enantiomeric excess was usually around 3 – 4 % ee. Interestingly, the activity towards acetic acid ethyl ester – the standard substrate to determine esterase activity by a pH-stat assay – was decreased considerably for various mutants. Whereas wild-type PFE showed an activity of 6 Umg^{-1}, the activity of the Leu181Val/Ala209Asp mutant was 2.3 Umg^{-1} and some variants showed only activities <1.0 Umg^{-1}.

The 3D-structure of PFE is unknown, however it was possible to perform homology modeling based on the solved 3D-structure of a haloperoxidase from *Streptomyces aureofaciens* [25]. This revealed that both mutations are not close to the binding pocket but solely at the periphery of the enzyme [24].

Although this method was useful to turn an inactive wild-type enzyme into an active esterase, the best mutant showed only modest enantioselectivity. Further work showed that the enantioselectivity of PFE could be increased from E = 3.5 to E = 5.2 – 6.6 in a single round of mutation [26]. Libraries were created by epPCR as well as by using the

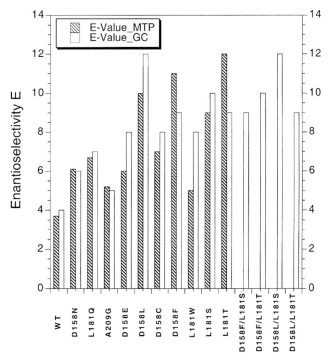

Fig. 14.2. Mutants obtained by directed evolution and saturation mutagenesis showing enhanced enantioselectivity in the resolution of 3-phenylbutyric acid derivatives. MTP: E-values (E_{app}) determined in microtiter plates using the corresponding (*R*)- and (*S*)-resorufin ester; GC: E-values (E_{true}) calculated using the equation E = [ln(1-c)*(1-ee$_S$)]/[ln(1-c)*(1+ ee$_S$)] from data determined by gas chromatographic analysis on a chiral column from samples obtained after esterase-catalyzed hydrolysis of (*R,S*)-3-phenyl butyric acid ethyl ester [24].

mutator strain from *Epicurian coli*. An extremely accurate determination of the enantioselectivity was achieved by using resorufin esters of (*R*)- or (*S*)-3-phenylbutyric acid, which allowed measurement of fluorescence in microtiter plates (MTP) avoiding problems with interfering compounds present in the culture medium. A further increase in enantioselectivity up to E = 12 was achieved by saturation mutagenesis at all three positions identified for the first generation mutants (E. Henke and U.T. Bornscheuer, unpublished). However, combinations of best variants by site-directed mutagenesis gave no further improvement (Fig. 14.2). It should be noted, that E_{app} values determined in MTPs were quite close to E_{true} values, which have been determined by gas chromatographic analysis of samples from resolutions of racemic mixtures of the corresponding ethylesters of 3-phenylbutyric acids. Here, also competition between the two enantiomers is included, whereas the MTP-assay delivers only apparent E values. Significantly higher increases in enantioselectivity were achieved for the resolution of methyl 3-bromo-2-methylpropionate. Using the homology model of PFE, Trp_{29} and Phe_{199} were identified as positions for random mutagenesis. Screening of libraries using Quick E [27] identified a mutant Trp29Leu exhibiting an E = 90 compared to E = 12 for wild-type PFE. This is the highest selectivity ever achieved up to now by directed evolution techniques (U.T. Bornscheuer and R.J. Kazlauskas, unpublished), and allows the synthesis of optically pure substrate and product of this compound.

The knowledge about mutations affecting enantioselectivity and altered substrate specificity derived from these directed evolution studies of PFE is currently used as a basis to design variants combining both properties.

In an excellent contribution, Fersht and co-workers evolved phosphoribosylanthranilate isomerase (PRAI) activity from the $(\beta a)_8$-barrel scaffold of indole-3-glycerol phosphate synthase (IGPS) [28]. Their success provides a striking example for testing the" conserved scaffold" hypothesis of enzyme evolution. Careful design of the basic PRAI loop system into IGPS created the scaffold for directed evolution by DNA-shuffling and StEP. Active variants were identified by an agar plate assay using an *E. coli* strain that does not grow in the absence of tryptophan, which in turn is the product of PRAI activity. The mutant obtained had 28 % identity to PRAI and 90 % identity to IGPS. The same hypothesis that several different enzymes having the $(\beta a)_8$-barrel scaffold evolved from a common ancestor was used as a basis for the alteration of the substrate specificity of another isomerase having only 10 % sequence homology [29]. This showed that a single round of gene-shuffling turned ProFARI (*N'*-[5-phosphoribosyl)formimino]-5-aminoimidazole-4-carboxamide ribonucleotide isomerase) – involved in histidine biosynthesis – into a mutant with PRAI activity catalyzing the same reaction in tryptophan biosynthesis (Scheme 14.4). *In vivo* selection of mutants in an *E. coli* strain lacking PRAI activity yielded two variants each with only a single amino acid substitution. One mutant catalyzed both reactions, although with only low

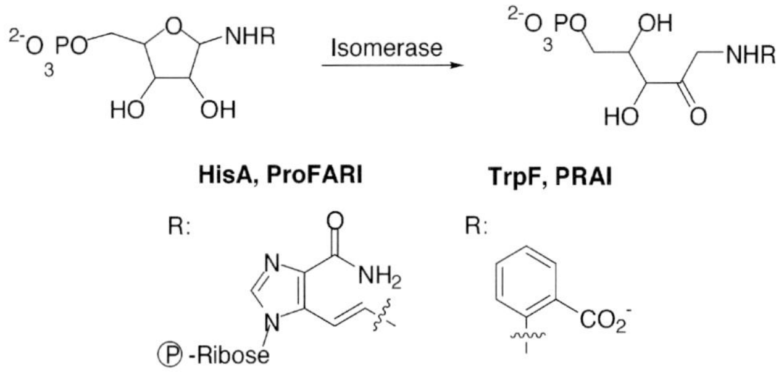

Scheme 14.4. Substrate specificities of ProFARI (involved in histidine biosynthesis) and PRAI (involved in tryptophan biosynthesis).

residual wild-type activity (<4 %). Both examples have been already discussed extensively in a recent review [30].

14.3.2
Group 2

DNA-shuffling was successfully used to turn an *E. coli lacZ* β-galactosidase into a fucosidase [31]. Screening of ~10 000 colonies per round based on chromogenic fucose substrates resulted in the identification of evolved fucosidases with 10 – 20 fold increased k_{cat}/K_m for the fucose substrate compared to the wild-type. Compared to the native β-galactosidase, the final evolved enzyme showed a 1000-fold increased substrate specificity for *o*-nitrophenyl fucopyranoside versus *o*-nitrophenyl galactopyranoside and a 300-fold increase for the corresponding *p*-nitrophenol derivatives. In a related manner, attempts were made to convert the *E. coli lacZ* β-galactosidase into a β-glucosidase, however so far only minor changes in the substrate specificity have been reported [32].

In another example, some mutants of the P450-monooxygenase from *Bacillus megaterium* (P450 BM-3, expressed in *E. coli*) obtained by directed evolution using epPCR were found to hydroxylate indole at two different positions (Scheme 14.2, **2a, b**) which upon air oxidation and dimerization yield indigo and indirubin [33]. So far, both compounds were known to be formed by the action of naphthalene dioxygenases or styrene monooxygenases, but are not produced by wild-type P450 BM3 and are thus a result of altered substrate specificity. Sequencing of several individual clones revealed that three positions (Phe87, Leu 188, Ala74) contained mutations. Next, simultaneous saturation mutagenesis at all three positions was performed to identify best "replacements" which led to the identification of a triple mutant (Phe87Val, Leu188Gln, Ala74Gly) enabling formation of indigo and indirubin in milligram amounts.

In a very recent paper the activity of a glutaryl acylase from *Pseudomonas* SY77 towards adipyl-7-ADCA – an important precursor in the synthesis of semi-synthetic cephalosporins – was altered by directed evolution [34]. Random mutagenesis of the a-subunit, growth selection of active variants followed by saturation mutagenesis identified three mutants showing decreased V_{max}/K_m values for glutaryl substrates and higher values towards the adipyl analog. The authors used a growth selection based on adipyl-leucine or adipyl serine to identify acylase mutants acting on adipic acid instead of glutaric acid side-chains in cephalosporins.

14.3.3
Group 3

A P450 monooxygenase from *Pseudomonas putida* primarily catalyzes the hydroxylation of camphor (P450$_{cam}$) utilizing the cofactor NADPH. In contrast, hydroxylation of naphthalene (Scheme 14.2, **3**) occurs only to a minor extent. Coexpression of P450$_{cam}$ with horseradish peroxidase in *E. coli* in the presence of hydrogen peroxide allowed hydroxylation in the absence of NADPH via a "peroxide shunt" pathway. Moreover, under these reaction conditions, hydroxylated naphthalenes form fluorescent dimers amenable to digital image screening [35]. Thus, approximately 200 000 mutants obtained by epPCR and the staggered extension process (StEP) [36] were easily screened for increased fluorescence suggesting enhanced activity of P450$_{cam}$. In addition, mutations could be identified having altered regiospecificities. Thus variants of P450$_{cam}$ were generated showing ~20-fold higher activity compared to the wild-type as well as mutants catalyzing the formation of dihydroxy compounds [37].

Phospholipase activity was introduced into a *Staphylococcus aureus* lipase mutant previously created by site-directed mutagenesis, which displayed very low phospholipase activity. Next, directed evolution using epPCR and gene shuffling led to a substantial increase in the phospholipase/lipase activity ratio [38]. Mutants were first screened in a qualitative agar plate assay followed by high-throughput chromogenic assay based on a thio-derivative of a phospholipid. Upon phospholipase activity, the free SH-group was reacted with 5,5'-dithiobis(2-nitrobenzoic acid) yielding 5-thio-2-nitrobenzoic acid amenable to spectrophotometric quantification. The best variant contained six mutations and displayed a 11.6-fold increase in phospholipase activity and a 11.5-fold increased phospholipase/lipase ratio compared to the starting mutant. Recently, similar findings were reported for a lipase from *Bacillus thermocatenulatus* expressed in *E. coli* [39]. A single round of random mutagenesis followed by screening of 6000 transformants on egg-yolk identified three variants with ~10 – 12-fold phospholipase activity. Further rounds of random mutagenesis gave no improvement, but site-directed mutagenesis (Leu353Ser) gave a 17-fold enhanced phospholipase activity.

The substrate specificity of a cytochrom c peroxidase (CCP) from *Saccharomyces cerevisiae* towards guaiacol (Scheme 14.2, 4) was increased 300-fold by means of DNA-shuffling [40]. Screening was performed by spreading colonies onto nitrocellulose membranes. Mutants showing enhanced activity towards guaiacol were identified by visual inspection of fast-staining colonies due to a brownish color based on the formation of tetraguaiacol. Interestingly, sequencing revealed that the – in the superfamily of peroxidase – fully conversed residue Arg48 was changed to His. The authors believe that this mutation plays a key role in the increase in activity toward the phenolic substrate guaiacol.

Wild-type aldolase from *E. coli* poorly accepts non-phosphorylated D-2-keto-3-deoxygluconate (KDG). Several rounds of epPCR and shuffling identified mutants with significantly enhanced activity towards KDG [41]. The best mutant showed a 70-fold increased catalytic efficiency with concomitant decrease in the activity towards the phosphorylated analog. Assaying was performed in microtiter plates based on reduction of released pyruvate by lactate dehydrogenase yielding a decrease in NADH concentration. Furthermore, library screening identified mutants accepting L-glyceraldehyde as substrate, which is converted by the wild-type aldolase to only a negligible extent.

14.4
Conclusions

The examples provided in this Chapter demonstrate that directed evolution resembles a very useful tool to create enzyme activities hardly accessible by means of rational protein design (Table 14.1). Even if the desired substrate specificity is known from other biocatalysts – e.g. phospholipase A$_1$ activity – the advantage of the directed evolution approach resides in the already achieved functional expression of a particular protein. Thus bottlenecks arising from the identification of enzymes by traditional screening and cultivation methods can be circumvented. In addition, directed evolution can dramatically reduce the time required for the provision of a suitable tailor-made enzyme, also because cloning and functional expression of the biocatalyst has already been achieved.

The striking examples by the Fersht and Sterner groups for the evolution of $(\beta a)_8$-barrel enzymes also demonstrate that even functional and structural barriers between different enzymes can be overcome by means of directed evolution.

Acknowledgements
I am especially grateful to Prof. Romas Kazlauskas (McGill University, Canada) and Dipl.-Chem. Erik Henke (Greifswald University) for their support.

References

[1] Bornscheuer, U.T., Kazlauskas, R.J. (1999), *Hydrolases in organic synthesis – regio- and stereoselective biotransformations*, Weinheim: Wiley-VCH.

[2] Bornscheuer, U.T. (2000), *Enzymes in Lipid Modification*, Weinheim: Wiley-VCH.

[3] Drauz, K., Waldmann, H. (1995), *Enzyme catalysis in organic synthesis*, Vol. 1 & 2 Weinheim: VCH.

[4] Faber, K. (2000), *Biotransformations in Organic Chemistry*, Berlin: Springer.

[5] Patel, R.N. (2000), *Stereoselective Biocatalysis*, New York: Marcel Dekker.

[6] Wong, C.-H., Whitesides, G.M. (1994), *Enzymes in synthetic organic chemistry*, Oxford: Pergamon Press.

[7] Liese, A., Seelbach, K., Wandrey, C. (2000), *Industrial Biotransformations*, Weinheim: Wiley-VCH.

[8] Miller, C.A. (2000), Advances in enzyme discovery, *Inform* **11**, 489–495.

[9] Bornscheuer, U.T., Pohl, M. (2001), Improved biocatalysts by directed evolution and rational protein design, *Curr. Opin. Chem. Biol.* **5**, 137–143.

[10] Arnold, F.H. (1998), Design by directed evolution, *Acc. Chem. Res.* **31**, 125–131.

[11] Arnold, F.H., Volkov, A.A. (1999), Directed evolution of biocatalysts, *Curr. Opin. Chem. Biol.* **3**, 54–59.

[12] Arnold, F.H. (1999), Unnatural selection: molecular sex for fun and profit, *Engineering Science* 40–50.

[13] Bornscheuer, U.T. (1998), Directed evolution of enzymes, *Angew. Chem. Int. Ed.* **37**, 3105–3108.

[14] Arnold, F.H., Moore, J.C. (1997), Optimizing industrial enzymes by directed evolution, *Adv. Biochem. Eng./Biotechnol.* **58**, 1–14.

[15] Schmidt-Dannert, C., Arnold, F.H. (1999), Directed evolution of industrial enzymes, *Trends Biotechnol.* **17**, 135–136.

[16] Reetz, M.T., Jaeger, K.-E. (1999), Superior biocatalysts by directed evolution, *Topic Curr. Chem.* **200**, 31–57.

[17] Petrounia, I.P., Arnold, F.H. (2000), Designed evolution of enzymatic properties, *Curr. Opin. Biotechnol.* **11**, 325–330.

[18] Tobin, M.B., Gustafsson, C., Huisman, G.W. (2000), Directed evolution: the "rational" basis for "irrational" design, *Curr. Opin. Struct. Biol.* **10**, 421–427.

[19] Sutherland, J.D. (2000), Evolutionary optimisation of enzymes, *Curr. Opin. Chem. Biol.* **4**, 263–269.

[20] Jaeger, K.E., Reetz, M.T. (2000), Directed evolution of enantioselective enzymes for organic chemistry, *Curr. Opin. Chem. Biol.* **4**, 68–73.

[21] Reetz, M.T. (2001), Combinatorial and evolution-based methods in the creation of enantioselective catalysts, *Angew. Chem. Int. Ed.* **40**, 284–310.

[22] Bornscheuer, U.T., Altenbuchner, J., Meyer, H.H. (1998), Directed evolution of an esterase for the stereoselective resolution of a key intermediate in the synthesis of Epothilones, *Biotechnol. Bioeng.* **58**, 554–559.

[23] Bornscheuer, U., Herar, A., Kreye, L., Wendel, V., Capewell, A., Meyer, H.H., Scheper, T., Kolisis, F.N. (1993), Factors affecting the lipase catalyzed transesterification reactions of 3-hydroxy esters in organic solvents, *Tetrahedron: Asymmetry* **4**, 1007–1016.

[24] Bornscheuer, U.T., Altenbuchner, J., Meyer, H.H. (1999), Directed evolution of an esterase: Screening of enzyme libraries based on pH-indicators and a growth assay, *Bioorg. Med. Chem.* **7**, 2169–2173.

[25] Hecht, H.J., Sobek, H., Haag, T., Pfeifer, O., Pée, K.H.v. (1994), The metal-ion-free oxidoreductase from *Streptomyces aureofaciens* has an α/β hydrolase fold, *Nature Struct. Biol.* **1**, 532–537.

[26] Henke, E. , Bornscheuer, U.T. (1999), Directed evolution of an esterase from *Pseudomonas fluorescens*. Random mutagenesis by error-prone PCR or a mutator strain and identification of mutants showing enhanced enantioselectivity by a resorufin-based fluorescence assay, *Biol. Chem.* **380**, 1029–1033.

[27] Janes, L.E., Kazlauskas, R.J. (1997), Quick E. A fast spectrophotometric method to measure the enantioselectivity of hydrolases, *J. Org. Chem.* **62**, 4560–4561.

[28] Altamirano, M.M., Blackburn, J.M., Aguayo, C., Fersht, A.R. (2000), Directed evolution of new catalytic activity using the α/β-barrel scaffold, *Nature* **403**, 617–622.

[29] Jürgens, C., Strom, A., Wegener, D., Hettwer, S., Wilmanns, M. , Sterner, R. (2000), Directed evolution of a (βα)$_8$-barrel enzyme to catalyze related reactions in two different metabolic pathways, *Proc. Natl. Acad. Sci. USA* **97**, 9925–9930.

[30] Stevenson, J.D., Lutz, S., Benkovic, S.J. (2001), Retracing enzyme evolution in the (βα)$_8$-barrel scaffold, *Angew. Chem. Int. Ed.* **113**, 1906–1908.

[31] Zhang, J.H., Dawes, G., Stemmer, W.P. (1997), Directed evolution of a fucosidase from a galactosidase by DNA shuffling and screening, *Proc. Natl. Acad. Sci. USA* **94**, 4504–4509.

[32] Stefan, A., Radeghieri, A., Gonzalez Vara y Rodriguez, A., Hochkoeppler, A. (2001), Directed evolution of β-galactosidase from *Escherichia coli* by mutator strains defective in the 3'→5' exonuclease activity of DNA polymerase III, *FEBS Lett.* **493**, 139–143.

[33] Li, Q.-S., Schwaneberg, U., Fischer, P., Schmid, R.D. (2000), Directed evolution of the fatty acid hydroxylase P450 BM-3 into an indole-hydroxylating catalyst, *Chem. Eur. J.* **6**, 1531–1536.

[34] Sio, C.F., Laan, J.M.v.d., Riemens, A.M., Quax, W.J. (2000), Altering substrate specificity of *Pseudomonas* SY77 glutaryl acylase towards adipyl-7-ADCA, *Biotechnology 2000 meeting, Berlin, in press*.

[35] Joo, H., Arisawa, A., Lin, Z. , Arnold, F.H. (1999), A high-throughput digital imaging screen for the discovery and directed evolution of oxygenases, *Chem. Biol.* **6**, 699–706.

[36] Zhao, H., Giver, L., Affholter, J.A., Arnold, F.H. (1998), Molecular evolution by staggered extension process (StEP) in vitro recombination, *Nature Biotechnol.* **16**, 258–261.

[37] Joo, H., Lin, Z., Arnold, F.H. (1999), Laboratory evolution of peroxide-mediated cytochrome P450 hydroxylation, *Nature* **399**, 670–673.

[38] Kampen, M.D.v. , Egmond, M.R. (2000), Directed evolution: From a staphylococcal lipase to a phospholipase, *Eur. J. Lipid Sci. Technol.* **102**, 717–726.

[39] Kauffmann, I., Schmidt-Dannert, C. (2001), Conversion of *Bacillus stearothermophilus* lipase into an efficient phospholipase with increased activity towards long chain fatty acyl substrates by directed evolution and rational design, *Protein Eng.*, in press.

[40] Iffland, A., Tafelmeyer, P., Saudan, C., Johnsson, K. (2000), Directed molecular evolution of cytochrome c peroxidase, *Biochemistry* **39**, 10790–10798.

[41] Fong, S., Machajewski, T.D., Mak, C.C., Wong, C.-H. (2000), Directed evolution of D-2-keto-3-deoxy-6-phosphogluconate aldolase to new variants for the efficient synthesis of D- and L-sugars, *Chem. Biol.* **7**, 873–883.

Subject Index